CONTROLLING THE
WATER

CONTROLLING THE
WATER

Matching Technology
and Institutions in
Irrigation Management
in India and Nepal

edited by
Dik Roth
Linden Vincent

OXFORD
UNIVERSITY PRESS

OXFORD
UNIVERSITY PRESS

Oxford University Press is a department of the University of Oxford.
It furthers the University's objective of excellence in research, scholarship,
and education by publishing worldwide. Oxford is a registered trademark of
Oxford University Press in the UK and in certain other countries

Published in India by
Oxford University Press
YMCA Library Building, 1 Jai Singh Road, New Delhi 110 001, India

First Edition published in 2013

ISBN-13: 978-0-19-808292-7
ISBN-10: 0-19-808292-4

Typeset in Galliard 10/13
by The Graphic Solutions, New Delhi 110 092
Printed in India by G.H. Prints Pvt Ltd, New Delhi 110 020

CONTENTS

TABLES, FIGURES, AND BOXES

ACKNOWLEDGEMENTS

This volume, and the research programme 'Matching Technology and Institutions' on which it is based, would not have been possible without the participation, support, and commitment of many people in India, Nepal, and elsewhere. First of all, the financial support for this programme by the Ford Foundation, New Delhi, is kindly acknowledged. We are especially grateful for the support to the programme by Ujjwal Pradhan and Vasant Saberwal. A crucial role has been played by Franz von Benda-Beckmann and Peter Mollinga, who were among the scientific initiators of this Wageningen University research programme. On the Indian and Nepalese sides, we would like to thank the many institutions hosting and persons facilitating the programme's workshops held in India (Ahmedabad, Bangalore, Hyderabad, Thrissur) and Nepal (Kathmandu). Both the workshops and the individual research projects have greatly benefited from the commitment and participation of several people who acted as local supervisors and contributed to the workshops: Vishwa Ballabh, Ajaya Dixit, Gopal Naik, R. Parthasarathy, Late S.T. Somashekara Reddy, and Nirmal Sengupta. Thanks to the enthusiastic participation of many, the workshops became a lively and dynamic place for the exchange of ideas, experiences, and opinions, and the discussion of research plans, progress, and results. Aside from the (then) PhD researchers who have contributed to this volume, the following people need to be specifically mentioned: Olivia Aubriot, Smita Gate, Chitra Krishnan, Preeta Lall, G. Mini, Preeti Misra, Shuku Pun, J.D. Sophia, Sriprakashsingh Rajput, and Diana Suhardiman. Finally, there would not have been any book, workshop, or PhD defence without the administrative support by Mona Challu (Ford Foundation) and Rein Alink, Gerda De Fauw, and Gerrit van Vuren (Wageningen University).

ABBREVIATIONS

ADB	Asian Development Bank
ADBN	Agricultural Development Bank of Nepal
AED	Agricultural Engineering Department
AEPC	Alternative Energy Promotion Centre
AMIS	Agency-Managed Irrigation System
AO	Association Organizer
AP	Andhra Pradesh
APC	Agricultural Price Committee
APFMIS	Andhra Pradesh Farmers Management of Irrigation Systems Act
APM	Adjustable Proportional Module
APROSC	Agriculture Project Services Center
BC	Branch Committee
BIS	Bachcha Irrigation System
BLGWIP	Bhairahawa Lumbini Groundwater Irrigation Project
BPL	Below Poverty Line
BTI	Butwol Training Institute
BYS	Balaju Yantra Shala
CACP	Committee on Agricultural Costs and Prices
CADA	Command Area Development Authority
CADP	Command Area Development Project
CCA	Culturable Command Area
CIP	Chitwan Irrigation Project
CMIS	Chattis Mauja Irrigation System
DAO	District Agriculture Office
DC	Distributary Committee
DCS	Development and Consultancy Services

DDC	District Development Committee
DEE	Deputy Executive Engineer
DIO	District Irrigation Office
DoI	Department of Irrigation
DRMHPEC	Daunne River Micro Hydro Power Executive Committee
DTW	Deep Tube Well
EDC	Economic Development Council
EDI	Economic Development Institute
EG&S	Ecosystem Goods and Services
EMC	Energy Management Committee
EU	European Union
EUG	Energy Users Group
FAO	Food and Agriculture Organization
FG	Functional Group
FMIS	Farmer-managed Irrigation System
FSL	Full Supply Level
FTL	Full Tank Level
GA	General Assembly
GIS	Gayatripuzha Irrigation System
GoAP	Government of Andhra Pradesh
GoI	Government of India
HYV	High-yielding Variety
ICA	Irrigable Command Area
IIMA	Indian Institute of Management Ahmedabad
IIMI	International Irrigation Management Institute
ILC	Irrigation Line of Credit
IMP	Irrigation Management Project
IMT	Irrigation Management Transfer
IMTP	Irrigation Management Transfer Programme
INCID	Indian National Committee on Irrigation and Drainage
INGO	International Non-governmental Organization
INR	Indian Rupee
IRRI	International Rice Research Institute
ISF	Irrigation Service Fee
ISP	Irrigation Sector Programme
IWMI	International Water Management Institute

IWRM	Integrated Water Resources Management
JA	Joint Azmoish/Joint Inspection
KIS	Khageri Irrigation System
KMI	Kathmandu Metal Industries
KRRS	Karnataka Rajya Raitha Sangha (Karnataka State Farmers' Association)
LBC	Left Bank Canal
MASSCOTE	Mapping System and Services for Canal Operation Techniques
MBC	Madhira Branch Canal
MC	Management Committee
MEA	Millennium Ecosystem Assessment
MHES	Micro-Hydel Energy Systems
MHFG	Micro Hydro Functional Group
MID	Minor Irrigation Department
MLBC	Mula Left Bank Canal
MLD	Ministry of Local Development
MoA	Memory of Agreement
MoU	Memorandum of Understanding
MoWR	Ministry of Water Resources
MPPU	Multipurpose Power Units
MRBC	Mula Right Bank Canal
MSL	Mean Sea Level
MSP	Minimum Support Price
MVDC	Madhaulia Village Development Committee
NEA	Nepal Electricity Authority
NFIWUAN	National Federation of Irrigation Water Users of Nepal
NGO	Non-governmental Organization
NIS	New Institutionalist School
NPC	National Planning Commission
NR	Nepalese Rupee
NSLC	Nagarjuna Sagar Left Bank Canal
NTS	New Traditionalist School
NWGIS	Nepal West Gandak Irrigation System
O&M	Operation and Maintenance
PC	Project Committee
PDDP	Participatory District Development Programme

PDS	Public Distribution System
PIM	Participatory Irrigation Management
PIS	Panchakanya Irrigation Scheme
PTD	Participatory Technology Development
PWD	Public Works Department
R&D	Research and Development
RADC	Remote Area Development Committee
RBC	Right Bank Canal
RD	Revenue Department
RDO	Revenue Division Officer
REDB	Rural Energy Development Board
REDP	Rural Energy Development Programme
REDS	Rural Energy Development Section
REMREC	Resource Management and Rural Empowerment Centre
RKMHC	Roshi Khola Micro Hydro Cooperative
RRHMC	Roshi River Micro Hydropower Management Committee
RRMHP	Roshi River Micro Hydropower Project
SATA	Swiss Association for Technical Assistance
SIDA	Swedish International Development Authority
SMC	Sub-project Management Committee
SMIS	Sorha Mauja Irrigation System
SOPPECOM	Society for Promoting Participative Ecosystem Management
SRD	State Revenue Department
STW	Shallow Tube Well
TAFCORN	Tamil Nadu Forest Plantation Corporation
TDP	Telugu Desam Party
TVDC	Tikuligarh Village Development Committee
UML	United Marxist-Leninist Party
UMN	United Mission to Nepal
UNDP	United Nations Development Programme
UNIDO	United Nations Industrial Development Organisation
UP	Uttar Pradesh
USAID	United States Agency for International Development

USD	United States Dollar
VDC	Village Development Committee
WALMI	Water and Land Management Institute
WECS	Water and Energy Commission Secretariat
WI	Works Inspector
WRCP	Water Resources Consolidation Programme
WUA	Water Users' Association
WUC	Water Users' Committee
WUG	Water Users' Group
ZP	Zilla Parishad

INTRODUCTION: TECHNOLOGY AND INSTITUTIONS IN IRRIGATION WATER MANAGEMENT — MATCH OR MISMATCH?

Dik Roth and *Linden Vincent*

This book explores the means of water control in irrigation management in India and Nepal. The contributing authors analyse relationships seldom explored in an in-depth way in irrigation studies: between water sources, technologies, and institutions in relation with wider agrarian, socio-political, and environmental changes. Our main points of departure are a contextualized understanding and an interdisciplinary approach towards water control in irrigation. Water control involves the interactive dynamics between technical, organizational-institutional, and socio-political control, and shows the match or mismatch between technology and institutions. An interdisciplinary perspective shows the interlocking dynamics of social choices in irrigation development, their social conditions of use, and their social effects.

The contributors to this book describe and analyse locations of research of which the hydrological, political, and agricultural boundaries often also determine the development of technological, regulatory, and managerial systems. Contrary to assumptions about control of designs, organizations, and human behaviour that guide the 'hydraulic mission' of the irrigation sector (Molle *et al.* 2009; Wester *et al.* 2009), the contributors show how irrigators and operators use and contest social and political relations and control of water and landscape in diverse irrigation settings: large agency-managed or jointly managed canal systems, small diversion and tank-fed gravity systems that are farmer-managed, and groundwater-based irrigation. These physical settings define spaces within which the relationships between social, technological, and legal systems can be studied. Public agencies have even seen them as requiring different programmes of intervention for the promotion of technology and management reform (Bottrall

1978). In fact, misfits between institutions and technological control in these domains are often a source of conflict.

This volume has developed from PhD research undertaken in the research programme 'Matching Technology and Institutions in Land and Water Management' supported by the Ford Foundation and Wageningen University, The Netherlands, between 1997 and 2007. The focus on India and Nepal reflects the core importance of irrigation in these countries. We hope this volume can inform future rural development and water reform policies, and inspire new interdisciplinary field research on the interrelationships of technology, societal change, and water control. Relatively little work has been done that explores these relationships for the diverse scales of irrigation technologies in India and Nepal. This research can also help in understanding new technological, social, and institutional trajectories emerging under agrarian and technological change. These are the main scientific justifications for this volume as well as for the original research programme.

Access to water requires technologies and institutions that regulate human use of the environment. There is a vital interface between technology and institutions for allocation and delivery of irrigation water at all levels. However, relatively little attention is paid to misfits in the evolution of these systems. Ironically, these gaps and contradictions occur precisely because of the long-term strategic importance of irrigation in many countries. This reflects a history of changing water interventions and the gaps and discontinuities of knowledge systems around water regimes and water engineering. Economic and environmental changes often put pressure on relations that have evolved between technologies and the institutions around them so that new relations can be mediated. This problematic is especially relevant where uncertainty and scarcity of water are fuelled by social and economic changes. Studying matches and mismatches in these relationships is important because water continues to be a key resource for livelihood security, used as an explicit intervention tool for economic growth and poverty alleviation. As pressure on resources increases, the position of small-scale and vulnerable producers is a major concern.

This introduction provides an overview of the contributions it makes to the study of water control. It also opens up discussion about the match or mismatch between technology and institutions. In

keeping with our perspective of water control, we use a flexible definition of (irrigation) water management as 'a form of social interaction,
among different stakeholders, using different methods, resources and
strategies, about water use and distribution activities' (Gutierrez 2005:
21). This definition enables us to look at interactions of stakeholders
at multiple levels: in the system where local cultural norms may be
present, but also in regional and global interactions that are shaping
irrigation management reforms. The concept of 'institutions' allows
us to see both the diverse grouping of actors shaping water management and the principles behind existing or promoted practices. This
enables us to keep a wide conceptualization of management including
both processes of governance of water and practical operational tasks.
This expresses that politics are an intrinsic part of water management
at different levels.[1] We return to this critical issue of understanding
the nature of technology, institutions, and governance in Chapter 1.

We complete this introduction with an overview of these interlocking processes and their effects for irrigation, showing the social
forces shaping irrigation development and irrigation as a social force
(Eggink and Ubels 1984). The discussion then moves through a short
overview of irrigation in India and Nepal, to a more detailed review
of contemporary water control issues in three key irrigation settings
that form the core of this volume: canal systems, groundwater-based
irrigation, and storage-based irrigation from tanks and ponds. We
end with a short chapter overview.

FORCES SHAPING 'MISMATCHES' IN WATER
CONTROL IN SOUTH ASIA

Irrigated agriculture has a long history in Asia, and has contributed
importantly to development in the continent. It has increased food
security, sustained rural livelihoods, and made staple foods like rice
available for a rapidly growing and increasingly urban population. It
has created rural employment, stabilized outputs, and lowered food
prices (Barker and Molle 2005; Merrey *et al.* 2007). In the twentieth
century, radical changes were brought about by the intensification of
irrigated agriculture using production-increasing technologies—the
Green Revolution—in combination with the expansion of irrigated
area (Bayliss-Smith and Wanmali 1984; Farmer 1986). In Asian

countries with large and growing populations like China, India, and Indonesia, massive food shortages and starvation now belong to the past. This is one of the greatest development achievements of the last century (Shivakoti *et al.* 2005).

However, a high price has also been paid environmentally, socially, and economically. In irrigation development, sustainability and equity have often been subordinated to increase in production. Irrigated agriculture causes environmental problems like waterlogging, salinization, overexploitation of (ground)water resources, pollution, and health problems (Ballabh 2005, 2008; Lahiri-Dutt and Wasson 2008). While often promoted for its contribution to poverty alleviation (through production and employment generation; see Chambers *et al.* 1989; Hasnip *et al.* 1999), irrigation has also changed access to water (and land) and economic risks in production. In some cases, irrigation development has increased inequality and marginalization of the poor, and broader socio-economic developments have shown that the number of people structurally suffering from hunger is growing again, especially in South Asia and Africa.

Other new challenges are also emerging: climate change, urbanization and industrialization, privatization of water (for example, tube wells), and growing water scarcity and competition (Barker and Molle 2005; Joy *et al.* 2008). Bottrall (1992) warned of the problems emerging in South Asia's irrigation networks due to the failure to adapt technologies and institutions over time (to changes in the socio-political environment) or in space (from one kind of physical environment to another). He called for a greater understanding of options for change. This volume takes up his debate in the light of a new understanding of technology, institutions, and reforms taking place in the region.

Irrigated agriculture is and will remain vital to national and rural food security, agro-industries, and export. It will have to keep pace with ongoing population growth and a skyrocketing global demand for food, feed and fuel crops. Major drivers in irrigated agriculture are the growing middle classes with purchasing power (rice and meat consumption) and the attempt worldwide to reduce dependence on fossil energy. However, growing water scarcity and competition will also hit the agricultural sector, as the transfer of water to other uses seems unavoidable (Ballabh 2005; Reddy and Dev 2006).

In India, for instance, the enormous expansion of irrigation in the last century was initially driven by the development of surface water resources under various governance, management, and property regimes (public, common, private; farmer-managed, jointly-managed, state-managed) in various agro-ecological and other settings (Sengupta 1991). However, Indian agriculture has recently become increasingly dependent on groundwater (Shah 2009). The bad physical and managerial conditions of many surface irrigation systems and the scarcity of surface water sources have driven millions of farmers into groundwater exploitation. In the process, overexploitation of aquifers has turned regions with substantial groundwater resources into 'dark zones' (Bhatia 1992; Prakash 2005; also see Prakash, this volume).[2]

Such developments have taken place with a narrow focus on production increase through water resources development. Important lessons from farmer-managed irrigation systems (FMISs), tank irrigation, and other water harvesting and storage systems based on local technology have not been learnt (Ambler 1994). New policies and reforms have reshaped cropping patterns, transformed designs, and reformed water management systems in the name of efficiency, cost reduction, equity, or water savings. Other important social drivers are the wider agrarian changes, local or regional politics, and state and national policies now reshaping access to water for agriculture, together with urbanization and new capital-flows under globalization.

In recent years governments have adopted policies to transform their bureaucracies and experiment with new forms of local governance. Across schemes we can see the influence of policies of irrigation management transfer (IMT), decentralization of responsibilities, and new governance institutions (Mollinga and Bolding 2004). The resulting 'politics of policy', as irrigators and other social actors adapt to, reinterpret, and reshape policy guidelines in their actions to obtain water, defend established interests, or pursue other goals, are analysed here (see Bala Raju Nikku, this volume). These new forms of governance, together with changing agrarian relations, have increased the focus on political action around irrigation systems. Irrigation systems have become important political platforms for bureaucrats in search of social legitimacy, politicians mobilizing voters, and farmer movements aiming to influence tariffs and markets (Briscoe and Malik 2007; see Shah, this volume). These processes also influence national politics. These dynamics have been discussed in various chapters of this volume.

There are both worrying and encouraging trends in how the new challenges are being dealt with. The last few decades have seen the continued ambitions of water bureaucracies to solve all problems by pursuing their hydraulic mission of controlling water by building large infrastructural works (for example, river linkages and dams; see Dhungel and Pun 2009; D'Souza, D. 2002; D'Souza, R. 2008; Whitehead 2008) or continued dependence on other blueprint engineering solutions. Equally worrying is the idealization of 'community' and 'local knowledge' embodied in 'traditional' irrigation technology as *the* solution to all water problems.[3] In India and elsewhere, these trends have polarized debates about large water infrastructure (Molle et al. 2009; Mollinga 2008b).

More encouraging is the growing recognition of the importance of specific contexts of water control (Baviskar 2007; Donahue and Johnston 1998; Lahiri-Dutt and Wasson 2008). Notwithstanding all recent talk about a 'global water crisis' and the need for 'global water governance' (for example, Pahl-Wostl et al. 2008), water problems and solutions always relate to specific agro-ecological, socio-political, economic, and cultural contexts and histories (Mollinga 2008a). South Asia, with its huge diversity of water histories, cultures, and local uses (Sengupta 1985; Singh 2002) illustrates this. Therefore, water challenges need a contextualized understanding of water control before new, more productive, equitable, and sustainable water futures can be explored.

The in-depth field studies in this volume (see Figure 1) show that prescriptive policies and technologies are often part of the problem rather than the solution. Especially the role of technology at the interface of ecology and water resources, landscape, and society, with its organizational and institutional arrangements, requires more theorizing and understanding. Rather than new blueprint recipes for change or 'nirvana concepts' (Molle 2008), more attention to the complexities of water environments and systems in irrigation is needed. The case studies show the diversity of technological and institutional trajectories and choices made by water users, engineers, and policymakers, adding new analyses of irrigation to existing water studies (for example, Shivakoti et al. 2005; Singh 2002).

FIGURE 1   India, Nepal, and the Research Locations under Study

Source: Authors.

TAKING AN IRRIGATION FOCUS

In recent years we have seen a marked shift from sectoral to integrated water resources management (IWRM) approaches. Therefore, the irrigation focus of this volume (except for Chapter 11 by Amreeta Regmi and Linden Vincent) requires explanation. First, irrigation is still making enormous contributions to rural livelihoods and food provision to urban populations (Molden 2007; Shah 2009). However, under the influence of rapid socio-economic and other transformations, irrigation-based livelihoods are under threat. Therefore, their protection, taking into account productivity, equity, and sustainability, is a key challenge.

Second, the irrigation sector still accounts for more than 70 per cent of freshwater use globally (Molden 2007). Agriculture will be the main victim of any future inter-sectoral water transfers. These redistributive measures are, by definition, sensitive. Challenging established claims to water, livelihoods, and economic and political interests, they will unavoidably generate tensions and conflicts.[4] Hence, the need is to find negotiated solutions, paying attention to existing water rights and water-based livelihoods (Bruns and Meinzen-Dick 2000; Bruns et al. 2005; Meinzen-Dick and Pradhan 2005). There are no easy 'win-win' solutions, so water issues tend to be highly politicized and politically contested (Mollinga 2003).

Third, important lessons can be learnt from irrigated agriculture (Merrey et al. 2007). Through its long history and development in a variety of localities, irrigation may best show us the linkages between land, water, technologies, and institutions, and provide a greater understanding of their historical trajectories of change under specific conditions. In whatever contexts of technical and social change—water transfers or institutional reforms—greater knowledge and understanding of these processes and conditions is necessary (Bruns et al. 2005).

IRRIGATION IN INDIA AND NEPAL: A SHORT OVERVIEW

The agro-ecology of India and Nepal makes irrigation an important basis of agriculture. It has shaped a diverse range of designs of systems for extracting, storing, and delivering water. Dependent on monsoon circulation patterns, much of the region has a seasonal climate that also shapes river flows. Core areas of India and parts of western Nepal are under semi-arid climates, where the rainy season is highly variable, giving rise to drought-prone areas and deserts (Jairath and Ballabh 2008). Irrigation can intensify crop production to two or more crops per year, expand the irrigated area, overcome dry spells, and help timely land preparation, planting, and harvesting to maximize land use and minimize damage from weather risks.

These benefits have been known since long, giving rise to a range of locally developed technologies for collecting, storing, and distributing water (Ambler 1994; Sengupta 1985, 1991; Singh 2002). Several authors highlight the importance of understanding water resource systems in their ecological dimensions and hydrological complexity and not just as sources for water abstraction and transfers (Bandyopadhyay 2009; Biswas *et al.* 2009; Gupta 2008; Gyawali 2001; Jain *et al.* 2007). Bottrall (1992) has highlighted four main physical environments in South Asia. These are the sites of locally adapted irrigation development, but also of inappropriate 'blue-print style' interventions by public agencies. This volume provides examples from these contrasting environments: the drought-prone uplands of central and south India (see chapters by Jyothi Krishnan, R. Manimohan, Vishal Narain, Bala Raju Nikku, Anjal Prakash, and Esha Shah); the low-rainfall alluvial plains of northwest India (one of the cases in the chapter by Narain); the high rainfall floodplains with good groundwater options (see chapters by Suman Rimal Gautam and Puspa Raj Khanal); and the Himalayan region (see chapters by Umesh Nath Parajuli and Amreeta Regmi and Linden Vincent).

Since the colonial period, irrigation and water engineers have regarded the development of infrastructure as their critical mission, thereby showing a supply augmentation rather than demand management orientation. In the recent past, for instance, opportunities for regulating water, lessening flood risk, and storing water for irrigation and hydroelectric power generation in river basins have stimulated dam building. The number of large dams in India grew from less than 300 in 1947 to over 4,000 by 2000. Most of these were built between 1971 and 1989 (Barker and Molle 2005).[5] Irrigation is either the primary purpose or one of the multiple purposes of 96 per cent of these dams. Dams have become an immensely contentious topic because of displacement of people[6] and environmental disruption.[7] Large-scale irrigation development has caused problems of over-development and over-allocation of rivers, turning once perennial rivers into seasonal ones, and fuelling the demand for more storage infrastructure. Moreover, large canal systems fed by dams are extremely complex technically and managerially, making effective water delivery a challenge (see Narain, Nikku, and Krishnan, this volume).

India and Nepal have a wide range of aquifers—confined and unconfined, shallow and deep—recharged by monsoon rains, rivers, irrigation systems, and other seepage sources. These range from the large-scale, high-yielding aquifers of the Indo-Gangetic Basin in the Terai of India and Nepal (Mukherji *et al.* 2009) to the increasingly over-exploited and low-yielding hard rock aquifers of central and southern India (Knegt and Vincent 2001; Kumar 2007; Prakash 2005). In India, groundwater has overtaken canal irrigation as the main source for irrigation (Shah 2009). This evidences a shift away from farmer reliance on public sector investments towards private investments in the era of 'atomistic irrigation' (Shah 2009; see Ballabh 2005). The groundwater boom and problems of overdraft are well documented, and there is growing understanding of the socio-ecologies that shape farmer responses and institutional development in different locations (Shah 2009).[8] Less documented, but a theme in this volume, is the crucial role of agrarian relations around ground-water use in determining winners and losers, and the organizational and institutional dimensions of farmers' choices between surface and groundwater sources in settings of 'conjunctive use' (see Prakash and Gautam, this volume).

Irrigation concerns forms of technical, economic, social, and political control. Irrigated agriculture developed in societies for which elite control over people, territories, and natural resources was a crucial determinant of political power, status, and wealth. In India and Nepal, expansion of control and patronage by states and ruling kingdoms through grants of land and other resources played an important role in irrigation development (Cullet and Gupta 2009; Pradhan *et al.* 2000; Singh 2002).[9] Apart from control of people and territory, revenue (for example, on products, land tax) was an important driver of irrigation development.

Large-scale irrigation development based on western engineering in India dates from the colonial period. British rule established government control over water. After the annexation of Punjab in 1849 (Barker and Molle 2005) and the Great Famine of 1876–8 (Gulati *et al.* 2005) the British colonial administration began to construct larger systems in the alluvial plains of north India and the deltas of south India (Molle *et al.* 2009; Singh 2002; Stone 1984; Whitcombe 1972). These transformed the hydrology, economy, and political rule of these areas (D'Souza 2006a; Gilmartin 1994). The 1873 Northern India Canal and Drainage Act increased state control and weakened local and customary rights (Cullet and Gupta 2009). Colonial irrigation development was strongly connected to broader (sometimes contradictory) interests and objectives like famine prevention, avoidance of political unrest, and generation of revenue. Specific concerns with the equitable spreading of benefits from irrigation developed through policies of 'protection' and 'localization'. These shaped the technology design and water allocation principles (Barker and Molle 2005; Mollinga 2003).

Irrigated agriculture in Nepal is known for its strong tradition of FMIS in the hills and southern plains (Lam 1996; Martin and Yoder 1987; Pradhan *et al*. 2000; Shivakoti and Ostrom 2002). These systems are technically and institutionally highly adapted to local climate, topography, and hydrology (Parajuli, this volume; Pradhan, P. 1989; Pradhan, U. 1988; Yoder and Upadhyaya 1987; Yoder *et al*. 1987). Until the 1950s, patterns of control and resource allocation of the FMIS were recognized as important, except for sporadic large-scale civil engineering works.[10] Later, greater state involvement and international donor support created a shift in focus to large-scale irrigation development (Khanal 2003; Pradhan 1996). Historically, the focus of state control shifted from land to water resources, related to broader transformations of the Nepalese state from the nineteenth century onwards. This has led to a gradual dissociation of land and water rights, and a weakening of local water rights relative to state claims (Pradhan *et al*. 2000).

The biggest changes in both countries were brought about by the postcolonial expansion of professional water bureaucracies, and associated routines of policy-making, planning, and implementation through civil engineering and construction: the 'post-colonial hydraulic mission' (Molle *et al*. 2009: 335). International donor funding was a major driver of the increased powers and intervention capacity of these bureaucracies. Irrigation development became the domain of civil and agricultural engineers and economists, whose priority was production increase. Irrigation and water bureaucracies developed a predilection for the large-scale and modern. This was often detrimental to existing technologies or institutions (Donahue and Johnston 1998; Gyawali 2001; Singh 2002), as little or no attention was paid to the social, political or ecological dimensions of irrigation and water management (Moench *et al*. 2003). Such biases can be explained from deeply engrained institutional incentives, primarily the nexus between the choice for large-scale infrastructure and the existence of 'iron triangles' between bureaucrats, politicians, and business interests (Molle *et al.* 2009: 336–7). This supply-side focus has proven quite resistant to change, even in the face of growing civil society movement protests (Mollinga 2008b).

The chapters in this volume focus on three major domains of irrigation, represented by three types of systems: canal (surface) irrigation (see chapters by Narain, Nikku, Khanal, Parajuli, and Regmi and Vincent), groundwater irrigation (see chapters by Gautam, Prakash), and 'traditional' irrigation systems based on tanks and ponds (see chapters by Krishnan, Shah, Manimohan). In the following paragraphs we give a concise overview of the major issues in each of these domains.

CANAL IRRIGATION: REFORMS IN NEED OF REFORM

Canal systems are designed physical systems involving many functional components that shape the 'service' and performance provided in water delivery. There are multiple options for works to acquire water (dams, diversions) and convey and distribute it (canals, division structures, drainage). Yet there are also hybrid systems (van Halsema 2002): physical system and social organization must evolve together to function. At the same time they transform local hydrological systems through abstraction, storage, and transfers of water. Thus, irrigation systems emerge from the interplay of political choices, scientific (engineering) endeavour, and chosen management options (D'Souza 2006a; Gilmartin 1994; van Halsema and Vincent 2006). This complex and historically changing interplay of society, engineering science, and hydrology in canal irrigation is illustrated in the chapters on canal technology by Nikku, Narain, Khanal, and Parajuli.

Through public agency development in colonial India norms of design and allocation became embedded in large-scale irrigation development. Government agencies became increasingly involved in the operation of large-scale systems, right down to outlets or the operation of gates linking minor canals with the watercourses that delivered water to farmers' fields.[11] A critical principle that shaped technology and water allocation is the difference between the so-called 'protective' and 'productive' irrigation design, the former strongly linked with famine prevention. Protective irrigation systems, based on 'scarcity by design' (Jurriëns *et al.* 1996; Mollinga 2003), supply water sufficient for only part of the irrigation area and on the basis of one crop a year. The idea is to avoid crop failure by thinly spreading the available water over as large an area as possible. Such schemes are completely 'supply driven', and not related to actual and fluctuating crop demand, although turns are made as regular and reliable as possible. Their structures have a low management intensity (without operators and as interference-proof as possible). An example are the *warabandi* or 'fixed turn' systems in India (and Pakistan), discussed by Narain, which serve an estimated 24 million hectares (Barker and Molle 2005).

By independence in 1947, 16 per cent of the irrigated area in India was based on protective designs. However, since then, crop production has changed to more water-demanding crops, political and organizational controls have changed, and enforcement has become less strong. Head-end farmers, for instance, tend to take more water than allocated to them. In addition, the restrictions on surface water supply have been a major driver of the development of wells in irrigated command areas and conjunctive use systems. Many systems have been designed for productive irrigation. Hence there is more effort to supply the amount of water needed by specific crops. Nevertheless, the public agency will still 'localize' or define the area to be irrigated (see Mollinga 2003). In fact, with insufficient water supply and funds for completion of construction, many large schemes have not achieved their original design areas and are less productive than is often assumed.

Protective and productive irrigation, upstream–downstream struggles, interference in canal operations through division technology, and reforms through IMT are discussed in the chapters on Indian canal irrigation by Narain and Nikku, and by Khanal for the FMIS of Nepal. Linkages of groundwater and canal irrigation systems of the Nepalese FMIS in relation to the conjunctive use are discussed by Gautam. Krishnan also shows the disruption to local pond-based irrigation systems and property rights by the development of large-scale canal systems and new technologies.[12]

The expansion of large-scale, agency-managed systems, legitimized by the need for increasing staple food production, ran on apace through the twentieth century. However, realization grew that a construction and supply focus worked at the expense of the managerial–institutional dimensions of irrigation. High construction and employment costs in state bureaucracies meant that little money was invested in the maintenance of large-scale systems, which often degraded. Revenue from long-established charges on irrigated land made hardly any contribution to cost recovery (see Ballabh 2005). The decision of funding nations and agencies to finance irrigation development by loans has increased the influence of international policies for development and reform in the region.

These processes have led to a gradual change in irrigation policies from resource development and expansion to management and control (Barker and Molle 2005). A first change came in the 1970s with the concept of Command Area Development, to develop the full potential of irrigated areas and raise productivity, especially in major and medium irrigation systems (Ballabh 2005). Here agencies first began to focus on local organizations. This initiated a trend to examine whether improvements could come from farmers themselves distributing water below the structures still maintained by agency staff. These programmes gave much more attention to the development of infrastructure and institutions at the lower (tertiary) level of large schemes. However, in these efforts to work with the needs of farmers sometimes investments were made in new infrastructure and institutions that proved hard to sustain after project support was withdrawn. The effects of such intervention are discussed in the case study from Nepal by Khanal.

A second major reform has been the promotion of decentralized irrigation management since the 1990s. This policy is known variously as participatory irrigation management (PIM) and IMT, involving the transfer of management powers to local organizations. It has been backed strongly in both the countries: in India with strong state involvement (Mollinga and Bolding 2004), in Nepal through new laws supporting change and international facilitation of training and support for newly developing water users' associations (WUAs) (Khanal 2003). These policies recognized that system rehabilitation may be necessary to improve water delivery before systems can be handed over to farmers. Farmer involvement in rehabilitation has also been part of policies to help train and empower the new WUAs.

However, experiences around irrigation reform in many countries, including India and Nepal, show that established interests and strategies to maintain them tend to be stronger than the ambitions of the reformers (Mollinga and Bolding 2004). The chapter by Nikku shows the realities of such reforms as experienced by water users and front-line operators, how these changes have given new opportunities and challenges, and how certain actor groups can capture or transform efforts for change. The chapter on Nepal by Khanal studies these reforms from a different perspective, looking inside the system of support for transformations in WUAs. To avoid blueprint approaches, programmes in Nepal used participatory methodologies and social organizers to work with farmers to find what changes in technology and institutions they prioritized.

The shift from agency management to joint management between state irrigation departments and groups of farmers, alongside efforts to renovate the systems and deal with inadequacies of water supply, is a critical feature of large-scale systems. The chapters mentioned discuss these interfaces between irrigation departments and farmers, showing both material and social dimensions of struggles between them and giving detailed insights into the reform processes. These interactions, and their mediation by technologies and institutions, are a core contribution to this volume.

Growing state involvement and reforms are also present in the histories of small run-of-river canal systems, which were often farmer-managed in the past. However, other dynamics are shaping their technology and institutions in the face of societal and environmental changes. They face new challenges of maintenance and resource mobilization, or new infrastructural designs in headworks and com-mand areas by public intervention. These struggles and search for financial support have occurred at a time when local politics changed and opened up to new actors, making these locally managed systems a focus for the interests of politicians. The catchments where these systems are located have often been affected by changing hydrology, land use and ecology. Socio-political changes often bring new require-ments like registration of water allocations to gain formal rights over them, and defence of water rights against upstream development and downstream demands (Khanal 2003; Pant 2000; Pradhan 1989; Yoder *et al.* 1987).

There has been a long debate about how states can support small-scale irrigation systems better by recognizing their existence, enabling farmer authority for management, being open to farmer-initiated technological developments, and being critically aware of new designs state agencies bring into systems. However, the bureaucratic understanding of small-scale systems is still rather poor, and needs improvement through inventory and research (Ambler 1994; Sowerine *et al.* 1994; Yoder and Upadhyaya 1987).

GROUNDWATER: DISASTER IN DISGUISE?

All over the world use of groundwater for agriculture has grown exponentially (Giordano and Villhoith 2007).[13] This growth is largely concentrated in Asia, primarily in India; Nepal has a relatively small but rapidly growing groundwater economy. In both countries, groundwater is primarily important for smallholders. India ranks first on a worldwide list of groundwater-exploiting countries, both in terms of irrigated area and share of the global total (Molden 2007).[14] There are marked differences between regions, ecosystems and states in availability and importance of groundwater, degree and types of scarcity, societal setting and impact, policies, and options and possibilities for expansion and regulation (Kumar *et al.* 2005; Shah 2009). The technological assemblage needed to access groundwater—wells, pumps, energy, and distribution devices—has itself a history and sociology of changing options, and so does the technology for drilling and configuring artificial recharge. Both of these are shaped by and have an impact on the evolution of institutions and organizations for system operation, water access, and control in changing local aquifers, ecologies, and agrarian societies.

The use of shallow wells and traditional lifting devices, often combined with other sources, have a long history throughout India (Sengupta 1985; Shah 2009). In the 1940s, well irrigation became recognized by the British as an alternative to surface systems (Singh 2002). When the government started propagating and stimulating mechanized pump and tube well technology through credits and subsidies from the 1970s, tube well irrigation boomed.[15] With currently more than 19 million structures, groundwater has become crucial for irrigated agriculture (Barker and Molle 2005; Roy and Shah 2003). The number of tube wells is increasing so fast (0.8 to 1 million new wells every year) that Shah (2005: 19; see Shah 2009) speaks of a 'veritable anarchy'.

Groundwater development in Nepal has been steadily growing since the 1970s, but on a much smaller scale than in India. It is most important for both domestic and agricultural use in the Terai (Mukherji et al. 2009). Groundwater resources are relatively abundant there, and recharge seems to make more intensive but sustainable exploitation for agricultural development possible. The resource is still mostly exploited for drinking water through individually owned and manually operated shallow tube wells (STWs).[16] It is increasingly used in agriculture, the backbone of rural livelihoods in the Terai (Kansakar 2005). Deep tube well (DTW) development for irrigation has primarily taken place through donor programmes for irrigation and agricultural development, often related to broader infrastructural construction programmes. Gautam (2006) has done path-breaking research on the complex ways in which the technologies delivered through such programmes interact technically and institutionally with existing forms of water use from various surface and groundwater sources. Complex linkages between various systems (for example, FMIS and DTW systems) have never been taken into account, thus turning 'conjunctive use' into an abstract ideal rather than an established practice (see Gautam, this volume).

Groundwater development has contributed considerably to production increase and expansion of irrigated agriculture. It can also increase equity and food security (for example, shallow groundwater for small-scale farming), and strengthen rural livelihoods, more than what large surface systems have done (Molden 2007). Thanks to groundwater irrigation, many small farmers in India have escaped from marginality and poverty in a way impossible with surface irrigation or rain-fed agriculture (Kahnert and Levine 1993; Moench et al. 2003).[17]

Groundwater irrigation does not require intervention of a bureaucracy, and stimulates demand-driven individual or group initiatives. It creates the possibility of irrigated farming outside the commands of surface systems, and greater flexibility and water security for farmers within such—often bureaucratically managed and unreliable—systems.[18] Thus, 'conjunctive private sector investment' (groundwater exploitation; tube wells, pumping) has come to the aid of many farmers (Barker and Molle 2005: 56). Its on-demand, on-site availability and reliability makes groundwater an important tool for reducing drought loss risk (Moench et al. 2003; Shah 1993).

Groundwater is more productive than other forms of irrigation, with important livelihood consequences (Molden 2007; Roy and Shah 2003). Some authors have stressed the possibility of water markets developing around groundwater exploitation, contributing to efficient and equitable resource allocation. However, there is also considerable concern about socio-ecologies where inequity and oligarchies may occur (Chambers *et al.* 1989; Shah 1993, 2009).

Groundwater characteristics (co-)determine how it can be governed, managed, and exploited (Burke and Moench 2000). Its subsoil and complex location in shallow or deep aquifers makes it invisible, harder to define its boundaries, and difficult to limit access to. In many countries, including India and Nepal, those who control land from which groundwater can be accessed are considered its lawful owners (see Vani 2009). In view of the resulting highly privatized, informal, and uncontrolled nature of exploitation, restricting groundwater use is difficult, causing a 'tragedy of open access' (Meinzen-Dick and Pradhan 2005: 191; Knegt and Vincent 2001).

Using the resource requires investments in well construction (boreholes) and lifting or pumping technology, which is increasingly costly as the water table falls (van Steenbergen and Shah 2003). In this volume, Prakash explores the relationships between social differentiation and groundwater exploitation in Gujarat. As the author shows, groundwater distribution and control crucially depend on the ability of farmers to control land and invest in groundwater technology as the water table goes down. Thus, the well-to-do and powerful classes can monopolize control (Prakash 2005; Prakash, this volume; see also Dubash 2002; Janakarajan 1994). Moreover, groundwater development has often taken place at the expense of livelihoods based on shallow wells and lifting devices, which have suffered from decline of the groundwater table. Other systems, like tanks, declined with the spread of tube wells in combination with other factors like changing property rights and the spread of high-yielding varieties (HYVs) (Janakarajan 2004).

Thus, especially in India, social and class concerns are the subject of intense debates about groundwater: does it benefit the poor or reproduce existing inequalities in landownership? If left to 'the market', groundwater can create, perpetuate or sharpen inequalities. Groundwater markets do not serve a level playing field. They inequitably allocate benefits and negative effects, especially if only the wealthy can afford deeper drilling to keep up (or rather 'keep down') with the groundwater table (see Moench *et al.* 2003; Prakash, this volume). Therefore, equitable groundwater use would at least require specific policies to counter its biases (Ballabh *et al.* 2008).

Groundwater use can also negatively impact ecosystems, with further economic and social consequences (Bhatia 1992; Roy and Shah 2003). The negative environmental impact of decades of government stimulation is visible in India. Where withdrawal exceeds recharge (either natural or from surface storage or harvesting systems), decline and depletion (groundwater, deep aquifers) occur. A general distinction must be made between shallow and deep aquifers: while shallow aquifers are often rapidly recharged from surface sources, recharge of deep aquifers is much slower (Barker and Molle 2005). In states with intensive and growing groundwater use like Punjab, Haryana, and Gujarat, excessive exploitation is creating huge environmental problems (Prakash, this volume; Roy and Shah 2003). In the 'dark zones'[19] of Punjab, for instance, the situation is critical: in 1993 almost 94 per cent of its utilizable groundwater for irrigation was used (Roy and Shah 2003). Declining water tables and quality problems like saline intrusion and pollution (fertilizer and pesticide use; see Ballabh 2005) are major threats to domestic uses of groundwater and to irrigation systems based on non-mechanized lifting devices. Other threats are land subsidence and damage to ecosystems (see Molden 2007).

Groundwater has long been regarded as a domain needing technical, hydrological and economic expertise only (Moench 1992; Vincent 2003). However, there is a growing awareness of the possible linkages between exploitation, ecological damage, poverty, and issues of access and equity. The main challenges are: reversing over-exploitation by supply and demand management rather than just resource development (Ballabh 2005; Molden 2007; Roy and Shah 2003); a reversal of macroeconomic incentives (Kumar *et al.* 2005; Singh 2002); and a shift of focus from merely higher-level legal regulation towards the development of locally accepted and enforceable norms and rules. As van Steenbergen and Shah argue, groundwater laws 'have not translated into anything that approaches real life' (2003: 255). Recent central government regulation in India (the 2005 Model Bill to Regulate and Control the Development and Management of Groundwater) still largely recognizes the basis of legitimate access in landownership (Cullet and Gupta 2009). Shah (2005) pleads for using the energy–irrigation nexus for demand management, using inter-basin transfers to recharge unconfined alluvial aquifers, and stimulating a mass-based recharge movement to revitalize groundwater socio-ecologies.

TANKS AND PONDS: PART OF AN ALTERNATIVE TO LARGE-SCALE SOLUTIONS?

The history of small-scale irrigated agriculture in India and Nepal, supported by local technologies for water harvesting, has often been treated as belonging to the past. This distorted picture of traditional irrigated agriculture is rightly criticized. A variety of local techniques to store rainfall and runoff have been documented (Sengupta 1985; Shankari and Shah 1993). In this text we focus on the contemporary use of tank and pond technologies. Important lessons can be learned from such technologies and associated management arrangements. This volume documents these technologies as 'social designs' (Shah 2003) in how their component elements have been constructed and modified, tying in with the changing nexus of social power and state policies, with important effects in changing access, rights, and local ecology.

Historically, in a bureaucratic environment increasingly opting for 'scientific irrigation' (Molle *et al.* 2009), and concerned to extract more revenue, rent and production, these technologies were discarded as unscientific and backward from the colonial period (in India) onwards (D'Souza 2006a; Sengupta 1985).[20] According to Barker and Molle (2005), a factor that has contributed to the marginalization of Asian FMIS in general is the dualistic history of irrigation development, which caused these systems to be ignored by donors and irrigation agencies. An important element of traditional water control that thus disappeared was its diversity and local adaptation, in which irrigation was seldom an end in itself but part of a much broader set of societal and livelihood concerns (Sengupta 1985).

India offers a variety of locally embedded water diversion, harvesting, lifting, transport and storage technologies (dams, tanks, ponds, wells, lifting devices, canals), with the monsoon rainfall patterns as a major determinant. With springs and other traditional harvesting systems, tanks, and ponds (*kulam*; Krishnan, this volume) are among the oldest and most important runoff surface water harvesting and storage technologies (Janakarajan 2004). Sengupta (1985) estimated the then tank-irrigated area at 4.5 million hectares.[21] Tanks and ponds are basically reservoirs for capture and storage of runoff water, enclosed by valley, catchment slopes, and constructed embankment or bund (Krishnan 2009; Shah 2003; van Oppen and Subba Rao 1980). Tank cascades, where the water resources and releases of one tank are connected by releases or overflow with others, are found in India and Sri Lanka (Sakthivadivel and Gomathinayagam 2006). Such systems are discussed by Manimohan in this volume.

Tanks and other traditional technologies mainly belong to the category of minor irrigation which, in view of serious management problems, received increasing government attention from the First Irrigation Commission (1901–3) onwards (Sengupta 1985). Nevertheless, variable support for tank irrigation and transfer between diverse public authorities in local government and irrigation departments has left tanks misunderstood and poorly managed (Agarwal and Narain 1997). The changes in tank systems under the influence of agrarian changes have been documented, but are not yet fully understood and therefore still debated (Mosse 1999). Janakarajan (2004), citing a study of 15 tanks in Tamil Nadu, mentions changes in landownership (from upper castes to cultivating castes) and the growing importance of groundwater (well irrigation, mechanized lift irrigation) in relation with the spread of the water-sensitive HYV technology as the most important factors. Sengupta (1985) mentions the destructive influence on tank irrigation of the expansion of canal irrigation systems and, even more, of the booming of groundwater irrigation. All have contributed to maintenance and rehabilitation problems for tanks (Sakthivadivel and Gomathinayagam 2006). Another social force has been the imposition of more standard designs for irrigation structures and water delivery, without attention to older design principles that reflected knowledge of local ecology and equity principles.

Mosse (1997b, 1999; see also Shah, Manimohan, this volume), in particular, is critical of simple story lines that assume a clear trajectory of degradation, decline and collapse of autonomous traditional management that can be explained in dichotomized frameworks (for example, traditional/modern; community/state). Often, such explanations reflect governmental and bureaucratic organizational assumptions, needs and expediencies rather than actual processes of change. According to Mosse, tank systems 'have, in fact, been interpreted as being in a state of decline, neglect and disrepair wherever they have been described' (1999: 307). The time perspective from which 'decline' was observed largely determined the explanations developed. Thus, narratives of decline are ideological statements rather than historical accounts. Such tank ideologies, in their turn, legitimize new regimes for tank development or restoration. Some of these efforts for reviving tanks (Palanisami 2000) form a backdrop to the studies in this volume.

Three authors deal with changes around tank and pond irrigation. Esha Shah contributes to debates about tank irrigation by exploring state–society relationships and their development in a context of agrarian change, focusing on the emergence of farmers' movements and 'new agrarianism'. Taking a 'social shaping of technology' perspective she uses this analysis in her explanation of changes in the technological designs of tanks. Krishnan explores processes of ecological degradation of water sources focusing on ponds. She links these changes to broader socio-economic and institutional transformations, especially of property rights regimes for land, water, and forests. Manimohan analyses how both agrarian and ecological change have shaped the evolution and maintenance of infrastructure in tank cascades in a 'wet' and a 'dry' zone.

There is growing awareness of the shortcomings of current supply-oriented water resources development through large-scale infrastructural works like dams, storage reservoirs, and river linkage schemes (D'Souza, D. 2002; D'Souza, R. 2008). Though these have been fiercely criticized for their ecological and social impact, the continued policy focus on such 'solutions' hampers a critical appreciation of their effects (Iyer 2008; Singh 2002). In the highly polarized debates about water infrastructure, the mega-projects of the state's hydraulic mission are increasingly pitted against the solutions of those who plead for a return to an often idealized 'traditional' community and its local knowledge, technologies, and management practices

(Shah 2003; Singh 2002). However, as we have seen, dichotomies like 'traditional' versus 'modern' are not very useful (Mosse 1997b). Several authors have rightly stressed the need for more balanced approaches that explore the options and possibilities of local technologies and management practices without excluding the possibility that interventions on a larger scale may sometimes be needed (Gulati *et al.* 2005; Iyer 2008).

THE CHAPTERS IN THIS VOLUME

After this introduction, in Chapter 1 Linden Vincent and Dik Roth present a conceptual–theoretical framework for analysing water control. The chapter discusses two issues relevant to the study of (mis-) matches of technology and institutions. The first is interdisciplinarity and its integrating concept, water control. These have shaped the socio-technical approach taken in all chapters. It argues for an approach to technology as both a material and a social construct that can be used to study creation, operation and governance of irrigation systems. This chapter also provides a critical perspective on 'institutional thinking' that increasingly shapes policies on water technologies and reforms, with a focus on the role of property rights to land, water, and infrastructure.

In Chapter 2, Vishal Narain discusses options for institutional reforms in the warabandi and *shejpali* systems of water allocation in large-scale irrigation systems in northwest and western India. Irrigation reforms tend to pay scant attention to water allocation, canal technology, and their interrelationships. The author shows that the potential of such reforms is crucially shaped by these factors. Warabandi and shejpali represent different forms of water allocation, corresponding to different technologies or designs. The chapter explores their varying potential for market creation, redefinition of water rights and decentralization. Though often presented as alternative approaches to reform, they are rarely considered together or in relation to each other. Narain makes a case for mainstreaming considerations of technology in proposals for institutional reform in irrigation.

In Chapter 3, Bala Raju Nikku discusses irrigation reforms in Andhra Pradesh (AP), India. These reforms, known for their 'big bang' approach, have attracted much attention. The 'Andhra Model' is based on the neoliberal belief that scaling down the irrigation bureaucracy, increasing water charges, and transferring maintenance and water distribution to elected WUAs can effectively address irrigation management problems and lead to growth of the sector. The author critically analyses the workings of this model of IMT and PIM. Using a 'politics of policy' framework, he provides an alternative perspective on such reforms. He shows that the process was captured by the irrigation bureaucracy, representatives of WUAs, and political leaders, among others.

In Chapter 4, Puspa Raj Khanal discusses IMT–PIM policies in Nepal and the use of participatory technology development (PTD) to transform irrigation management. Using evidence from three case studies of reforms in systems of different scales and complexity, the author shows the changing dynamics in and between technology and institutions. A major weakness of the reforms is that they do not provide enough incentives for improved management performance. Khanal argues that future programmes should build on, but move beyond, contemporary models of PIM and PTD, to bring a new realism in attempts to reform technology and institutions together. The author proposes an alternative 'irrigation modernization' approach that encompasses local performance priorities, to achieve service-oriented management in accordance with the resource base of the irrigation systems.

In Chapter 5, Umesh Nath Parajuli focuses on technologies and their links with management in FMIS in Nepal. Infrastructure design in FMIS is strongly influenced by local environmental conditions and management needs. As these tend to be dynamic, infrastructure has a strong adaptive capacity. Using case studies of three FMIS, the author documents a variety of principles that guide design. The viability of various types of water distribution systems and division structures differs in any given area. Choice is largely dictated by hydrology and flow hydraulics, agro-ecological conditions, and locally existing institutions, which in turn shape irrigation management functions. The author suggests that recognition of these linkages in the design of hill irrigation systems can enhance system performance and improve the livelihoods of local communities.

In Chapter 6, Suman Rimal Gautam discusses institutional changes around groundwater technology in the Tinau river basin in the Nepalese Terai. Interventions to promote deep and STWs did not take into account existing water use practices, nor social conditions and institutional arrangements around them. The study brings to light the dynamic character of the management institutions that emerge in the interaction of canal and pump technologies and use of different water sources. Though irrigation policy stresses 'conjunctive management', interventions in groundwater and surface water tend to be completely isolated from each other. The author argues that institutions evolve from these complex interactions of technologies and sources of water. To understand changing institutions, resource use, and technological performance it is necessary to look beyond the formal objectives and arrangements of interventions, and to focus on local action processes.

In Chapter 7, Anjal Prakash takes the reader to water-scarce Gujarat, India, where groundwater markets have been developing in an agricultural area with a rapidly declining groundwater table. The author investigates the factors shaping unrestrained groundwater use through DTWs, and the responses of various social groups. Focusing on the politics of groundwater markets and its interrelation with social differentiation and class–caste relations, he analyses issues of access and control over land and groundwater, laying bare the role of the dominant classes in determining access to water. In this context of increasingly constrained groundwater access, groundwater markets and sharecropping arrangements crucially mediate changes in agrarian relations.

In Chapter 8, Esha Shah analyses transformations of tank design and management in Karnataka, India. Criticizing assumptions about the superiority of community management, the author argues that tank technology reproduces and reinforces agrarian power relations. Technical artefacts articulate with social tensions, relations of power and ideologies, and are thus transformed historically. Tank technology and policies have changed as a result of the mutually transforming roles played by the state and society. The village community fabric changed with the rise of farmers' politics, heralding a shift in tank technology, use and management. Shifts in designs have perpetuated and reproduced existing social relations of power, or created new ones. The author argues that those who approach artefacts and knowledge systems as objects of values and virtues fail to capture their social and political scripting, thus denying technology a historical perspective.

In Chapter 9, Jyothi Krishnan discusses access to water in a paddy growing area in Kerala, India. She describes how the existing property rights regime around water disregards the ecological properties of water and the integrated nature of land and water use, creating unequal access and unsustainable use. In the use of energized water lifting devices, little attention is paid to the interrelationships between surface and groundwater. This results in unsustainable and inequitable use and management of tanks and tube wells. The author argues that the ecological characteristics of land and water should become an inherent component of property rights. Existing regimes should be modified to take into account the integrated nature of land and water, and the hydrological interconnections between components of the water cycle. Further, technologies should be subject to ecological considerations that prioritize sustainable and equitable use of water.

In Chapter 10, R. Manimohan analyses struggles in tank cascade landscapes in south India. Changing land use has caused environmental problems like soil erosion and sedimentation. These have transformed the social and material character of the agrarian landscape. The author argues that soil erosion and sedimentation are geographical expressions of political economy relations, implying new forms of social differentiation in access to and control over land and water. Two distinct landscapes—wet and dry zones—have emerged, characterizing different patterns of water management and agrarian change. While subsistence farming characterizes the dry zone tank cascade, the wet zone moves towards commercial farming. These characteristics shape water rights in a manner that suits the requirements of the new agrarian regime. Such changes have also resulted in a shift in the design of irrigation artefacts in tank cascades.

In Chapter 11, Amreeta Regmi and Linden Vincent analyse micro-hydel designs developed within irrigation canal systems in Nepal. They study two systems with contrasting design approaches and outcomes in evolution of power supply and organizational management. The authors develop several concepts to analyse technology and institutions, and their co-evolution. Technological systems need to evolve from hydraulic assemblages into transformative units capable of evolving with changing power and water supply demands of a village. Good performance across technical, political, and financial accountability is important for the local management system. Constitutional accountability is also vital. In this respect, management of systems combining micro-hydel and irrigation is not different from management of irrigation systems alone.

In Chapter 12, Peter P. Mollinga moves debates about interdisciplinarity in irrigation studies forward by suggesting starting points for a more comprehensive interdisciplinary social theory on irrigation. Mollinga reviews the boundary concepts that have emerged in interdisciplinary irrigation studies. The focus is on concepts that capture the hybridity of irrigation systems as complex systems, and cross-disciplinary boundaries. The author discusses concepts capturing the materialization of rights, design-management relations, and the social construction of technology, landesque capital, and ecosystem goods and services (EG&S). Finally, the broader issues of space–time relations and cultural politics of water are explored. On the side of formal theory, a focus on a combination of the concept of hydro-social cycle with structure–agency theorization as morphogenesis is proposed. On the side of substantive theory, three avenues for investigating the materiality of the social process of irrigation are proposed: the commodity form of water, a materialist institutionalism, and the embodiment of agency. The chapter concludes by listing five domains for further research activities.

Notes

[1] Narrow views of irrigation management stress the physical tasks of water delivery. Broader definitions focus on multiple activities in system maintenance, governance, and preservation (Coward 1979; Uphoff 1986). The risk of these definitions is that the politics of institutions and rule-making are not addressed. Ostrom (1992) distinguishes between management and governance to differentiate between the regular tasks of ensuring water delivery and the making of rules about rights and responsibilities. We keep a broad perspective of management including governance, to keep understanding technology and institutions together, and to recognize the role of politics around irrigation artefacts.

[2] 'Dark zones' are areas characterized by a high rate of groundwater development (85–90 per cent).

[3] We do not idealize or essentialize the meaning of terms like 'traditional' or 'indigenous'.

[4] Meinzen-Dick and Pradhan (2005) discern four major mechanisms for water reallocation: administrative, market-based, collective negotiation, and 'other means' (for example, force, theft).

[5] Dam building increased from the 1950s, peaked between the 1960s and 1980s, and declined from the mid-1980s due to low grain prices, rising construction and land development costs, and growing environmental opposition (Barker and Molle 2005).

[6] Many irrigation systems based on multipurpose water storage not only have a command area but also an 'obey area', where local people suffer the damage caused by dam building (Sengupta 1985).

[7] For Nepal, see Gyawali (2001).

[8] Socio-ecologies describe the interactions and influences between ecological factors, and social structure and organization. Typologies have been developed for groundwater in South Asia in particular, to describe cycles of resource development, trends in self-regulation of resource use, and livelihood/production systems (see Shah 2005, 2009).

[9] For example, birta and jagir lands in Nepal: grants to individuals, empowering them to control land and develop water resources by labour mobilization, and facilitating revenue collection (Pradhan et al. 2000; Khanal, this volume).

[10] See Dahal (1997) for their emergence in Nepal.

[11] Outlets are important structures, where water moves from minor canals into watercourses taking water to farmers' fields, and management responsibility often moves from the public agency to the farmers (see Narain, this volume). For design, technology choices and farmer–agency interactions, see Bolding et al. 1995; Diemer and Slabbers 1992.

[12] In India, major or large-scale systems have a culturable command area (CCA) greater than 10,000 hectares. Medium schemes are between 2,000 to 10,000 hectares, and minor or small-scale systems are under 2,000 hectares.

[13] Molden (2007: 395) mentions an increase in groundwater use from 100–150 cubic kilometres in 1950 to 950–1,000 cubic kilometres in 2000, and a contribution to irrigated area growing from some 30 million hectares in the 1950s to between 70 and 100 million hectares in 2000. Shah (2005: 18) estimates the area depending on groundwater irrigation at 85–95 million hectares.

[14] The area in India under irrigation in 2007 was estimated at around 26.5 million hectares. Its share of groundwater irrigation in the global total irrigated area is 38.6 per cent. Estimated groundwater withdrawal in India would total around 250 cubic kilometres per year by 2010 (Molden 2007).

[15] Tube well irrigation grew from 0.6 million hectares in 1960–1 to 30.6 million hectares in 1987–8. It delivers water to more than half of the net irrigated area (Singh 2002).

[16] According to Kansakar, around 800,000 STWs are operated for this purpose in the Terai. Groundwater development for irrigation lagged behind,

with 206,000 hectares (out of a potential of 612,000 hectares) being irrigated through around 60,000 STWs and 1,050 DTWs (Kansakar 2005: 96).

[17] This evidence concerns both regional (state) level and micro-level (see Moench *et al.* 2003).

[18] There are important hydrological linkages—refill through seepage from canal systems—between canal irrigation and groundwater irrigation (Dhawan 1993; see Barker and Molle 2005). Ballabh (2005) also notes the flipside of this linkage: waterlogging and salinity.

[19] See Note 2.

[20] Singh concludes for India that 'a mature irrigation science existed in India, before the advent of the British colonizers' (2002: 34). See also Sengupta (1985).

[21] Janakarajan (2004) gives an estimation of 200,000–250,000 tanks (for India) with a total water spread area of 3 million hectares and command area of 4.5 million hectares.

1 ANALYSING WATER CONTROL: INTERDISCIPLINARITY, SOCIO-TECHNICAL APPROACH, AND INSTITUTIONS IN WATER MANAGEMENT

Linden Vincent and *Dik Roth*

The introduction to this volume has shown that the main challenges in irrigation are not primarily technical, but involve complex relationships between technical, socio-political, and institutional factors and processes. However, policies dealing with these problems often reduce such complex processes and relationships to simple routines of technical intervention combined with the use of socio-legal engineering instruments, presented as replicable and context-independent (Molle 2008), politically neutral, and uncontested. While such policy biases continue to be reproduced by mainstream domains of knowledge and expertise and established bureaucratic routines, the crucial linkages and relationships in transformations around water resources are not taken into account. In this chapter we provide an interdisciplinary framework for analysing water control that fills such gaps.

We first explore the socio-technical approach, used by the authors in this volume, as an interdisciplinary approach to water provision and use. The first part of the chapter explores this interdisciplinary approach from a focus around technology and reviews its integrating concept: water control. It focuses on two embodiments of technology in society. The first, and easiest to recognize, is technology as physical works and artefacts like dams, canals, and gates. Though clearly material objects, they are also social objects in how they have been constituted, are used, and have effects on society. The second embodiment is that of human knowledge and skills. These must also be understood as social constructs used for the creation, operation, and governance of irrigation systems. Often, different knowledge systems—for example, of engineers and farmers—may be involved. These issues are dealt with in the second section of this chapter.

A second set of discussions, presented in the third section, deals with institutions, specifically focusing on property rights to land, water, and physical infrastructure. This section aims to provide a critical perspective on the kind of 'institutional thinking' that increasingly shapes water policies. We first discuss how institutions are often approached in resources management, and then focus on the important role of property institutions in water governance and management. We argue that attention to legal-institutional complexity and property rights is crucial for understanding and changing water control realities. In the conclusion we return to the value of interdisciplinary approaches for understanding contemporary and future water management developments.

INTERDISCIPLINARITY: ITS NECESSITY, DYNAMISM, AND THE SOCIO-TECHNICAL APPROACH

Irrigated agriculture involves various institutions and social relationships around use and command of land, water, labour, and inputs. Water use has a dimension of infrastructure and tools, and of techniques and knowledge to use them. Society designs, develops, operates, manages, and maintains irrigation systems. Irrigation systems have been studied in many ways, to explore one or more of these dimensions (Vincent 1997). The interdisciplinary socio-technical approach aims to understand why and how systems are designed and operated in certain ways under particular spatial, institutional, and political configurations and choices. It can show how and why these processes and artefacts may be amenable to change, and why people—individuals, groups of farmers, state and non-state organizations—intervene in, contest, or struggle to reform irrigation and care for irrigation systems that have meaning for them, for reasons extending beyond the purely economic dimension of production. Irrigation systems are not only physical-technical settings in which water management and crop production evolve. They are also social settings and arenas of struggle, where technology influences access, water and land rights, and management arrangements. Diverse actors with different knowledge and strategies for obtaining and using water meet there. Emerging in the mid-1980s (Kloezen and Mollinga 1992; Uphoff 1986), this approach has evolved to research different problem contexts. These have included learning from farmer management (Coward and Levine 1987), understanding irrigation as a 'social force' transforming production relations, resource access, and system management (Eggink and Ubels 1984), challenging the assumed neutrality of irrigation and scientific knowledge (Diemer and Slabbers 1992), and, the interest of this volume, the dynamics between technology and institutions.

The need for interdisciplinarity in research comes from the critical importance of understanding how society controls and uses nature through the mediation of technology and institutions, requiring understanding of multiple linkages and dynamics. There are many interdisciplinary research approaches to irrigation and water management with different emphases and models of the interaction of social, agro-ecological, and physical systems.[1] However, the socio-technical approach brings a focus on, and a better understanding of, important local relationships and interactions. These include: the analysis of the interrelations of technology and management conditions and needs, and technology and property relations; design paradigms and processes; communication and interaction between irrigators, engineers, and administrators; and adaptation and impact of irrigation technology. The approach is relevant for seeing how broader policy reforms in land and water impact water delivery and access, and how concerns of scarcity, sustainability, equity, and justice are addressed (Vincent 1997, 2001).

The socio-technical approach enables the integrated analysis of the social construction of irrigation technology in its design choices (selection of artefacts and components, and the output and benefit sharing it enables), social conditions of use (operation and management), and social effects (outputs and transformations of social relations, and impact on people and habitats). The focus of this volume is on how water supply of and access to water are mediated by technology and institutions operating at various levels. This core relationship is shaped by wider forces of state and non-state institutions, agrarian structure, and agro-ecology, including the shaping effects of landscape and other technologies, and practices of production and resource management (Mollinga 2003).

WATER CONTROL

The interdisciplinary concept of water control has been developing since the mid-1990s (Bolding *et al.* 1995; Mollinga 2003), with concerns not only for a new understanding of irrigation realities to inform social action, but also to develop more social theory around the study of materiality that helps wider reflection on technology and development. It has drawn upon social theories of practice, actor-oriented sociology, concepts of agency, and actor-network theory. These analyses developed out of an early concern to better understand water distribution and system performance through the interactions of irrigators and engineers, and to understand how crops, water, and people interact, using concepts like actor, practice, and knowledge (Diemer and Huibers 1996; Diemer and Slabbers 1992). A search for understanding new water development questions and policies shaping social transformations was also prominent. Some of these efforts developed from older critiques of irrigation, notably about the behaviour of irrigation bureaucracies and inappropriate technology policies in diverse water resource contexts (Ambler 1994; Bottrall 1978).

The concept of water control draws on two complementary conceptualizations of technology. The first is technology as a mediation between society and nature, highlighting how social choices, capabilities, and practices shape irrigation systems. It shows how society shapes technology in its capacity to intervene in hydrology and ecosystems. We can also understand technology design and outcomes as a social process in which different actors interact continuously to shape technology outcomes. The second is technology as a capacity to transform goods into desired things, manifested as material objects, and in the invention and employment of artefacts that move water and irrigate crops. Focusing on the design of an artefact, one can study its requirements of use, how its form is socially constructed, and how it has social effects (Khanal 2003: 12; Bolding *et al.* 1995).

A focus on water control allows attention to the movement and use of water, while also exploring a variety of relations and transformations. To do this, water control is analysed for different but interrelated domains of interaction where different forms of control are contested: hydraulic or technical water control (design and operation of infrastructure and water delivery); organizational control (the organizational forms interacting to get water delivered and crops produced); and social and political control (the social and political relations that structure irrigation) (Bolding *et al.* 1995; Mollinga 2003). Some authors distinguish socio-legal control, especially in situations of legal pluralism (discussed ahead), with interacting local, national, and global norms; and cultural-metaphysical control, which takes into account alternative cosmo-visions (Boelens and Davila 1998).

Broadening the scope to ecology and landscape control, this approach can enable a focus on socio-technical regimes shaping agro-ecology in land and watershed management practices and landscape use linked with irrigation. It can analyse mechanisms developed by irrigators to achieve desired water distribution, given their knowledge of and control over ecological and social dynamics (see chapters by Gautam, Khanal, Parajuli, this volume). This enables studies of opportunities and dilemmas that arise with changing conditions of water technology and institutions (Vincent 1997: 13). The socio-technical approach can thus interact with the growing field of study of socio-ecology in South Asia (see Shah, this volume). This can give in-depth insights into struggles over collective and private water rights under changing technologies, evolving governance over land and water, and continued adaptive design and operational sustainability. The chapters by Shah, Prakash, and Krishnan in this volume show how irrigation technology reshapes rights, access, and control over water. Command over water becomes a factor in the changing power base of agrarian relations between farmers, and between farmers and the state.

The chapters that follow take up more specific questions about water control, and introduce complementary concepts to inform the research. Thus we can look at technologies in multiple ways. Shah (this volume) discusses artefacts as delegated values, ethics, and property rights embodying a technical code for their functioning. Design is not only an application of knowledge, but also a social ordering (discipline, moral obedience), an invisible yet embedded requirement to make an abstract system function. When we locate technologies in real geographical and social space, we start to see the physical settings in which people must operate and maintain a system, and the social settings where people govern and allocate resources. This has to be done while integrating water management in the wider governance system, cooperating to make systems run, utilizing knowledge for design, renovation and modernization of systems, and contesting the distribution of production and benefits.

EXPLORING WATER CONTROL AND THE MATCH OF TECHNOLOGY AND INSTITUTIONS

This interdisciplinary approach recognizes that the complexities of water management and frameworks of analysis change with time. Thus, while we present the concept of water control and the socio-technical framework used in our analyses, any book also delimits its scope of work in time and space. In this section, we briefly outline the key relationships and questions explored in this volume to increase understanding of contemporary irrigation and water management questions in South Asia, looking at technology from 'the social' around it.

WATER CONTROL AND PHYSICAL SETTINGS: DESIGN-MANAGEMENT INTERACTIONS

Recognition of the physical setting created by irrigation technologies is an important theme of this volume, emphasizing their social construction and the diversity of situations. There is an older history of debate about the understanding of physical settings and how they operate, and the 'mismatch' of new technological and institutional interventions (Bottrall 1992; Coward and Levine 1987; Diemer and Slabbers 1992; IIMI/WECS 1987). These debates yielded a new understanding of the hydraulic levels and domains of irrigation systems, where gates and outlet structures were important points of interaction and contestation between water users and agency staff supplying water. They showed how artefacts shaped water supply and were socially constructed (Bolding *et al.* 1995). They also showed the disruption caused by government control of organization without clear thinking about the division of rights and responsibilities between governments, communities, and individuals, or about appropriate choices of water control (Bottrall 1992; Coward and Levine 1987).

This volume continues this debate, adding an exploration of the effects of new policies and problems in these settings. An important new focus is understanding the effects of irrigation management transfer (IMT) policies that have sought to delegate responsibilities to new local organizations determined by physical settings. Renovating and modernizing technologies can be part of these policies. Further, policies of participatory technology development (PTD) and participatory irrigation management (PIM) claim to be based on farmers' preferences. This volume documents local dynamics that are reshaping these physical settings, particularly the effects of agrarian change in irrigated areas and transforming political control. It also investigates the effects of resource scarcities triggered by uncontrolled development of resources and declining land and water productivity.

The critical influence of physical settings is most obvious in surface canal irrigation systems, especially large-scale systems. Narain (this volume) approaches irrigation systems as socio-technical systems, whose physical layout corresponds to certain preferred infrastructure, property rights, and organizational designs, determining options and constraints for institutional transformation. Technical configurations shape the service characteristics of water delivery, as in the 'protective irrigation' concept. Narain documents the *warabandi* and *shejpali* systems, the management of which is affected by decentralized management and farmers' efforts to gain more water for productive rather than protective irrigation. He argues that reforms can only be successful if they are realistically adapted to existing technologies and institutions, and not based on blueprint model design.

Khanal approaches this issue through the PIM and PTD perspectives. He shows how serious efforts to develop participatory, decentralized management of the large-scale West Gandak system according to a 'command area development' model in fact promoted a system still unable to pay for itself and cope with its complex and difficult water management problems. Later experiences in PTD enabled the users and representatives of other systems to decide on technological improvements within their management capabilities. They were also better able to create a sustainable management structure that could coordinate delivery across physical settings and was 'affordable' in terms of time and money.

Looking further into farmer-managed irrigation system (FMIS), the chapters by Parajuli, Khanal, and Regmi and Vincent document the diverse technologies found and their developments in response to social and environmental conditions and transformations. Parajuli emphasizes the need to take into account agro-ecological and socio-cultural contexts to understand technology outcomes. The proportioning weir, for instance, cannot be introduced everywhere. In some locations very simple technologies are used. Whole systems may have field deliveries made by one functionary only, if this is best for the agro-ecology and accepted by farmers.

Another theme is 'technological blindness' as a source of mismatch. This may cause inappropriate technology development or unsustainable development of resources. Gautam shows how farmers in the Terai of Nepal develop various water sources conjunctively. They shift between surface and groundwater technology options, depending on the costs and options for linking up to existing institutional arrangements, and reliability of water offered. However, state organizations brought in new technologies without paying attention to other water sources. Dialogue between agencies and farmers about resources and technology in use were lacking, leading to the underutilization of group borewell investments.

Krishnan's analysis shows the disruptive effects of new canal development on older systems of collective tanks and ponds. It has changed hydrology, access, and property rights, and triggered private groundwater development. Current land property categories create an inadequate understanding of the ecological properties of resources. Thus, land use leads to ecologically unsustainable practices. Expansion of private property regimes at the expense of common property promotes over-exploitation and creates inequalities. The resources under state control are reproducing unequal and unsustainable patterns of groundwater utilization.

Our understanding of transformations of physical settings under public intervention and agrarian change also needs to extend down to the field level. In their studies of irrigation tanks and access to land and water, both Manimohan's and Shah's chapters in this volume analyse tanks as sites of interaction between different social groups and the state, and different knowledge traditions. Diverse forces of agrarian change, the shift into transplanted paddy, and the imposition of standardized design interventions have brought into the tank systems risks of greater inequality and exclusion. Shah documents how these have driven changes in the infrastructure of water control in the tank bund (for example, disappearance of the plug and pole control in the sluice) and in the layout of irrigation canals in the command area. Older principles of water allocation disappear, and there is no longer a service to the whole command area. Manimohan analyses the socio-technical landscape of tank cascades to show the effects of deforestation, erosion, and sedimentation. These changing environmental dynamics also reshape the strategies of tank users to access water from tanks or groundwater, and choices of production.

For groundwater use in Gujarat, Prakash reminds us to keep a broad understanding of how a tube well shapes productive capacity and ability to generate surplus income. This surplus (and access to it) is affected not only by falling water tables. There is also the critical issue of management of yield risk from declining soil fertility. He reminds us that we need to think about the new 'politics of decline' operating in irrigated agriculture and responses to this, one of which is the exit strategy of smallholder irrigators.

WATER CONTROL AND KNOWLEDGE:
MANAGEMENT TASKS AND SKILLS

The question of what management tasks are needed and by whom they are to be performed to make technology function has been an important research topic. Large-scale irrigation is still widely considered to require either full control by a public agency or joint management between an engineer-based agency to control major operations and local farmers' organizations at the outlet or tertiary levels. In this volume, Nikku refers to the works of Coward and Uphoff concerning the range of management tasks in irrigation and organization of rights and obligations (Coward 1979; Uphoff 1986). The chapters on FMIS in Nepal use the concept of 'hydraulic property', conveying that water rights are related to involvement in the creation and maintenance of hydraulic infrastructure (Coward 1986a). Shah, Parajuli, and Manimohan show how technology itself embeds institutions and management needs. Irrigation artefacts can ensure storage, control, and division of water supply according to socially agreed upon rules; in the case of the proportioning weir without needing a human operator.

The chapters develop a new concern regarding the structure of irrigation management, especially in view of reforms and widening interests in approaches integrating land, water, and forest management. Focusing on management reforms in large systems, Nikku and Narain bring in new frameworks that recognize higher state levels and their influence on irrigation management. In studying policy as process, Nikku illustrates how irrigation realities are shaped by four levels of power and political action. Narain shows how new participatory approaches actually reflect global discourses about what objectives should drive the design of new management institutions rather than what tasks the organizations could realistically undertake. Ideas dominating new management reforms were often promoted separately and without reference to local contexts. Nikku shows how new policy formulations to transform local irrigation management under new water users' associations (WUAs) were weakened by a failure to base these plans on realistic expectations of the behaviour of key stakeholders. Policy and field domains were often presented as separate and undifferentiated, and change as unidirectional. However, state-local interactions were characterized by blurred boundaries and struggles between various actors using their agency to transform, interpret, selectively embrace, or resist the policies. This shows the importance of understanding politics when thinking about matching technology and institutions. This requires an understanding of power and control, to analyse how actor strategies influence irrigation technologies and institutions. Khanal describes this shift in analysis as moving from functional to political models for WUAs, seeing these as two different approaches in thinking about organizational design. Understanding politics is important for thinking realistically about new governance options, which also depend on possibilities to build local social and political legitimacy. He critically analyses the structure of new management organizations and how these can be encouraged to evolve into strong and capable organizations.

Another new focus in this management domain of technology is the interest in management performance, both in terms of water delivery and performance of management tasks. There has been considerable expert discussion of the performance indicators for monitoring and comparing water deliveries, production output, and financial management (Molden *et al.* 1998). Instead, the authors in this volume focus on the performance criteria that local farmers and organizations prioritize, and how they try to achieve these. Regmi and Vincent use a framework of technical, political, and financial accountability to determine whether organizations prove capable or not to evolve as sustainable management organizations and to steer the evolution of new supply systems into transformative units capable of meeting changing needs. Manimohan looks into resources management in the tank catchment, in particular, management transformations and loss of collective institutional controls in managing land and forest that shape catchment runoff and siltation of tanks. He is also concerned about the confusion about responsibility for land and water management, which leaves catchments with very weak control of adverse changes in vegetation and soil management that also affect water resources.

WATER CONTROL AND AGENCY: SOCIAL POWER AND SOCIAL DESIGNS

In studying the relations between technology and institutions, we have to address the issue of social power: how it controls choices of technology and access to its benefits. Once again, we can see an older debate. Irrigation is not neutral; it will reach the richer farmers and can have negative effects on the poorest, unless there is concerted social action to counter this effect. The chapters in this volume continue to highlight these risks and the dynamics that shape this match or mismatch of technology and institutions.

Several chapters show a concern with issues of agency—not only in how state agencies maintain or shift capabilities to design, operate, and manage systems and resources, but in new governance forms in and between state and locality and in transformations of agrarian elites. Since 1985, shifts and transformations in political authority and legitimacy (Cooper 1998) at state, local, and international levels have brought new struggles into the management of technology and resources, with governance as a burgeoning topic from the 1990s.

The chapters not only report on the agencies that have shaped and maintained power in the large-scale transformations of regional hydrology in the name of development (see those by Narain, Nikku, and Khanal). They also explore the more conventional state programmes of technological intervention (those by Gautam, and Regmi and Vincent) and the new networks of agrarian power controlling local resources (those by Shah and Manimohan). We also see how local groups experiment with new technological and institutional practices (see chapters by Khanal, Gautam, and Manimohan). These chapters focus on the transforming agency of diverse collectivities: states, techno-scientific agencies, and community-based water organizations. This gives new insights in the social values and norms that keep technology and institutions together and functioning well, such as accountability, legitimacy, and representation.

Gautam highlights the dynamism of farmers in finding best technologies to access resources, both in selecting between well technologies and opting in and out of canal systems and conjunctive use. They use subsidy programmes for small farmers if available, but also shop around and collaborate to use pump equipment and other technology or water sources. Failure to understand farmer preferences and to make programmes accessible, and the rejection of new technologies are the causes of mismatch here.

Shah reflects upon the impact of 'new agrarianism' in shifting control over irrigation technology and generating new institutions that favour owner-cultivators. Changed relations between farmers and the state have influenced policies relating to production (for example, rice prices) and rehabilitation responsibilities, which have been devolved to the local level. New agrarian-based organizations have heavily mobilized owner-cultivators for support. These movements are actively driving agrarian change in a particular direction by making specific productive and knowledge choices, reshaping rice cultivation and water delivery. Thus, control over tank renovation and repair, and hence over their resources, shifts towards new dominant groups.

Study of the relation between theory, design, and action is also relevant. Several authors raise a debate on design, using its broad definition by Papanek (1985) as the imposition of meaningful order. They ask whose order determines the evolution of technical systems and management structures, and the transformation of whole ecosystems and social relations of production. This debate explores the performance criteria developed by farmers in their everyday worlds, rather than the norms of efficiency and productivity driving external interventions. Many reform programmes are linked with efforts to introduce 'modern' technical water control. These favour increased flexibility of water supply that can be better tuned to crop water demands and new systems for water pricing and charging, but often bring more complex technologies of water instead. Both Shah and Narain point out that, culturally, there is an alternative rationality to this economization of irrigation management. However, traditional technologies have largely failed to trigger state interest. Shah and Parajuli show the richness of design concepts in tanks and FMIS. However, these are hard to study: local knowledge is often implicit and tacit, embedded, and practical. Most public schemes, on the other hand, are based on more abstract scientifically derived knowledge, which is assumed to be generally applicable, but often does not produce better and more effectively managed systems.

Design also involves an understanding of communication procedures between actors with responsibilities and powers that make management more effective. Narain documents the changing procedures that gave local irrigation organizations in Maharashtra powers and responsibilities to request water for their farmers, and direction on how public agencies must supply this. These have led to improvements in water supply and organizational performance. The new evolution of joint irrigation management under IMT policies has to be followed by a more sensitive bottom-up design approach. Unfortunately, blueprint approaches to technology rehabilitation and organizational development under irrigation reform are a cause of mismatch and poor performance if local organizations cannot evolve or have no need for these particular models.

Khanal's analysis of PTD approaches shows that these demand sustained action and understanding, beyond superficial contacts between engineers and irrigators. He reviews the diverse levels and networks of contacts that committed engineers and farmer representatives need to work through to really build joint under-standing for new designs, and discusses the range of initiatives and commitments in negotiation and political representation this may involve. This goes far beyond a simple system 'walk-through' by farmers and engineers. Khanal tries to move the debate on PTD into a more sustained commitment to 'service-oriented water control'.

Not biased towards any particular technology, this approach tries to find technologies that match with local organizational capacity and management objectives.

Regmi and Vincent voice similar concerns over micro-hydel systems introduced alongside irrigation systems in Nepal. They give an example of a village that had to accept a preferred prototype of the designer rather than make its own choice. In such cases, systems remain a 'hydraulic ensemble' of parts rather than transforming as 'evolutionary systems' able to respond to changing power demands on them and to build the capabilities of their management organizations. They introduce the concept of 'technological democracy' to describe this ability to chose and support a technology that can evolve to support local needs, as a counterbalance to interventions that show authoritarianism in design. They propose that agencies promote constitutional accountability as well as technical, political, and financial accountability in new systems. Agencies promoting interventions can take responsibilities in building longer-term support for their designs, beyond just the 'turnkey' effects of installing new devices and prototype institutions.

These chapters show that changing agrarian and political relations have reshaped technology and institutions in their core artefacts—who manages irrigation and how, and who determines access to water and benefits. Mismatches become visible in how systems are operated and who gains access to its outputs. The promotion of new technologies and designs leads to the take-up of new resource exploitation opportunities. Nevertheless, these have not yet created socially and ecologically sustainable resource development and management. Therefore, the integrative design of irrigation systems within the broader agro-ecology and water cycle remains a challenge for the future, and the study of agency and power in transforming water control remains vital.

INSTITUTIONS IN IRRIGATION MANAGEMENT: PERSPECTIVES FROM INSTITUTIONAL AND LEGAL REALITIES

CHALLENGING MAINSTREAM APPROACHES TO INSTITUTIONS

Institutions are crucial in the governance and management of water resources. Many misfits between people, resources, and technology are institutional rather than technical, and, therefore, possible solutions will often be institutional as well (Merrey *et al*. 2007). Attention to the institutional dimensions of irrigated agriculture is a badly needed step away from technocratic water-supply-based policies. However, apart from this broad and superficial consensus, what remains are widely divergent views of institutions and the role they can play. This is not surprising: ways of using such concepts are often guided by broader policy or political agendas and objectives of the users (Bardhan and Ray 2008). What are institutions, then, how do we deal with them in our analysis of water control, and which pitfalls to avoid? This section deals with these questions, paying special attention to the domain of law and property rights.

As a point of departure for the analysis of institutions in water control, we need a broad definition of the institutional domain that does not contain implicit normative biases, and facilitates the analysis of institutional complexity. Several authors on institutions stress the regulation and control of human behaviour as a key element. Leach *et al*. define institutions as 'regularized patterns of behaviour between individuals and groups in society' (1999: 225). Scott (1995) analyses institutions as constituted by a complex of normative, cognitive, and regulative dimensions. Jentoft (2004) mentions two important elements: their durability beyond the lives of individuals, and the moral force in institutions with values forming their 'glue' (which does not mean, however, that institutions are free from contestation and conflict). Institutions may be more or less purposive, characterized by single or multiple purposes or functions, either explicitly or implicitly, as interpreted from their day-to-day working in society.

However, mainstream approaches to institutions in water policy tend to be more instrumental and based on a narrower conceptualization of the social dynamics that form the core of institutional processes. We find several aspects of the ways in which the role of institutions and processes of institutional change are often dealt with in policymaking quite problematic. They seem to hamper the kind of analysis proposed in this volume for the complex relationships between institutions and technologies in irrigation. What is required is a rethinking of the institutional domain and institutional processes. By their empirical richness, the chapters in this volume illustrate the need to be critical of the kind of simplistic and instrumental approaches discussed ahead.

AVOIDING NORMATIVE AND PRESCRIPTIVE BIASES

In much institutional thinking there is a strongly rule-centred, prescriptive bias. Such approaches focus on institutions regarded as 'good' and desirable from an often unspecified, normative framework (Harriss 2002). Starting from a priori ideas about 'good governance', 'robust institutions', or 'efficient management', shopping lists of 'shoulds' are formulated that influence policy agendas and interventions. Often the focus is one of socio-legal engineering and institutional design ('crafting institutions') (Ostrom 1992; for criticism see Cleaver 2002; Mehta *et al.* 1999). Thus, processes of institutional change are misinterpreted as the establishment of formal rules, laws, and organizational structures, reducing complex institutional processes to routines of rule-making with pre-defined purposes (for example, production, sustainable management), taking for granted their institutionalization in human behaviour. Complex socio-cultural processes thus become a 'resource bank' (Cleaver 2002) for institutional engineering and building 'social capital' (see Harriss 2002).

Such 'toolkit' approaches contain, for policymakers, attractive recipes for creating conditions and incentives for cooperation ('collective action') and reduction of transaction costs. Examples in irrigation are the routine establishment of WUAs and agendas for IMT. But what does such institutional engineering lead to in practice? Several chapters deal with complex processes of institutional transformation related to reforms. Their focus on institutional processes moves far beyond simplistic managerial and engineering notions of the institutional domain. Nikku shows for IMT in Andhra Pradesh (AP) that policies supporting institutional changes through irrigation reforms may be transformed or even blocked by actors with high and institutionalized stakes in maintaining the status quo. Plans for system rehabilitation in close cooperation with WUA members got stuck in what the author calls the 'contractor-bureaucrat-politician nexus'. Water distribution remained a domain in which the same interest groups as before had their way, at the expense of tail-end farmers. Similar institutional blockages influenced other domains like joint canal inspection and fee collection.

Gautam discusses this issue for Nepalese groundwater irrigation. Rather than following the prescriptions of groundwater development, water users explore various combinations of technical, organizational, and institutional options around multiple water sources. Institutional changes are not the product of the toolkit prescriptions of interventions, but of farmers' strategizing to gain, maintain, or strengthen access to water. In some cases, farmers backed out of tube well systems, leading to growing management, repair, and maintenance problems. In other cases, multiple options in groundwater and surface systems led to forms of institutionalization expressed in joint rules for both sources.

MOVING AWAY FROM A NARROW FOCUS ON WHAT SEEMS TO BE MANAGEABLE

Like 'social capital' and 'collective action', the term 'institutions' is often used with a preference for concentrating on those domains of social life that hold the promise (or illusion) of successful directed change through legal or policy instruments. As many local institutions are less visible for the outside observer, there is a tendency for mainstream institutional approaches to overvalue the tangible and the visible, the 'formalised modes of interaction and codified norms' (Cleaver 2002: 14). These tend to be the bureaucratic, single-purpose institutions rather than the more flexible, adaptive, and often multifunctional ones. However, this bias can lead to a neglect of important 'embedded' non-state social institutions like caste, kinship, and inheritance, or locally defined land and water rights, the relationships between them as well as complex linkages with the state institutional domain (Bruns and Meinzen-Dick 2000; Roth *et al*. 2005; also discussed ahead).[2]

Many case studies in this volume remind us that in real life there are many institutions, institutional processes, and forms of 'collective action' that seem to be quite effective but not per se conducive to policy goals like sustainable management, creation of markets, or equity. Or, as the examples from tank irrigation illustrate, institutional changes may make society as a whole more equitable, but tank management less sustainable. On the other hand, in many societies there exists an institutional richness and diversity that could, in principle, contribute to better land and water management, but is either not seen or not regarded as relevant by policymakers, planners, and engineers.

Examples abound of institutional processes that are not supportive of, or even at odds with, sustainable and equitable management, but these are not taken into account. For groundwater, Prakash lays bare the important role of caste as a key socio-political institution in determining the dividing line between winners and losers, those included and those excluded. While specific institutional approaches regard 'more market' (water markets) as a benefit, Prakash reminds us of wider institutional structures and processes (caste and class divisions) that should always be taken into account before subscribing to simple institutional recipes.

Both Shah and Manimohan trace broader political-institutional changes in society as an important causal factor behind problems with tank management. Shah points out that the emergence of 'new agrarianism' since the Green Revolution has created new economic opportunities for landowning classes, but has also set off radical changes in tank technology and the institutionalized relationships and practices of tank management and control. Adding to these transformations are the changing caste relationships and growing political representation of lower-caste groups in Indian society, which led to a breakdown of established patterns of labour allocation for tank maintenance. Manimohan comes to similar conclusions about the impact of processes of emancipation of agrarian labour as a consequence of the emergence of lower-caste or peasant political movements.

Several chapters illustrate the positive but often invisible role of institutions. Though these are usually not seen or regarded as relevant by policymakers, they contribute to the stability and local legitimacy of water distribution and management arrangements. Narain shows how local ties of family, kinship, and neighbourhood (*bhaichaara*) form the basis of time exchanges among farmers in warabandi systems. Such social institutions make deviation from the rather rigid warabandi system possible, increasing its local acceptance and legitimacy. Though the author does not draw this conclusion, introducing more 'market' in these systems could even cause a breakdown of such cohesive institutional elements. Gautam refers to similar institutionalized social norms leading to practices of sharing between neighbours and kin, making access to water easier and more flexible.

Keeping the Links with Local Water Use Contexts

There is a growing tendency to abstract away from local contexts of water use by defining water problems globally in terms of a crisis of the 'global water system' and devising institutional solutions in terms of generalized forms of rule-making and regulation. Thus, Pahl-Wostl *et al.*, stressing the need for 'global water governance', state that 'many anthropologists and others continue to argue that one needs to understand local rights, needs, and stakeholders in order to effectively address governance issues' (2008: 421), and conclude that 'some kind of formal global coordination is required in tandem with more decentralized network and market-based approaches' (2008: 432). Such an expectation that 'adaptive' governance can evolve using certain preferred forms of institutional set-ups hardly pays attention to local contexts. It misses the crucial point of the—by definition—locally embedded character of water use.

The volume provides overwhelming evidence of this importance of understanding water use in specific use contexts. Without idealizing 'the local', the case studies show that abstracted, decontextualized, or blueprinted institutional solutions are a recipe for trouble. This focus on 'the local' does not mean that the institutional linkages with higher-level processes of change, development of norms, and governance are not important. On the contrary, we see this as a crucial issue in research and policymaking. However, these linkages can only be seriously dealt with by taking an approach to institutions that does not make a priori choices on the basis of normative precepts about water governance institutions, what they should do, and where they should be located.[3]

Parajuli shows that the technical, organizational, and institutional characteristics of the FMIS in Nepal can only be sufficiently understood if they are analysed as embedded in their physiographic, hydrological, and socio-cultural contexts. Technological choices, within the conditions of possibility posed by the physical environment and the social criteria of equity, simplicity, and flexibility of farmers themselves, have important consequences for institutional arrangements and the definition of water rights. Institutional arrangements, in their turn, shape technical options for, for example, division works that should reflect institutionalized understandings of water rights in these systems. Such complex and locally embedded relationships constitute the very institutional 'glue' around technology in these FMIS.

Repoliticizing Institutional Processes

Many approaches to water management show a tendency to depoliticize interventions and institutional changes in, for instance, irrigation reforms (PIM, IMT), reallocation of water, and redefinition of water rights. These basically political processes characterized by a high degree of complexity and involving multiple and contradictory perceptions and interests are often presented as technical, rational-scientific, manageable, and uncontested 'implementation' routines with predictable outcomes (Harriss 2002; for IWRM, see Mollinga 2008a). Institutions are seen as causes of stagnation or instruments of change, as good or bad, robust or clumsy, but usually not as part of the broader socio-political landscape.

In the social-environmental interactions mediated by technologies and institutions we are concerned with here, such depoliticization leads to bad analysis and policies. Definition and analysis of complex social-environmental ('wicked') problems, and the devising of policy solutions by various parties inevitably bring in values and political contestations related to access, authority and rights, inclusion and exclusion, benefits and livelihoods, or wider processes of societal or agrarian change (see also Mollinga 2003, 2008a). The role of institutional processes under these conditions cannot be sufficiently understood from either a consensual or rule-focused approach to institutions. Rather, the institutional domain should itself be approached as the arena of social differentiation, power struggles, contestation and resistance, and conflicts about authority and legitimacy (Bardhan and Ray 2008; Sikor and Lund 2009). This requires full attention to the role of micropolitics in and around institutions and the role of human agency as a driver in processes of institutional change (see Agrawal 2003).

Several chapters in this volume analyse the role of political processes and institutions around water control. Nikku's 'politics of policy' is most explicit about 'the political' in policy processes. His analysis shows that the 'gaps' between formal policy intentions and field realities cannot just be explained as policy shortcomings to be repaired by 'better' policy. Rather, policy interventions are crucially mediated by actors with institutionalized political and economic interests. It is only through a 'politics of policy' framework that we can see these institutional processes in water reforms and interventions.

Shah discusses the politics of institutional processes around tank irrigation, and their impact on technologies, managerial arrangements, and institutions of the tank landscape. Critical of approaches that essentialize 'state' and 'community' and simplify their relationships, Shah discusses the political role of the new farmers' movement and its impact on agrarian and tank policies. She argues that the productive and knowledge choices made by politically organized landowning farmers are important drivers of agrarian change at various levels, including the communities of tank users. Thus, the changing relationships, interactions, and contestations between sections of communities, as well as between communities and the state, are played out and materialized in the tanks.

MOVING BEYOND STATE-BIASED APPROACHES TO INSTITUTIONS

Implicitly or explicitly, the term 'institutions' is often used restrictively for those arrangements belonging to the domain of the state, at least in the relative attention paid to them. Institutions are often equated with (state) organizations like government agencies, or associated in a functionalist and instrumentalist manner with specific state-defined or state-sanctioned purposes (Cleaver 2002; Jentoft 2004; Mehta *et al.* 1999). If non-state institutions are heeded at all, this usually happens on the basis of a rather artificial and biased distinction between 'formal' and 'informal', or 'modern' and 'traditional' institutions, the former often associated with state actors and the latter with non-state actors (Mehta *et al.* 1999). Such dichotomies have little value in analysing institutional processes from a perspective that takes into account both state and non-state institutions (Cleaver 2002).[4]

In such approaches, there is a strongly normative and ideological element involved, determining the objectives of institutional change or design and their linkages with technology, deciding on what 'good rules', 'good governance', or 'efficient management' are, and selecting the actors that should be allowed to play a role. However, in real life, actors like water users have a degree of agency, flexibility, and choice in the institutional arrangements they associate with, that guide their behaviour in specific situations, and which they regard as legitimate, whether these are state, non-state, or hybrid institutions emerging from processes of 'bricolage' (Cleaver 2002). The existence of multiple institutional arrangements should be taken into account in analysis and policy, without a priori preferences for specific ones based in normative and ideological arguments about the role of the state (for legal pluralism, see the following section).

Gautam's study illustrates the complex institutional processes around FMIS as sources of surface irrigation water. After installation of deep tube wells (DTWs), farmers in the area had given up their customary rights to surface water sources. However, when they had to pay for groundwater, they looked for possibilities to link up to the FMIS again and reactivate their water rights. Communities in the area entered into various political alliances and strategies to reach this objective, making use of either state programmes for the rehabilitation of FMIS or of alliances with neighbouring communities that would strengthen their water rights and enable them to build an intake dam. In this complex political-institutional environment, where farmers are constantly exploring opportunities to improve access or strengthen rights to water, boundaries between state and non-state institutional domains are not very relevant. In these processes, farmers also constantly negotiate and renegotiate their 'bundles' of rights and responsibilities attached to various water use options: their water rights.

FROM NARROW TO BROAD ANALYTICAL UNDERSTANDINGS OF PROPERTY RIGHTS AND 'THE LEGAL'

The use of a rights discourse in policies for and debates about resources management takes us into the domain of law and property rights. Property rights can be regarded as institutions as they are basically arrangements for regulation and control of human behaviour and incentives. The crucial role of property rights to land, water, and other resources (for example, physical infrastructure) as a mediating institution between resources, irrigation systems, technology, and (groups of) people has been long neglected in irrigation studies. A notable exception is Coward's (1986a) work on 'hydraulic property', which refers to the property rights to water, land, and infrastructure established, maintained, and reproduced through the investment of labour, capital, and building materials in irrigation systems (see also Mollinga, this volume). Property rights are now increasingly recognized as a crucial institution (Agrawal 2003; Leach *et al.* 1999), including in irrigation studies.[5]

As with institutions more generally, there are many ways of approaching property rights to land, water, and infrastructure conceptually and theoretically. Different approaches are characterized by underlying basic differences in worldviews, political ideologies, and disciplinary orientations. This has consequences for the scope of recognition of various institutional domains, for how water problems are defined, solutions conceptualized, and interventions and reforms devised (see Bardhan and Ray 2008).

Approaches to property rights with a strongly econom(ist)ic focus, such as neo-institutionalist or neo-liberal approaches, tend to narrowly conceptualize property by reducing it to the economic sphere, seeing it exclusively as an instrument for reducing transaction costs or a facilitator for market transactions guaranteeing the most efficient use of resources (Benda-Beckmann *et al*. 2006; Hann 2007; Roth *et al*. 2005). Explanations of human behaviour are often based on variants of rational choice theory, with individual utility maximization as the 'default assumption' (Bardhan and Ray 2008: 6). When applied to the domain of policy, such approaches tend to focus on rule enforcement, in institutions regarded as rule systems, needed to reach specific, predetermined managerial objectives.

However, little attention tends to be paid to the existence of various bodies of norms, rules, and rights in relation to the behaviour of real-life social actors with a degree of agency and choice. Nor is there serious appreciation of those values and meanings attached to natural resources that transcend the purely economic. In real life, institutional processes around water use typically involve social actors who are confronted with the uncertainties of multiple, ambiguous, and contradictory norms, rules, and rights, also involving issues of legitimacy, power, and authority, and meanings and values attached to water resources. When finding their way into policy, approaches that do not take into account institutional complexity tend to explain property-related mismatches in terms of gaps between policies or laws, on one hand, and 'implementation' or 'enforcement', on the other. Usually they do not go beyond the normative valuation of such gaps in terms of deviation from rules, propagating 'better policy', 'robust institutions', 'clear property rights', and so on.

The case studies in this volume take a less normative, more empirically, and analytically grounded approach to property rights. Narain stresses that property rights reforms and market creation are important dimensions of institutional reforms in canal irrigation. Creation of clear rights, trading, and pricing are part of a blueprint package, the applicability of which to specific systems is assumed rather than proven. Narain shows how definitions of water rights in protective warabandi systems are based on turns expressed in time shares regarded as property rights. Such water rights are institutionalized through the water division infrastructure, showing its basically socio-technical character. Thus, the physical layout of such systems, corresponding to local norms, property rights, and organizational design, is a major determinant of options and restrictions for institutional transformation.

Parajuli shows for Nepalese FMIS that water rights, embedded in and materialized through the physical infrastructure like the proportioning weir, are not only about the right to a share of water, but refer to a 'bundle' of rights and obligations that water users hold in these systems. These bundles are defined differently, with reference to different means of technical water control, and with a different degree of precision in FMIS positioned differently along rivers and in landscapes. While some systems must function under upstream control, others are demand-driven and give more precise irrigation under downstream control. Incidence of flash floods and presence of long canal stretches are important determinants of the load of management responsibilities (labour, resources), incorporated in the bundles of water rights and responsibilities in these systems.

Property rights also play an important role in Manimohan's case study of tank cascades in Tamil Nadu. Again, we can see how property rights are embedded in the tank cascade technology, primarily in the tank spillways or calingulas, which are 'material expressions of water rights between upstream and downstream irrigators in cascades'. Water rights were further defined locally in a system of rules and privileged rights between tanks in the cascade as a whole as well as at the individual tank level. However, with the emergence of new agrarian property relations (especially higher- and lower-caste property rights to land and water and related managerial rights and responsibilities), new technologies, and commoditized forms of agricultural production, these relatively well-adapted complexes of technology, tank management arrangements, and property rights have come under pressure. These transformations have led to changes in the tank technology, and have created new disjunctures between upstream and downstream areas, and wet and dry zones.

Krishnan discusses how the changing property rights around ponds, wells, streams, and canals contribute to ecological degradation of water resources. The complex and changing property statuses of various water sources that are not based on values of sustainable resource use, new water lifting technology (pumps), and a growing demand for water under the influence of new and intensified cropping preferences combine into new and unsustainable forms of water resources exploitation. Individualizing water control and making water use more flexible, these new technologies have become instruments for turning public or common property sources into open access sources. The water obtained from these sources becomes privatized by enclosing it in wells or ponds with a private property status. The author argues that these problems can only be solved by paying much more attention to the property status of water resources.

These and other case studies show important elements of the role of property rights that should be covered conceptually, theoretically, and in analysis.[6] First, water rights are not just the right to a quantity of water, but 'bundles' of rights and obligations. Second, water rights are dynamic and often deeply embedded in specific but changing socio-nature contexts of resource use. Third, there are important linkages between landscape and physical environment, technology, management, wider institutional environments, and the ways in which property rights to resources are defined. Fourth, rights to resources like water are often defined differently in different domains at different levels of governance, involving state and non-state political actors and institutions. The presence of multiple and competing institutions and definitions of property rights means that any one of these institutions does not have an all-determining influence on human behaviour. The plurality of institutional domains creates opportunities for institutional 'shopping' and the exertion of human agency by individual or group actors.

Mainstream approaches to law tend to narrow down 'the legal' to the domain of the state, either not seeing or not recognizing non-state bodies of norms, rules, and laws and their impact on resource management and governance. To effectively deal with the aspects of property rights mentioned above, we need a broad conceptualization of 'the legal' that includes bodies of values, norms, principles, and laws deriving from a variety of sources of validity, legitimacy, and authority. These may be located anywhere between the extremes of local socio-political groupings or entities (for example, customary systems) and international or transnational laws and principles. This existence of multiple, complexly interacting 'legal systems' is called legal pluralism or legal complexity (Benda-Beckmann *et al*. 2006; for irrigation, see Bruns and Meinzen-Dick 2000; Meinzen-Dick and Pradhan 2001; and Roth *et al*. 2005).

Attention to legal complexity makes it possible to take into account analytically not only state but also non-state domains. Rather than taking a normative position on the relative importance or validity of any of these, it focuses on the experiences, choices, and strategies of actors in their day-to-day dealings with natural resources (Benda-Beckmann *et al*. 2006). It regards 'the legal' as a domain of struggle and contestation in which laws and norms are used to legitimize competing claims. Thus, it shows what many resource struggles actually are: political processes at various levels rather than just 'management' in the narrow and often depoliticized sense it is often used in (Boelens *et al*. 2005). Through its focus on mobilization of competing sources of authorization and legitimization of claims, this also clarifies the linkages to broader struggles for political power and authority in society in relation to resource control (Sikor and Lund 2009).

Recognition of the existence of a plurality of legal systems and institutions more generally means that solutions to water problems have to be sought in negotiations between actors linked to a variety of legal-institutional domains. In several chapters of this volume, we can see the negotiations between domains of the state and customary law; those with different histories or knowledge characteristics (for example, in terms of professionalization), and those with different spatial and resource claims, or divergent bases of social or political legitimacy (Bruns and Meinzen-Dick 2000). This is an important insight not only for irrigation studies, but also for dealing with big challenges like water reallocations and transfers, and the associated renegotiations of water rights (Meinzen-Dick and Pradhan 2005). Taking into account legal complexity contributes to a better understanding of water rights and their consequences for institutional processes around water control.

Finally, these approaches that acknowledge legal and political complexity have consequences for how we deal with broader issues of 'water governance'. Its mainstream use is often loaded with assumptions about actors, relationships, steering capacities, and shared goals. Where use of the term governance allows for expression of the embeddedness of (water) resources in broader socio-political structures and processes—including contestation, negotiation of rights and interests, and power in society—it can contribute to the socio-political analysis on water control (Mollinga 2008a). When governance is defined as referring to 'processes of steering, ordering, ruling, and control' (Nuijten *et al*. 2004: 123), it creates even wider possibilities in the use of the concept of water control.

STAYING INTERDISCIPLINARY, LOOKING AHEAD

The critical role of irrigation and water management in the development of South Asia and the complex histories and challenges outlined in the Introduction demonstrate the need for interdisciplinarity in research. In this chapter, we have discussed how society controls water and the natural cycles and habitats linked with it through the mediation of technology, and have introduced some of the contestations and choices in this mediation explored further in subsequent chapters. We have highlighted relationships between technology and institutions, and approaches to study institutions shaping water use that can make a difference for understanding water control. These are critical areas where an interdisciplinary approach can be relevant for the work of engineers, policy actors, and local organizations, as well as researchers.

From the water control perspective, we have emphasized the value of a continued and evolving focus on the physical settings in which design-management interactions evolve (to recognize local specificities and avoid blueprints). It is also important to have a better understanding of the knowledge needed and available to make systems work as wanted, to be able to look in a more sensitive way at management and governance actions and options. We have also highlighted the benefits of recognizing the human agency present in water control: irrigation designs are social designs related to social power, and new institutional arrangements evolve within this relationship. We have also discussed some critical challenges for institutional thinking, particularly guarding against mainstream approaches and normative and prescriptive biases. Research has shown the importance of widening the perspective of what is included in 'management', of understanding the political character of institutional processes, and of looking beyond only 'state-based' approaches that specify what form, content, and purpose institutions should have. A critical attention to property rights, law in a broad sense, and legal pluralism in water and technology use help in understanding evolving relations of technology and institutions for irrigation and water management, also as related to the wider environment and landscape.

As argued, the interdisciplinary research used in this volume is not just made by integrating disciplines, and more than a methodological approach only. It is a collective effort to generate an understanding of the complexities in irrigation and water management through shared and expanding knowledge and theory. The final chapter in this volume returns to this effort, to review existing and emergent interdisciplinary concepts being used to study these relationships and approaches that we hope will be taken up critically by future water researchers, professionals, and users.

Notes

[1] These include approaches seeing irrigation as part of a natural resource system (focus on governance systems and collective action), of a farming or rural resources system (focus on the resource dynamics of agro-ecosystems and knowledge systems), or of a socio-physical system (stressing the physical behaviour of water flow and component technologies, and the management science of operations). The first tends to treat technology as a 'black box', so relationships between technology and management are poorly understood. The second tends to treat water supply as a 'given', leaving technology and politics of delivery unexplored. The third can be seen as reductionist, without awareness of social forces acting onto design and operations and with a narrow approach to knowledge, sometimes giving poor solutions to design dilemmas (Vincent 1997).

[2] In Ostrom's earlier work on institutional design (for example, the 1992 work), these often crucial, institutional dimensions were not much more than a footnote.

[3] Mollinga (2008a) rightly stresses the need to see water issues in context, but also to look beyond localities in both time and space. Conceptualization of water problems in terms of 'problem-sheds' and 'issue-networks' can establish such relevant linkages (see also Bruns and Meinzen-Dick 2000).

[4] Cleaver's distinction between 'bureaucratic' and 'socially embedded' institutions does not fully solve the problem. However, her focus on institutional 'bricolage' is relevant, stressing the dynamic character of institutions.

[5] Next to this focus on property rights, attention is needed to 'access'. Ribot and Peluso (2003) explore access in the boundary zone between the legal and the non-legal. Access, according to the authors, is about 'bundles of powers'. The structures and mechanisms through which access is denied or established, maintained, expanded, and controlled should be analysed as related to, but not identical with, the domain of property.

[6] For reasons of space we mention only a few important points here. For a fuller treatment, see, for example, Boelens and Davila 1998; Bruns and Meinzen-Dick 2000; and Roth *et al.* 2005.

2 DECENTRALIZATION, WATER RIGHTS, AND MARKETS: BRIDGING THE TECHNOLOGY–POLICY GAP IN INDIAN IRRIGATION

Vishal Narain

Globally, irrigation reforms have been a subject of much attention. Though irrigation is crucial in raising agricultural productivity and farm incomes, there is concern over the huge financial investments made in the sector, from which the returns have not been commensurate. It is estimated that the World Bank has lent some 35 billion dollars for irrigation development, equivalent to about 7 per cent of all its lending since the 1950s (World Bank and GoI 1999). Munoz *et al.* (2007) report that at present more than 60 countries in the world have implemented some type of irrigation sector reforms.

Typically, three approaches to institutional reform have been taken: decentralization, pricing, and property rights reforms and market creation. Decentralization is often assumed to reduce the managerial and financial burden on the state by giving a greater role to non-state actors, notably farmers. The argument of pricing water correctly is based on the premise that it would reflect the scarcity value of water and convey its economic significance, causing users to use it more judiciously. A similar rationale underlies efforts at property rights reform and market creation, namely, that in the absence of secure property rights, a sort of 'tragedy of the commons' would lead to overexploitation or misuse of the resource. This can be averted if secure property rights exist and water is freely traded to reflect its market price and scarcity value. Decentralization in the irrigation sector has taken the form of building user capacity for system management by forming user groups, often called water users' associations (WUAs). The trend towards state disengagement, giving a greater role especially to farmers, is known as irrigation management transfer (IMT) or participatory irrigation management (PIM) (Brewer *et al.* 1999; EDI 1998; Turral 1995; also see Nikku and Khanal, this volume).

Irrigation has played an important role in India's agrarian transition and development. Irrigated yields are much higher than those from rain-fed agriculture, and it is said that irrigation has lived up to both its protective as well as productive roles (Dhawan 1988). However, though huge investments have been made in large-scale canal irrigation systems, their functioning suffers from several weaknesses: a widening gap between the irrigation potential created and utilized; inequity in water distribution between head and tail reaches; waterlogging and salinity; poor recovery of irrigation charges and fees; and the deteriorating state of canal infrastructure.

Yet, while proposals for reforms in Indian irrigation have been debated for over two decades, their relationships with technology or design of canal irrigation have received little attention (Narain 2003a, 2008). Irrigation management reforms tend to be somewhat piecemeal, emphasizing one of the three approaches mentioned above, to the exclusion of others. Moreover, proposals for reform are often made without a clear understanding of field-level management practices.[1] This chapter examines how the potential for irrigation management reform through these different approaches varies with the design of canal irrigation, which shapes the institutional and normative practices with regard to water management. This is done with reference to two forms of irrigation prevalent in India: *warabandi* irrigation in northwestern India (and Pakistan), and *shejpali* irrigation in western India.

Research on warabandi irrigation was carried out at two locations on the Western Yamuna Canal—one each in Rohtak and Jind Districts—of the northwestern Indian state of Haryana. For shejpali, the site was located in the command of the Mula irrigation project in Ahmednagar district of the western Indian state of Maharashtra. All locations had seen the formation of WUAs. In the warabandi locations in Haryana, namely, the Rampur and Sitapur WUAs, they were formed under the World-Bank-supported Water Resources Consolidation Programme (WRCP). In the Mula project, the formation of the Lakshmi Narayan WUA had been led by a local non-governmental organization (NGO).

This chapter is divided into six sections. The first section provides an overview of the concept of protective irrigation. It is followed by a section describing warabandi and shejpali as two specific forms of protective irrigation and gives their basic principles of design and operation. The element of built-in water scarcity is elucidated. The third and fourth sections describe the water distribution process as actually observed in two WUAs studied in the warabandi and shejpali irrigation systems, respectively. The fifth section discusses the implications of these different forms for reforms based on the three measures mentioned earlier: IMT, property rights reform and market creation, and water pricing. The final section summarizes its key messages.

PROTECTIVE IRRIGATION

Protective irrigation is a specific form of large-scale canal irrigation found in the semi-arid, drought-prone regions of the Indian subcontinent (Mollinga 2003). These systems are designed and operated in such a way that the available water is spread equitably over a large area to reach a large number of farmers, thereby meeting only part of the crop water requirements. This is in contrast with the dominant design practice in irrigation engineering, which is to design irrigation systems in such a way that water supply covers the full crop water requirements, either completely by irrigation or in addition to rainfall (Jurriëns et al. 1996). Thus, an element of 'scarcity by design' is in-built in these systems. Further, the scarce water supplies are distributed among the farmers through some form of institutional rationing. This takes different forms in warabandi and shejpali irrigation, as we shall see ahead. So there is not only a 'scarcity by design' in the operation, but also a scarcity that is 'shared' among water users on a basis concretized through the principles of water allocation.

Protective irrigation originated in British colonial irrigation policy in the nineteenth century (Mollinga 2003). It then aimed at supplying limited quantities of water to subsistence-oriented farmers growing traditional food crops. This would protect crops and livelihoods from drought, prevent famine and social unrest, and secure colonial rule. Thus, 'protective works ... exist primarily for protection against famine, not to bring revenue to the state' (Narain 1922: 272). Similarly, as Attwood (1987) writes, when protective irrigation works were constructed in the Deccan in the early twentieth century, the 'protective nature' of Indian canal systems implied that, first, the canals were intended to protect food crop production against droughts over as wide an area as possible, and, second, they were not required to be self-financing. In other words, the authorities did not expect to recover the full interest on capital costs from irrigation charges.

Since independence, protective irrigation has remained a part of the government policy for agrarian development, emphasizing both growth and equity objectives: production and productivity increase, and the spread of the benefits of development over different sections of the population and regions (Mollinga 2003). It has guided the construction of canal irrigation systems in the phase of planned economic development. In 1987, the National Water Policy stated that 'the irrigation intensity should be such as to extend the benefits of irrigation to as large a number of farm families as possible, keeping in view the need to maximise production' (GoI/MoWR 1987: 9).

WARABANDI AND SHEJPALI AS FORMS OF PROTECTIVE IRRIGATION

THE WARABANDI SYSTEM

Warabandi is a specific form of protective irrigation found in northwest India and Pakistan. It can be understood as a socio-technical system with a composite technology and accompanying principles of social organization and definition of rights (Narain 2008). 'Wara' means turn and 'bandi' means fixation. Thus, warabandi means fixation of turns (Bandaragoda 1998; WALMI 1998a). It implies a rotational method of water distribution. The theory of warabandi is that each cultivator is assigned a turn, represented by a specific period of time—a time share—and that the volume of water available during that period is his to use. This time share becomes a property right legitimized by the state through the creation of a formal warabandi roster for the delivery channel in question. The warabandi share, as a property right, then serves to organize the social relations of irrigation among the cultivators, and between them and the irrigation agency (Coward 1986a).

The cardinal principle underlying warabandi is that the available water, whatever its quantum, should be allocated to cultivators in equal proportion to their culturable command area (CCA), and not only to some farmers in the command to meet their total demand. This is intended to impose principles of equitable division of water under general conditions of water scarcity in the system (Jurriëns et al. 1996).

Since the water supply is inadequate to run the whole system continuously, warabandi works through a system of rotation (Malhotra 1988; Malhotra et al. 1984). This system works at three levels. First, a main canal carries the water from the source (river, reservoir). The main canal feeds two or more branch canals which operate by rotation and may or may not run at full supply. This is the primary distribution system and runs throughout the irrigation season with varying supply. Branch canals supply water to a large number of distributaries, which must run at full supply level (FSL) by rotation. The distributaries operate by eight-day periods. This is the secondary distribution system.

Distributaries supply water to watercourses through ungated, fixed discharge outlets.[2] Watercourses are designed to run at full supply when the distributary is running full supply and its water is allocated to the farmers through the warabandi schedule. This is the tertiary water distribution system. A period of seven days (168 hours) is divided among the farmers in proportion to the size of their land-holdings. Water deliveries are controlled by time and are proportional to land. Below the outlet, the warabandi schedule is supposed to be followed by the farmers.[3] The government's main role is to assist in preparation of the irrigation roster and in settling disputes. Farmers are allowed free use of groundwater and freedom to plant whatever they want following their own assessments of the availability of water. There are no cropping restrictions.

SCARCITY BY DESIGN UNDER WARABANDI

As noted earlier, a shared 'scarcity by design' is an essential characteristic of protective irrigation systems. Scarcity conditions are created in these systems to divide water thinly over a large area and a number of farmers. Water allowance, capacity factor, and irrigation intensity are the expressions of this in-built scarcity. Each unit of the CCA is allocated a certain rate of flow called the water allowance. Its value is a compromise between demand and supply. For instance, in the Bhakra system in north India, the value of this allowance at the outlet to the watercourse is 2.4 cusecs[4] of water per 1,000 acres of CCA (0.17 litres per second per hectare) (Malhotra 1988; Malhotra *et al.* 1984).

No distributary operates for all days during the crop season: the ratio of the operating period of a distributary to the total period of the crop is called its capacity factor. This is again a compromise between demand and supply and is separately designed for each of the sub-periods. For example, in the Bhakra canal system, the designed mean capacity factors for *kharif* and *rabi* are 0.8 and 0.72, respectively, which means that during these seasons, each distributary may receive full water for about 144 and 129 days, respectively, that is, for about 273 days a year.[5] These values of water allowance and capacity factors do not ensure irrigation for 100 per cent of the CCA. The ratio of the irrigated area to the total CCA is known as the intensity of irrigation. Its value in the Bhakra canal is 62 per cent per year, or about 30 per cent per season.

When the distributary is running at FSL, the watercourse draws its full (authorized) discharge. The capacity of the watercourse varies from 1 to 3 cusecs, depending upon the command area (WALMI 1998b). The value of the water allowance at the watercourse head is 2.5 to 3 cusecs per 1,000 acres of the CCA.[6] To ensure that the stream size below the outlet is appropriate for handling by farmers (say 25–40 litres per second), outlet commands (*chaks*) are relatively large (100–300 hectares). If the chak is 100 hectares, and allowance is 0.2 litres per hectare per second, then the farmers receive 20 litres per second for 1.68 hours per hectare; if the chak size is 200 hectares, they receive 40 litres per second for 0.84 hrs.

Delivery capacity per hectare (allowance) is very low, about 10–15 litres per hectare per second at the outlet (300–400 acres per cusec). This is considered insufficient, even if given continuously, to meet the theoretical crop water requirements for more than perhaps 20–30 per cent of land in kharif and 35–45 per cent in rabi (Berkoff 1990). In a study of the Pabra distributary of the Bhakra system, in the Hissar district of Haryana, Malhotra *et al.* (1984) conclude that, for an average of 3–4 irrigations per field, the system of warabandi and the water available provide adequate irrigation to only about 25–30 per cent of the CCA; leaving 70 per cent of the CCA rain-fed. This, the authors argue, corresponds to the design objectives of warabandi.

On the western Yamuna canal, where the two research sites for this chapter were located, the allowance is 2.4 cusecs per 1,000 acres on the Bahlot sub-branch, 2.86 cusecs per 1,000 acres on the Butaana branch, and 2.57 cusecs per 1,000 acres on Rohtak distributary at the level of the outlet.[7] The designed irrigation intensity is 60 per cent for the entire year.[8]

THE SHEJPALI SYSTEM

Shejpali is partially a demand-based system in which the government agency allocates water in consideration to a farmer's request according to land and crops. The word 'shej' means turn and 'pali' means water; thus, shejpali is essentially a system of taking turns for water. Unlike the warabandi system, however, under shejpali, a farmer has the option in a season to take water or not, and beforehand he[9] has to file an application indicating area, crop, and number of required watering turns.

The rationing of scarce water is sought to be achieved through crop zoning and by sanctioning the types and areas of crops grown. It is thus a method of water control in which the demand for water is regulated. Farmers request water before the cropping season by presenting the irrigation department with their proposed cropping patterns. These are partly sanctioned by the irrigation department; the farmers are then entitled to irrigation supplies for these crops. The distribution of the sanctioned water is in rotation, taking into account the requirements of the sanctioned crops.[10]

An operational implication of the water distribution process arises from the characteristics of the gated pipe outlets from which water is released into the watercourses. Hydraulically, these outlets are non-modular structures (Bolding et al. 1995).[11] This implies that the discharge depends both on the upstream and the downstream water levels. Discharge is driven by the working head: the difference between the water level in the parent channel (in this case, the minor) and the watercourse. Thus, it varies as the level in either changes.

For instance, on the downstream side, when high fields are being irrigated, the watercourse heads up and the discharge is reduced (WALMI 1998a). When low fields are being irrigated, the water level goes down in the watercourse and the discharge is increased. If the watercourse silts up, the working head is reduced and hence discharge is reduced. Upstream, this has to do with the absence in protective irrigation systems of cross-regulators: these are elevations or other structures on the canal bed that stabilize water levels with varying discharge (Bolding *et al*. 1995).

Thus, shejpali is distinct from warabandi in terms of receiving and sanctioning applications for irrigation as well as operation and adjustment of the gated pipe outlets. As we shall see later, this has implications for the potential for management reform in these systems through IMT. Further, in both warabandi and shejpali, the institutional set-up is such that the irrigation fees collected go into the state exchequer rather than to the irrigation department.

SCARCITY BY DESIGN IN THE SHEJPALI SYSTEM

Design data on irrigation systems in Maharashtra, needed to assess their performance and gauge the magnitude of 'scarcity by design', are very scattered. The Second Irrigation Commission (GoI 1972) cites the case of the Jayakwadi dam. One of the modern post-independence dams of the state, it is designed to cover supply to a gross cropped area of 183,565 hectares and irrigable command area (ICA) of 141,640 hectares. The designed intensity of irrigation is 77 per cent.[12] Major crops to be irrigated are cotton and rabi crops, with rice on 5 per cent and sugarcane on 1.5 per cent.

For the Mula dam system, Lele and Patil (1994) cite the designed crop pattern as follows: 5 per cent is to be under perennials (mainly sugarcane); 20 per cent under two-season crops (cotton); 30 per cent under monsoon season (July to October) kharif crops; 42 per cent under rabi or winter crops (November to February); and 3 per cent under hot weather crops (March to May). These designs reflect a combination of assessment of soils and their required recharge to support a crop, and available water supply at different times of the year relative to the overall command area.

The CCA, ICA, the cropped area as designed, and the intensity of irrigation as per the design report of the Mula project (the research site for this chapter) is presented in Table 2.1.[13]

As can be seen, these irrigation intensities are lower than 100 per cent. The assumption is that some soils, like light or shallow soils, which will need more frequent irrigation, are more suitable for kharif cultivation when rainfall is also a factor. Whereas deep soils can hold more water between irrigations and so are more suitable to support

TABLE 2.1   Design Features of the Mula Project

Canal	CCA (in ha)	ICA (in ha)	Irrigation Intensity (ICA/CCA) (in %)	Cropped Area (CA) (in ha)	Cropping Intensity (CA/ICA) (in %)
Mula Right Bank Canal (MRBC)	85,167	59,292	70	62,257	105
Mula Left Bank Canal (MLBC)	15,182	10,121	66	10,627	105
Pathardi Branch	17,853	11,397	64	14,018	123

Source: Adapted from Narain (2003a).

rabi crops, when there is less or no rainfall and greater dependency on irrigation.

WALMI (1998b) presents the assumed irrigation duties in the upper Pravara Project, another important irrigation system in the state in Table 2.2.

In Maharashtra, the canals are designed taking into account the cropping pattern and the area to be irrigated in the peak period (Narain 2003a). The canals are usually designed with 0.7 litres per hectare per second in case of a non-paddy crop and 1.0 litres per hectare per second for paddy crops. The duty of various crops considered for designing canals are as in Table 2.3 (in acres per cusec; at the outlet level).

It must be noted that the figures of 0.7 litres per hectare per second and 1 litre per hectare per second are actually not far from crop water requirements. When considered in addition to rainfall this would, in fact, suggest that water availability in irrigated areas is not always as low as it is made out to be in Maharashtra.[14]

THE INTRODUCTION OF WATER USERS' ASSOCIATIONS

WUAs have been introduced both in the warabandi and shejpali systems. In the cases presented here, the Sitapur and Rampur WUAs in the warabandi system in northwestern Haryana were introduced basically as part of the World-Bank-supported WRCP initiative. The main role assigned to the WUAs was the maintenance of lined

TABLE 2.2 Duties (in acres) for Working Out Live Storage
Necessary for Irrigation in the Upper Pravara Project,
Maharashtra

Crop	Kharif Duty	Rabi Duty	Hot Weather Duty
Sugarcane Overlap	65	70	50
• Kharif	65		
• Rabi		70	
• Hot weather crops			50
Other Perennials	100	105	75
Long Staple Cotton	200	400	100
Two Season Crops	300	140	–
Green Manure	400	–	–
Kharif Groundnut	200	–	–
Bajari (Pearl Millet)	–	Not irrigated	–
Mung and Udud (Pulses)	–	Not irrigated	–
Rabi Jowar (Sorghum) after Mung, Udud, and Early Groundnut	–	200	–
Gram after Bajari	–	400	–
Hot weather Maize and Vegetables	–	–	100

Source: Adapted from Narain (2003a).
Note: The above figures are at the outlet level.

watercourses below the outlets. The Laxmi Narayan WUA, formed
in the shejpali system in Maharashtra, had a more significant role.
Its functions related to the management tasks performed earlier by
the irrigation department: operation of gated pipe outlets, collection
of applications for water, and releasing and distributing water below
the outlets. This WUA was formed as a result of lobbying by an
NGO with the irrigation department. The NGO also played a role
in organizing farmers for collective management. As a result of the
execution of the programme for IMT, the department entered into a
contract for volumetric supplies of water for each cropping season.

TABLE 2.3 Duties (in acres) for Various Crops for Canal Design

Crop	Kharif	Rabi	Hot Weather
Sugarcane	60	50	40
Sorghum	200	150	–
Bajari	200	–	–
Wheat	–	100	–
Vegetable	200	200	7.5
Pulses	200	200	100

Source: Adapted from Narain (2003a).

Box 2.1 presents some calculations, in a comparative perspective, of water scarcity in the WUAs studied. The calculations suggest that water was definitely scarcer in the case of the WUA in Haryana.

WATER DISTRIBUTION IN THE WARABANDI SYSTEM: THE RAMPUR AND SITAPUR WUAs

Since protective irrigation systems seek to maximize productivity per unit of water, while farmers seek to maximize productivity per unit of land, there is an inherent conflict between farmers' objectives and government objectives. Initially, the combination of the warabandi system with light crops like wheat in northwest India, allowed the systems to work as designed. With the introduction of high-yielding varieties (HYVs) and the commercialization of agriculture, this pattern became disrupted. The initial high design duties became inadequate. Farmers wanted to irrigate in two seasons instead of one. They also wanted to irrigate all their land, as against the planned low intensities. There was an increase in productive irrigation and rice cultivation (Jurriëns 1993).

Against this backdrop, two factors are significant in understanding the design characteristics of canal irrigation in northwest India's warabandi system. First, since irrigation is protective, the aim is to ration out water through a system of rotations through the warabandi system rather than to match supply with demand. Second, system control and operation are totally centralized. Thus, in principle, farmers have no control over the availability of water from the main system and cannot match water supply with crop requirements.

Box 2.1   Water Scarcity Calculations

Water Allocation in the Lakshmi Narayan WUA, Maharashtra[15]

Kharif: 0.434 million m³ to irrigate 94 hectares
434,000 m³ to irrigate 94 × 10⁴ m² 434,000 m³/94 × 10⁴ m² = 0.462 m = 462 mm

Rabi: 1.058 million m³ for 120 hectares
1.058,000 m³ for 120 × 10⁴ m² 1.058,000 m³/120 × 10⁴ m² = 0.882 m = 882 mm

Hot weather (summer): 0.283 million m³ for 62 hectares
283,000 m³ for 62 × 10⁴ m² 283,000 m³/62 × 10⁴ m² = 0.456 m = 456 mm

Total allocation = 462 + 882 + 456 = 1.800 mm

Irrigation depth for the year in the command of the WUA is 1,800 mm.[16] Adding the average annual rainfall of 594 mm, this gives about 2,400 mm, enough for, say, sugarcane.

Comparison with Designed Allocation on the MRBC:

Designed allocation on the MRBC:

Kharif: 196.31 million m³

Rabi: 256.53 million m³

Hot weather: 126.96 million m³

Total allocation: 579.8 million m³

[Figures obtained from the Operational Plan prepared for the MRBC in 1986 by Command Area Development Authority (CADA).]

Total ICA of the MRBC: 59,292 ha[17]

So, when this designed allocation is applied to the Lakshmi Narayan WUA, it would be as follows:

Kharif: (94/59,292) × 196.31 × 10⁶ m³ = 0.311 × 10⁶ m³
The actual allocation is 0.434 × 10⁶ m³. This is higher than the designed allocation:

Rabi: (120/59,292) × 256.53 × 10⁶ m³ = 0.519 × 10⁶ m³
Actual allocation: 1.058 × 106 m³. This is also higher than the designed allocation.

Hot weather: $(62/59,292) \times 126.96 \times 10^6$ m^3 = 0.133×10^6 m^3
The allocation is 0.283×10^6 m^3. This is also higher than the designed allocation.

Thus, in all the seasons, the WUA has succeeded in securing an actual allocation greater than the designed allocation.

Water Allocation in the Rampur WUA:

Capacity factor of the distributary: 165 days (90 days in the monsoon season + 75 days across four groups each for the rest of the year)

Designed discharge at FSL into the outlet: 1.56 cusecs (2.4 cusecs per 1,000 acres for 648 acres).

Area of outlet is 648 acres (260 hectares), all designed to get water from the canal, and under the warabandi schedule.

Total volume 165 days = $(1.56 \times 0.0283) \times 60 \times 60 \times 24 \times 165$ m^3 = 629,373.88 m^3

Irrigation depth = volume of water/command area of outlet = 629,374 m^3/ 260 hectares = 629,374 m^3/260 $\times 10^4$ = 0.242 m or 242 mm annually.

Adding an average annual rainfall of 500 mm,[18] we get an annual irrigation depth of 752 mm. Water is much scarcer by design in the Haryana case.[19]

Source: Adapted from Narain (2003a).

This research showed, however, that farmers employ strategies to increase control over water supplies (Narain 2003a, 2003b): they engage in time exchanges, supplement canal irrigation with tube well irrigation, tamper with outlets, insert siphons over the canal or distributary (water theft), and apply for a rice shoot (temporary pipe outlet).[20] The first, time exchange, improves flexibility of water application and may bring efficiency gains. The others enable an irrigator to get extra water.

TIME EXCHANGES

The most common response to a system of scarcity is a time exchange. A farmer who needs water outside his turn borrows from another farmer and returns it later. This deviation is justified as *bhaichaara*-based organization.[21] While the water right is constituted through state law, it is often realized through a different normative system: bhaichaara. This points to the existence of different normative frameworks (legal pluralism) in water distribution.

As noted earlier, under the warabandi system a period of one week (168 hours) is divided in proportion to the size of the land holdings falling under one chak, the area served by one *mogha* (outlet). Each piece of land, then, is supposed to receive water for a fixed time and day of the week. An irrigator's water right is determined in proportion to the size of his landholding.[22]

The so-called pucca warabandi schedule, which provides legitimacy to the water right so defined, determines water allocation below the outlet. It is framed by the irrigation department and has a legal basis in Section 55(A) of the Haryana Canal and Drainage Act of 1974. It specifies the time period and day for which an irrigator will irrigate his land. In local language, it is said to be *manjoor shuda* (approved, sanctioned by law).

In practice, within the warabandi schedule, farmers follow a distribution pattern based on their mutual understanding and cooperation shaped by bhaichaara. A bhaichaara committee is formed on the basis of mutual consent rather than through the force of law. When I asked the Secretary of the Rampur WUA to explain bhaichaara, he said 'it is when we give up our own self-interest for a greater common good'.

Thus, we could argue that time exchanges arise because of three factors: first, a well-defined water right, through the pucka warabandi schedule, that defines the basic unit of exchange; second, the protective nature of canal water supplies, creating a practical problem for field irrigation; and, third, the rigidity in the turn allocation and the small size of land-holdings that lead to short (hence unpractical and inefficient) irrigation durations; fourth, the existence of plural legal repertoires.

The characteristics of time exchanges can be summarized as follows: first, they cut across *thola*s (ancestral family units). Second, they are more common among members of adjacent pieces of land, but also take place among members whose fields are distant. Third, they cut across outlets. Farmer A, who irrigates from Outlet X, would exchange his time slot with Farmer B, who irrigates from Outlet Y.

Fourth, sometimes, negotiation and decision-making about the exchange takes place before the distributary is run; sometimes after irrigation has started; and at other times, it emerges spontaneously, at the spur of the moment. Fifth, time shares cut across seasons. The tail-enders accumulate their time shares by skipping irrigation in the kharif season and irrigating crops like sorghum, millet, and cotton from rainfall. They give their time shares to those who need water for, for example, paddy cultivation, and use their accumulated time shares to irrigate wheat in the rabi season.

OTHER TRANSACTIONS IN CANAL WATER

While water rights are tied to land, a transaction in a water right could take place along with or exclusive of the land. Some farmers give their land to others on a tenancy arrangement for a year.[23] This is done along with the warabandi share of water. The rate at which they give out the land includes the water share. This arrangement is known as giving land on *theka*. Smaller farmers often take land on theka. Larger farmers also give out their land on theka in smaller parcels to more than one smaller farmer, or simultaneously cultivate their own land and those of other farmers.

This status quo of the water right is maintained under so-called *saajedaari* (sharecropping) agreements. In this case, the landowner provides the land and the concomitant share of water. The landowner and the partner equally contribute other agricultural inputs, and the produce is shared equally as well. Sometimes, however, a farmer who does not depend on canal water may sell his share to another farmer, while he continues to irrigate his land from other sources like groundwater. The other kind of sale of a water right is for a specific turn, when a farmer does not need to irrigate during that turn. Some farmers point out that the rate depends upon the 'demand' for water. During the summers, water selling takes place for about INR 200 per hour; during the winter season, it could be about INR 100–150 per hour. This sale is not for the entire year, but only for one round of irrigation. In one case, transactions in surface and groundwater were combined. Farmer A took water from Farmer B's tube well and in return gave his surface water right for one irrigation turn.

Time exchange, however, is more popular than water sale. While there are water right sales, they are confined. A farmer prefers to lend his water share instead of selling it since lending creates a future claim to water. When a farmer borrows a time share of a fellow farmer, he is obliged to return it. However, a water right that is sold is either an irrigation turn given up, or the entitlement for an entire year foregone. Farmers also have a psychological resistance to selling their water right. Many irrigators expressed a psychological attachment to their water right. There is a feeling that it is not appropriate to sell a water share. A farmer's response to whether he would sell his water share was: 'How would I fill my stomach then?' This observation has practical implications for the creation of water markets in these irrigation systems, a subject discussed in detail ahead.

SUPPLEMENTING CANAL IRRIGATION WITH GROUNDWATER IRRIGATION

The second response to 'scarcity by design' is to supplement canal irrigation with groundwater irrigation. Farmers usually wait for the canal water, since it is much cheaper. They usually have information on when the distributary would be running the next time. If, however, they need to irrigate outside the schedule, they use tube well water. Similarly, when the distributary is 'on' and they feel that their share is inadequate to their requirement, they supplement it by tube well water delivered through the same watercourse.

Generally, when the canal is 'off' and irrigation is needed, farmers run their own tube well; if they do not own one, they buy groundwater. During the rabi season of 2000–1, the first wheat irrigation was from the canal, while the second and third irrigations used water from the tube well. Some farmers who used to buy groundwater dug their own tube wells for greater water supply security.

Transactions in groundwater depend on the relationship between buyer and seller. Groundwater is usually sold at INR 40 per hour.[24] Alternatively, the buyer provides the required amount of diesel. There is also a third form of arrangement, a contractual year basis: the seller provides groundwater at the rate of INR 1,000 per acre per annum.

Some tail-end farmers do not or hardly receive canal water, so they depend upon tube wells. If they do not have a tube well, they have to buy from others. Tail-enders who take tube well water, prefer to buy it from tube wells located near the head reaches. While tube wells near the tail reaches have saline water, in the head reaches the groundwater is of superior quality because of proximity to the canal.

WATER THEFTS ALONG THE CANAL

Another way of obtaining additional water is through water thefts. Water theft can take place by inserting a rubber siphon over the *hawdi* of the outlet (a circular embankment into which water flows from the outlet before flowing into the watercourse) over into the watercourse. Thus, more water is taken than flows from the hawdi into the watercourse. Another way is to break the brick along the side of the iron cast in the adjustable proportional module (APM). Both activities increase the discharge. Similarly, when an APM is installed without the hawdi, the brick can be broken from the side of the APM cast pipe.[25]

During my association with the Rampur WUA, water thefts did take place, and went unreported and/or unnoticed. During my contacts with the Sitapur WUA, a second WUA that I studied in the adjoining Jind district, I noticed water thefts along the minor. A farmer said: 'A rich farmer can bribe the *beldaar*[26] to allow him to insert a siphon, a small farmer simply steals water by inserting it.'

CREATING A WATER RIGHT: APPLYING FOR A RICE SHOOT

Under warabandi irrigation, a farmer can create a water right by applying for a rice shoot: a temporary pipe outlet, given for paddy irrigation between July and September, when surplus water is available following the monsoons. During the monsoon season of the year 2000, 16 rice shoots were sanctioned along the Sitapur minor.[27] At the end of the season, the irrigation department removes the rice shoots.

WATER DISTRIBUTION IN THE SHEJPALI SYSTEM: THE LAKSHMI NARAYAN WUA

There are three cropping seasons in Maharashtra: the rabi season from 15 October to 28 February, the hot weather season from 1 March to 30 June, and the kharif season from 1 July to 14 October. Rabi is the main irrigation season, kharif is the rainy season. In this season, farmers also use rain water; the role of irrigation is largely supplemental. In the rabi season, farmers irrigate wheat, *channa* (chickpea) and sugarcane. In kharif, they irrigate *bajri* (pearl millet), sugarcane, *jowar* (sorghum), cotton, lentils, and some fruits and vegetables. In the hot weather season, they irrigate mainly groundnut and sugarcane.[28] Typically, on the Mula minor no. X, there are three to four rotations in the rabi season, two in the hot weather season, and only one in the kharif season.

The WUA that had been formed there took over several irrigation management and distribution functions performed earlier by the irrigation department. The WUA calls for applications from the farmers (both members and non-members) for supplying water immediately after getting the information from the irrigation department on water availability for each season. This water availability is in terms of the sanctioned quota agreed upon in the memorandum of understanding (MoU) between the irrigation department and the WUA. When there is less water in the Mula dam, this quota is to be proportionately cut. The allocation at the time of this research was: 0.434 million cubic metres for kharif, 1.058 million cubic metres for rabi, and 0.283 million cubic metres for hot weather (see Box 2.1). These quotas are fixed to irrigate 94 hectares in kharif, 120 in rabi, and 62 in the hot weather season.

The secretary of the WUA issues a notice to all members and non-members indicating the date by which they are required to file their applications.[29] The farmers are required to indicate the crops that they wish to irrigate and the area under each. There are no crop restrictions applicable to the irrigators in this WUA case, unlike the cases with normal shejpali. This is an outcome of the terms and conditions expressed in the MoU between the WUA and the irrigation department.

After scrutiny of the applications, approvals are issued for water supply. The applications are arranged outlet-wise. The total quantity of water requested is compared with the sanctioned quota or the quantity of water agreed to be released by the irrigation department (in case of insufficient replenishment in the reservoir). If the quantity required can be accommodated from the available water, all applications are sanctioned, provided past dues have been paid. If, however, the demand is more, a proportional cut is applied.[30]

Demands are arranged outlet-wise and the period of supply for each outlet is worked out. The water is supplied at 1 cusec at the rate of 10 hours per hectare. The entire discharge from one outlet is allotted to one farmer at a time. The discharge for the minor is calculated from the number of outlets required to operate each day, plus a seepage allowance of 12 per cent or 1 cusec (whichever is more). This is indented with the irrigation department. Permitted applications of water (sanctions) are again arranged outlet-wise, and the outlet that has to run for the longest period is first identified. The minor must run for this period. The WUA requests the number of days and the exact dates for which the minor is to be run within the schedules of the main canal. The other outlets required to run for shorter periods are clubbed together such that their total running period is the same or less than the longest period, and the outlets to be running on each date are then identified.

The *paatkaris*[31] of the WUA then prepare individual schedules for water delivery for each outlet. These schedules show the name of the farmer, gat,[32] number, area sanctioned, area to be irrigated, time allotted, and the starting and completing time for irrigation. Time allocation is assessed from use of a constant discharge of 1 cusec. The final water distribution schedule, known as the *pali patrak*, is put on the WUA office notice board.[33] The outlet with the greatest demand is opened first, as it runs for the longest time.

Each farmer is issued an irrigation pass indicating the time from which he has to take water. On each outlet, distribution starts from the tail, and each farmer takes water, one at a time. So, for instance, if a farmer is issued a pass from 10.00 to 18.00 hours, the pass authorizes him to take water uninterrupted during this time. This is the basis of his 'water right' or entitlement. When his turn is over, he invites the next farmer to take water, as indicated in the pali patrak. Sometimes farmers exchange their turns. If, for example, a farmer's land is not ready for irrigation, he can exchange his slot with another farmer. This happens outside the formal distribution schedule prepared by the WUA and is based on mutual understanding between irrigators.[34]

The paatkari monitors water distribution and keeps the keys of the outlets along the minor. The keys of the minor head outlet stay with the canal inspector from the irrigation department. The paatkari is paid by the WUA. All water applications are submitted to him before the start of the season. He gives a notice of the water distribution schedule ten days in advance. The keys of the outlets are retained by the paatkaris.[35] They also supervise that farmers complete the irrigation in the allotted time and keep a watch on irrigation offences which should be reported to the WUA.

As water distribution proceeds, the outlets are opened to accommodate a discharge of one cusec, and a red mark is put on the parshall flume that serves as a benchmark for measuring the discharge.[36] When the irrigation rotations are over, the paatkaris shut down the gate and put a lock to prevent tampering. This technical innovation of installing measuring devices and the institutional innovation of the WUA paatkari to monitor these discharges essentially overcomes an operational weakness of the system, namely, that the outlets require constant adjustment and measurement for maintaining discharges.

I did not notice incidences of breaking the lock on the outlets or of water thefts during my association with the WUA. According to my field observations, incidences of breaking the locks or of manipulating the discharge of water did not occur after the formation of the WUA. There was a greater sense of discipline among the WUA members. The act of measuring the discharges by the paatkari eliminated efforts to manipulate the discharge, either at the head of the minor or along it. One of the reasons for this was the greater sense of ownership. If the farmers or WUA members broke the pipe, they would have to repair it. So it would pose a financial burden on the farmers and the WUA.

Thus, three functionaries of the WUA play an important role in the water distribution process as it proceeds: the secretary, and the two watermen or paatkaris. The secretary essentially performs a liaison function with the irrigation department, receives water applications from farmers and issues water passes and sanctions. He informs farmers about rejections and sanctions. He also performs other functions such as keeping records of the meetings of the WUA and entertaining visitors.

The paatkaris prepare the water distribution schedules for each rotation and outlet. This is transmitted to the irrigation department through the secretary. They also open the outlets for 1 cusec flow, monitor the water distribution process, and report irrigation offences.[37]

The role of the canal inspector from the irrigation department is to release water at the minor head. His role ends there. As noted earlier, a standing wave flume is installed at the minor head to measure discharges.[38]

A *panchanama*[39] should be issued on any additional area irrigated over and above that sanctioned, any unauthorized irrigated area, and area that could have been irrigated from wastage. However, in practice, it is not usually issued, since it is considered 'psychologically damaging'.[40] Those who take water out of turn or without applying for it are charged double the normal rate.

As indicated in the MoU, the WUA has been allotted quantities of water for each season. During scarce periods, the irrigation department proportionately cuts the amount. The Lakshmi Narayan WUA usually demands the entire volume of water allocated according to the MoU (IIMA–IWMI 1999). Within the overall rotational plan and its allocation, the WUA requests water delivery according to its own schedule during the season. The WUA can request the discharge and days of delivery of water in each rotation. This ability to control deliveries allows them to match the crop water requirements more closely with water supplies. The canal inspector releases water to the canal when requested by the WUA waterman. This facility is not available to members of the Mula minor no. Y or the other non-transferred canals. The IIMA–IWMI (1999) study showed that the duration of the rotation and the discharge at the head of the minor showed greater variation and range on the Mula minor no. X than on the minor no. Y. This shows that the WUA had taken advantage of the greater flexibility that management turnover conferred.

Another provision is that the water not taken by the WUA during the rabi season may be used in the hot weather season after a 30 per cent deduction for evaporation losses. This is useful for the irrigation of sugarcane, particularly if groundwater levels are low and for the cultivation of high value crops. The WUA has taken advantage of this facility several times beginning in 1991–2 (IIMA–IWMI 1999). This facility, again, does not exist for the non-transferred canals.

As in the warabandi cases, one way of coping with 'scarcity by design' is to meet the gap through groundwater irrigation. This is also one of the conditions of possibility in the WUA: one of the reasons that this system works is that all demands do not depend upon it. Similarly, for pearl millet in the kharif season, farmers typically wait for the rains. The crop needs water 15–20 days after sowing in June-July; they irrigate from the canal only when the rains fail. It usually needs two waterings before its harvest in November or December.

STRATEGIES FOR IRRIGATION MANAGEMENT REFORM: POTENTIAL AND POSSIBILITIES IN THESE FORMS OF PROTECTIVE IRRIGATION

This discussion of water distribution has important implications for the potential of irrigation management reform strategies. As described, the water distribution process in warabandi and shejpali is very different, both in how it is conceptualized and how it is implemented. I now examine the implications of these differences (in technology or design and in the accompanying social, normative, and institutional systems) for reform through the three sets of measures mentioned earlier: decentralization, pricing, and water rights reform and market creation.

DECENTRALIZATION

The warabandi and shejpali systems have different potentials for reform through IMT. Shejpali has operational implications in terms of making and processing applications, and operating gated pipe outlets, which could be handed over to the Lakshmi Narayan WUA. There is a potential for change through the formation of user groups for taking over these functions. As shown earlier, WUA formation made possible the taking over of several functions involved in operating the shejpali system, which brought about positive changes in water distribution among the farmers.

This potential remained limited with warabandi, where there already was a form of water allocation and distribution. In this system, water distribution is relatively simple with fewer levels of operation below the outlet. Water allocation is defined through the warabandi schedule, and water is distributed among the farmers through this schedule, mediated by social relationships based on bhaichaara. The formation of WUAs at that level makes relatively little difference to the water distribution processes.

Essentially, warabandi and shejpali, as alternative water allocation systems, have different operational implications that define the potential for reform through WUA establishment differently. Thus, discussions about the potential for reform through IMT must start from an understanding of the technology for water allocation, its institutional and normative systems, and the kind of reform that is appropriate to that context.

In shejpali, WUAs can take over the water distribution function, improve water control, and curb opportunities for illicit and extra water payments. There is also a possibility of changing control relations between the users and the bureaucracy, and of putting in place a set of mutual accountability relationships by instituting volumetric water supplies and introducing volumetric payments, as already discussed. The operation of the shejpali system requires an organizational structure for the performance of water management and distribution functions, and the operation of gated pipe outlets that can be provided by WUAs. In the warabandi areas, on the other hand, the scope for transfer of powers to users or for changing control relations between the users and the irrigation department is much more limited. There is already a mechanism for sharing and distributing water at the field level among the farmers. WUA formation will have very limited effects on water management and distribution practices, except if allocation and distribution at higher levels would come under farmer governance.

In Haryana's warabandi system, WUAs have been formed below the outlets, mainly under the World-Bank-supported WRCP. At that level, there is already a form of social organization engaged in water management and distribution. As shown here, the real management challenges are located above the outlets: water thefts, poor main system management, and tampering with outlets. At this level, and where no dent has been made yet through efforts at irrigation reform, reforms are needed.

PRICING

Pricing irrigation water to reflect its scarcity value through the elimination of subsidies has been a major theme in Indian canal irrigation reform (GoI/MoWR 1987; Gulati and Chadha 1999). It is argued that underpricing of agricultural inputs creates wasteful use. However, if the policy goal is efficiency and water saving, as it usually is professed to be, pricing will be ineffective in the warabandi system. Here, a farmer draws a fixed water share, regardless of the price he pays. The farmer's water right is defined in terms of time rather than volume. Raising the price of water will not induce a farmer to use less. Similarly, if the goal is resource generation for the upkeep of systems, then, under the present institutional set-up, it would be ineffective.

As noted earlier, resources go into the state exchequer rather than the irrigation department. Thus, the pricing of canal irrigation will be an ineffective tool for reform in northwest India.

As documented earlier, more effective strategies for water saving and equity would have to focus on preventing farmers from taking more than their protective share of water by tampering with outlets or inserting siphons along the canals, so that tail-end farmers can receive their entitlement. This is an accountability issue that calls for responsible local governance.

Further, the switch from a crop/area basis of pricing to volumetric pricing at the user level may not be practical in the warabandi system. Water is delivered through fixed discharge outlets like open flumes and the adjustable proportionate module. However, it may be an option for the shejpali system, accompanied by volumetric payment systems. With gated pipe outlets, the installation of measuring devices, such as the V-notch and the standing wave flume, make it possible to implement volumetric deliveries and pricing to user groups below the outlets. Volumetric supplies have been successfully experimented with in the shejpali system in Maharashtra, as has been already discussed.

PROPERTY RIGHTS REFORM AND THE CREATION OF WATER MARKETS

The creation of water markets through the institution of well-defined, tradable property rights is often advocated. The premise is inspired by fundamental neo-classical economics: well-defined, secure property rights in water will lead to a situation where water is allocated to the highest valued uses, at a market clearing equilibrium price. Furthermore, this price, when it is constituted through the interface of the forces of demand and supply, will convey the scarcity value of water (Anderson and Snyder 1997; Rosegrant and Binswanger 1994).

While theoretically the argument in favour of market creation may sound appealing, it acquires a new dimension in the context of the design characteristics of canal irrigation. Under protective irrigation like warabandi, where water is 'scarce by design', the possibilities for water markets are limited. My observations suggest that while there are some sales, they are confined geographically and very localized. The basis of a water sale is a surplus. A farmer chooses to sell his water right only after he has met his own requirements. When a farmer's water right is inadequate relative to his requirement, as is the case here, there is no saleable surplus. I also found that farmers have a psychological resistance to selling their water share. When a farmer does not need his water share, he chooses to lend it instead of selling it. Lending creates a basis for a future claim, since the borrower is obliged to return it.

In the shejpali system, the creation of water markets is at least technically feasible, given that it is possible to deliver volumetric supplies to farmers and outlets. However, given that water rights get defined in a strict sense only after the farmers have made applications and after these have been sanctioned, it is hard to think of a long-term institutionalization of water markets. In fact, during my field research, no sales of water rights among individual farmers were noticed in the shejpali system.

* * *

Proposals for reform in Indian canal irrigation put emphasis on correcting the incentives facing users and providers of water through pricing, financial accountability, autonomy and privatization, and water rights reform and market creation (Rao and Gulati 1994). Other recurring themes are securing greater coordination and integration in the water sector and its mainstreaming into overall development programmes; making technical improvements in system management; and devolution of functions and powers to user groups to secure decentralization (Mollinga 2000; Narain 1998, 2000; Vaidyanathan 1999). These proposals tend to be rather piece-meal and isolated, and the relationships among them as well as with the technology or design of canal irrigation are hardly considered.

This chapter has shown that the potential for reform in irrigation systems varies with the design of canal irrigation and the concomitant institutional framework for water allocation. The warabandi and shejpali systems of irrigation need to be understood as composite socio-technical systems, comprising a specific physical layout and an accompanying institutional framework for irrigation management, involving an organizational set-up for irrigation functions and specific notions of rights and entitlements. Their physical characteristics impose certain organizational requirements that shape the extent of user involvement possible in management, thereby influencing the potential of decentralization. Water rights are embedded in these alternative technologies; they represent certain configurations of design and social relationships. The definition of rights, whether volumetric or in terms of time or turn, influences the role that pricing may have on water use decisions.

While pricing, rights reform and market creation, and decentralization are commonly understood as alternative approaches to institutional reform in canal irrigation, they should perhaps be seen as interrelated components of overall reform packages. Further, their fit with the technological design characteristics needs greater appreciation. A prerequisite is that we see irrigation systems as composite socio-technical systems, wherein the physical layout corresponds to certain norms, property rights, organizational designs, and options for institutional development and transformation.

Notes

[1] Empirical studies have hardly influenced public policymaking in irrigation.

[2] These are semi-modular outlets: discharge varies with the upstream water level, for example, the open flume and the APM.

[3] However, it is rarely followed as prescribed.

[4] A cusec is a discharge of one cubic foot per second (28.3 litres per second). Many irrigation schemes in India still use the older imperial units of discharge and flow. They are also used here, but converted to metric units or given as the converted amount in brackets where relevant. One acre equals to 0.4047 hectares.

[5] Kharif and rabi are the two major cropping seasons of northwest India. Rabi is the winter season, while kharif is the rainy or monsoon season. Western India has a third season, called the hot weather season.

[6] Outlet discharges are based on the water duty used for a scheme, or the area (in acres) that can be irrigated by a flow of one cusec. For instance in this case, an outlet of 28.3 litres per second (1 cusec) would serve 120–160 hectares.

[7] Personal communication, Mr Mattoo, CADA, Rohtak, 11 December 2001.

[8] The limited description of technical design characteristics often limits the study of these systems (Jurriëns et al. 1996). One problem with the documentation of design data is that it is not stated to what level it applies;

similarly, when intensities are given, we do not know how they vary across the seasons (Jurriëns *et al*. 1996).

[9] Note that 'he' stands for (s)he.

[10] Further details are provided in WALMI (1998a).

[11] This should be contrasted with outlets in the warabandi system that are semi-modular, wherein the discharge varies only with the upstream water level.

[12] This intensity is split up as 60 per cent for rabi, 10 per cent for hot weather, and 7 per cent for kharif. Personal communication, Dr S.A. Kulkarni, Indian National Committee on Irrigation and Drainage (INCID), New Delhi.

[13] Personal communication, Dr S.N. Lele, Society for Promoting Participative Ecosystem Management (SOPPECOM), Pune.

[14] In the command of the Mula project (my research site) the average annual rainfall is 594 millimetre (Lele and Patil 1994). However, this is much less than the modal rainfall occurring 50 per cent of the time. Allowances need also to cover for conveyance and distribution losses in the system.

[15] These figures were arrived at through negotiation between the irrigation department, the WUA, and an NGO.

[16] There are pieces of land that receive canal water in all three seasons.

[17] This assumes that the water is spread over the entire 59,292 hectares.

[18] This is an estimate of average rainfall in Rohtak district based on data obtained from the agriculture department, Rohtak district, Haryana.

[19] The possibility of lateral inflow through canal seepage is not taken into account here. In both cases, there is some recharge from canal waters.

[20] Wahaj (2001) discusses similar responses in Pakistan.

[21] 'Bhai' means brother; bhaichaara indicates a feeling of brotherhood, friendship, and affection. Farmers explain bhaichaara as 'being born from the same womb', familial ties, friendship, and brotherhood.

[22] In the chak under the Rampur outlet that I studied, each *kila* (acre) of land was allocated 14.5 minutes of irrigation.

[23] In Rampur village, this rate varies anywhere from INR 4,000 to INR 10,000 per kila per year, depending on the quality of the land. Land near the canal commands the highest price, as much as INR 10,000 per kila.

[24] This amount is an hourly rate independent of quantity.

[25] These acts are liable to a fine up to 30 times the water rate.

[26] Lower-level irrigation department worker who patrols the canals.

[27] Interview with Gauge Reader at the head of the Sitapur Minor, 19 September 2000.

[28] Jowar (sorghum) and bajri (pearl millet) are staple food crops in Maharashtra grown for subsistence. Wheat is grown for subsistence and the market. Sugarcane and groundnut are commercial crops.

[29] Unlike in warabandi, farmers have an option to apply or not to apply for water.

[30] Normally, smaller demands of up to 0.5 hectare are admitted in full, and cuts are applied to the demands for larger areas.

[31] Waterman in charge of water distribution for about 800 to 1,000 hectares in the shejpali system. He is the lowest rung of the irrigation department, in direct contact with the irrigators; he opens the outlet gates and releases the water for distribution among the irrigators.

[32] Gat is the revenue record number.

[33] Earlier a notice was also put on the individual outlets, but was often torn down by village children. So the practice was discontinued.

[34] But it is interesting to note that with the specification of the time for which a farmer is to take water in this system, it comes very close to how the warabandi system works on the field (below the outlet) in northwest India. In fact, during my interviews, when I asked a farmer how he took water, he pointed to his watch and said 'according to time'.

[35] When the WUA is not formed, this is done by the Paatkari.

[36] This parshall flume is located on the downstream side of the pipe outlets.

[37] They also repair and maintain the field channels, including desilting and deweeding, keep daily water accounts and, after the rotation, prepare a completion report on water used. The WUA hires labour for cleaning the watercourses. See Lele and Patil (1994).

[38] Gauge patti in the local language.

[39] A panchanama is a statement of penalty.

[40] When I mentioned the panchanama in an interview with a Paatkari of the WUA, there was a furor. It seemed to be a real act of shame and dishonour to be issued a panchanama.

3 THE LOCAL POLITICS OF POLICY IN THE ANDHRA PRADESH IRRIGATION REFORMS

Bala Raju Nikku

Neo-liberal policies promoting financial reform and state disengagement have had a large impact on the irrigation sector in developing countries. There has been a growing interest in the impact and performance of irrigation management transfer (IMT) or participatory irrigation management (PIM) models (Groenfeldt and Svendsen 1997; Mollinga and Bolding 2004).[1] The Mexican and Philippine irrigation reform models were widely debated and disseminated as successful examples of, and models for, management transfer policies. At the same time, the lack of critical analysis of the IMT models is also evident (Mollinga and Bolding 2004; Nikku 2006; Oorthuizen 2003; Rap 2004).

Andhra Pradesh (AP), a federal state in southeastern India, is known as the 'rice bowl' of India.[2] With a population of 76 million (in 2010), it is the fifth largest state.[3] It presents a wide spectrum of climatic and other variations. Rainfall varies from 500 millimetres in the southwest to about 1,200 millimetres in the north. Its economy is still based primarily on agriculture. AP benefited from the development of irrigation infrastructure before, during, and after the colonial period, and was a leading participant in the Green Revolution. However, a decline in public investments and a lack of user participation in irrigation system management have led to the deterioration of irrigation infrastructure. To address this issue, AP has embarked on irrigation management reforms.

The AP reform policy is based on the neo-liberal assumption that scaling down the irrigation bureaucracy, increasing water charges, and transferring maintenance of irrigation infrastructure and water distribution to elected water users' associations (WUAs) can address the ills of irrigation management and lead to growth of the irrigation sector. The AP irrigation reform of 1996, implemented with a 'big bang' approach, has attracted much national and international attention; it became known as the 'Andhra Model' (Nikku 2002, 2003).[4]

This chapter critically examines the Andhra model of irrigation reform and its proclaimed success, and shows how various policy actors attribute different meanings to the reform policy and contest its implementation. It focuses on local interactions at the level of WUAs as they were developed and empowered by the state. Thus, this chapter contributes to discussions on the interfaces between policy, irrigation technology, and institutions in these irrigation reforms. With this aim, this chapter next outlines the conceptual framework of 'politics of policy' used in the analysis, and the subsequent sections examine the main characteristics of the AP irrigation reform and the contestations of different policy actors. These contestations and processes of negotiation are discussed for important arenas of irrigation: operation and maintenance (O&M) and rehabilitation works, irrigation water distribution, joint inspection (*azmoish*; JA) and irrigation fee collection, and the position of the *luskars* (the lowest-level irrigation department staff).[5] Together, these sections provide critical insights into the limitations of blueprint approaches and the proclaimed success of the AP irrigation reform model.

The findings presented here are based on empirical research undertaken in the Nagarjuna Sagar irrigation project in AP from March 2001 to July 2002, and additional short fieldwork periods between 2002 and 2004. The case study site was the Madhira Branch Canal (MBC), a secondary canal of the Nagarjuna Sagar Left Bank Canal (NSLC) system, irrigating 39,387 hectares of command area. Some 44 per cent of this area is localized for producing 'wet' irrigated crops like paddy, with the rest prescribed for irrigated dry crops; this crop plan and wider scarcities emerging in the NSLC system have created competition and head–tail end struggles for water. The MBC is 33 kilometres long, extending further as the Nidanapuram Major Canal for another 50 kilometres. The research adopted an interdisciplinary approach, using a combination of qualitative research tools including participant observation at the village and WUA levels, semi-structured interviews with irrigation officials and WUA leaders, and group and key-informant discussions with WUA members.

POLICY PROCESSES IN IRRIGATION: A CONCEPTUAL FRAMEWORK

IMT has become a major strategy in irrigation. All over the world governments are reducing their roles in irrigation management while farmer groups or private organizations are expected to replace them (Vermillion 1992). Governments often pursue management transfer programmes to reduce their expenditures on irrigation, improve productivity, and stabilize deteriorating systems (Vermillion 1997: 1). 'Autonomy with accountability' seems the core idea of turnover policies (Kloezen and Samad 1995). The instrumental and mechanistic character of the IMT model and the tendency to see it as a neutral planning and implementation routine, instead of a complex socio-political and policy process, needs further analysis.

POLICY AS PROCESS

Policy is a process of 'becoming' rather than 'being' (Dye 2001).[6] Ham and Hill note that policy is dynamic rather than static and that we need to be aware of shifting definitions of issues (1984: 12). Conceptualizing policy as a process means an explicit acknowledge-ment of the importance of the social and historical context in which policy is shaped and implemented (Mooij 2003). Mooij and de Vos (2003) argue that the idea of policy as process prevents assumptions that policies are 'natural phenomena' or 'automatic solutions'. A policy is never fixed, but is always evolving and changing. Policies are embedded in social relations, power structures, negotiations, and cooperation practices in society.[7] This makes policy very com-plex. Some scholars see policy as essentially a process of bargaining and competition between different groups in society (Dahl 1961). Policy actors can be individuals or a collective influencing a policy to achieve their own or collective material and non-material interests. The interaction between actors and policy has been explained with concepts like 'policy networks', 'pluralist interest groups', or 'policy pressure groups'[8] (Keeley 1997).

A process-oriented framework starts from the observation that the outcomes of policy implementation are highly variable. Implementation is an ongoing, complex, and interactive process of decision-making by interest groups involved: governments, bureaucrats, 'beneficiaries', funding agencies, and so on. Policy implementation is an example of strategic action in which (usually) a government agenda becomes articulated with local interests, the policy content is renegotiated and transformed, and particular intended and unintended outcomes are produced.

POLICY AS POLITICS

The 'politics of policy' framework provides a further tool for analysing individuals' actions in different policy arenas. Policy is a contested resource, shaped by various stakeholders (Mollinga 2001). Mollinga and Bolding remark that the word 'politics' is virtually absent from the formal policy discourse on irrigation reform. It sometimes appears in euphemistic terms or as a black box, but 'explicit analysis of the political dimensions of irrigation and irrigation reform is rare' (2004: 4). Grindle (1977) sees 'the politics of policy' as including 'politics of implementation', referring to the political nature of the policy formulation and implementation process. Different interest groups and individual actors contest policy at all stages and try to shape it in particular ways.

POLICY AS AN ARENA FOR CONTESTATION

Policy contestation[9] is a form of collective and individual action in space and time. Policy actors mobilize individual and collective resources and networks in their location to influence the policy process. If power is always constituted and exercised in social relationships, policy contestation by actors is directly proportional to their ability to use and mobilize power.

This chapter focuses on the execution of irrigation reform policy at the local level of the system. Day-to-day implementation is seen as an arena in which actors responsible for allocating resources engage in influencing resource allocation decisions related to critical fields of 'water, wages and works' (Oorthuizen 2003). This involves, in particular, politicians and bureaucrats at the state level, the implementing bureaucracy at the local level, and recipients and 'beneficiaries' of reforms, particularly water users, at the local level. Much of the irrigation administration literature focuses on the higher-level bureaucrats. The upper levels are certainly important, but so are frontline workers in the field who daily confront the problems about which the upper officials make policy (for the role of lower-level bureaucrats, see Lipsky 1980; Rap 2004; and Tendler 1997). Egeberg (1999: 167) concludes that most students of public administration seem to have focused on behaviour and attitudes without relating them explicitly to bureaucratic structure.

UNDERSTANDING THE NATURE OF THE INDIAN IRRIGATION AND CANAL BUREAUCRACY

Independent India inherited many capital-intensive large irrigation systems from the colonial period. The management of these systems fell largely in the domain of the states and the national government. Construction-oriented development continued under successive governments, and irrigation acts like the Bengal Irrigation Act, the North Indian Irrigation Act, and the Bombay Irrigation Act further strengthened state control over the large-scale irrigation sector.

After independence, the Indian canal bureaucracy retained much power. The irrigation sector has been monopolized by state irrigation departments. The irrigation bureaucracy gained control over the management of irrigation systems, and management practices were changed to suit its convenience (Sengupta 1997). Wade (1982) also stresses that canal officials enjoy great discretionary powers: their decisions impinge heavily on the political prospects of politicians and on the economic well-being of local communities. A report of the World Bank and the Government of India (GoI) states that 'the irrigation departments are generally viewed as centralized, bureaucratic, isolated, and financially dependent on state government funds. The departments are not accountable to the end beneficiaries: the farmers. Their emphasis tends to be on system construction rather than O&M of existing projects' (World Bank and GoI 1999).

There is considerable literature on the relationships between bureaucrats and politicians (Kothari and Roy 1969). Bhalerao (1973) notes that bureaucracies have to function under democratic political leadership. Though states need bureaucracies, often both the political executive and the bureaucrats seek to protect and strengthen their powers. While doing so, they may accommodate each others' interests, resulting in a politician–bureaucrat nexus.

THE MISMATCH OF TECHNOLOGY, INSTITUTIONS, AND POLICY

Being an important input for irrigated agriculture and other uses, water is heavily contested at various levels by (groups of) farmers, irrigation engineers, and other interest groups like state-level politicians. Many scholars have shown that irrigation infrastructures can become signposts of struggle, their design being altered and re-altered under the influence of the changing balances of power around the policy process (Mollinga and Bolding 1996; Narain 2003a). The Andhra experience of irrigation reforms is unique in terms of its sheer size, spanning over 4 million hectares of farm land and about 6 million farmers organized in 10,800 WUAs, 323 distributary committees (DCs), 23 major, and 60 medium project committees (PCs). Once the institutional structures and procedures related to reform policies have been established, it is crucial to understand how these policies influence actual institutional processes around irrigation and vice versa. The outcomes of the reform are shaped by different actors and their ability to influence policy.

THE MAIN CHARACTERISTICS OF THE ANDHRA PRADESH IRRIGATION REFORMS

Salient features of the AP irrigation reform are its legislative support, transfer of certain responsibilities from the irrigation bureaucracy to WUAs, and the rapid approach to implementation that earned it the name of 'big bang' approach. In addition, the World Bank and other agencies extended their financial support to PIM in the state during the leadership of Mr Chandra Babu Naidu, chief minister of the state between 1995 and 2004.[10] The World Bank played an important role by providing not only funds but also advisory input and by increasing the visibility of the AP reforms (see Nikku 2006; Reddy 2002). To implement the reforms successfully the state enacted the Andhra Pradesh Farmers Management of Irrigation Systems Act (APFMIS Act) No. 11 of 1997.[11] The reforms provided a 'window of opportunity' for advocates of IMT policies seeking to address the technical, organizational, and socio-political dimensions of irrigation management.

MOTIVATIONS FOR IRRIGATION REFORMS IN ANDHRA PRADESH

About 40 per cent of the gross cropped area in AP is irrigated and contributes 60 per cent of total agricultural production. However, agricultural growth declined to less than 2 per cent per annum in 1995. Therefore, rehabilitating and sustaining irrigation and enhancing its agricultural productivity are of paramount importance to AP (GoAP 1996). Thus, the declining performance of existing irrigation systems led to the introduction of reforms. Peter (2002) argues the need for reforms, pointing to the dilapidated condition of systems, bad maintenance, shrinking command areas, low irrigation efficiencies, and tail-end problems. He also notes the 'anarchic situation' in command areas, where head-enders appropriate most water, and farmers tampering with the irrigation structures cause further damage to the systems. A further problem, according to this author, is a lack of coordination between the departments of irrigation, agriculture, and revenue. As a consequence, a gap has grown between the potential irrigated area created and the area actually utilized. This 'gap area' has been estimated at around 0.5 million hectares. The situation is further aggravated by, among others, a lack of established O&M procedures and a lack of funding for O&M; most O&M funds were used to pay staff salaries rather than undertake maintenance.

Other scholars have argued that the AP irrigation reforms are the outcome of an alliance between the ruling party political heads and international donor agencies like the World Bank[12] (Reddy 2002; Venkateswarulu 1999). Yet others are of the opinion that the APFMIS Act is meant as a first step in the direction of a comprehensive Irrigation Act applicable to the entire state of AP. Furthermore, the existence of a large irrigation bureaucracy (12,000 engineers and 42,000 other staff) in AP raises the issue of downsizing the bureaucracy under reforms. Capacity-building and attitudinal change of the irrigation bureaucracy are also given importance under the reform programme.

Thus, it is clear that a host of political, economic, technical, and ideological motives have led to the irrigation reforms. It was, moreover, a combination of a genuine need for reforms and a favourable political environment. It was evident that irrigation sector performance was deteriorating, and, at the same time, the Telugu Desam Party (TDP) government headed by Mr Naidu wanted to bring changes in the administration to gain an identity. The Chief Minister himself seemed to be a strong supporter of the reforms. Thus, the low level of performance of irrigation systems, political support, and timely availability of external funds together led to the irrigation reforms.

THE APFMIS ACT OF 1997 AND INSTITUTIONAL ARRANGEMENTS

The APFMIS Act of 1997[13] was prepared as comprehensively as possible, to be applicable across the entire state. The Act was passed in the Legislative Assembly of AP on 27 March 1997. Consequently, in the *State Gazette*, it was published as: 'An act to provide for farmers' participation in the management of irrigation systems and for matters connected therewith or incidental thereto.' As a result, state-wide elections were held in June 1997 for more than 10,000 WUAs and 174 DCs covering major, medium, and minor irrigation systems.

The APFMIS Act recognizes that the 'scientific and systematic development and maintenance of irrigation infrastructure is considered best possible through farmers' organizations'. The Act also states that 'such farmers' organizations have to be given an effective role in the management and maintenance of the irrigation system for effective and reliable supply and distribution of water' (APFMIS Act 1997).

Under the Act, each WUA has a right to receive irrigation water as per agreed allocations and information related to water supplies (periods and quantity) and canal opening and closing dates. The members enjoy a right to receive water as per a specified quota for use, and are allowed to grow any crop other than those prohibited by law, adjusted to the water allocated to them. They also have the right to sell or transfer the water share to other water users in the WUA (rules 5 and 7 of the APFMIS Act 1997). These are crucial provisions for both WUAs and users, especially given how irrigation management is otherwise in the hands of the irrigation department. None of the earlier irrigation acts in the state had these provisions.

The reform policy envisages a change in the attitude of the irrigation bureaucracy. It should become a facilitator rather than remaining a traditional provider. The Act provided additional powers to irrigation bureaucrats to approve financial and technical matters. It also assumes that the collective role of the water users makes the Irrigation Department more sensitive to the service role it should be playing. A major intention of the Act is to facilitate collective action by users so that they exert pressure in times of need, which should make the irrigation department service-oriented and accountable to users.

The Act aims at forging linkages between the different levels of farmer organizations. In major irrigation systems, for example, WUAs are responsible at tertiary canal level, a DC at the secondary canal level, a PC at the system level, and an apex committee at the state level. The representation of WUA leaders at each level was viewed as a strategy to strengthen the decision-making process. Another important feature is that the Act authorizes the WUA to manage conflicts and impose fines. Thus, the Act provided scope for creating incentives and rewards for individuals and institutions, leading to better management of the irrigation systems. It also envisages two-way communication between users' organizations (WUAs, DCs, and PCs) and the irrigation agency at all levels (see Figure 3.1).[14]

IRRIGATION OPERATION AND MAINTENANCE WORKS

ADDRESSING THE MAINTENANCE CRISIS

Gulati *et al.* (1999) argued that 'silently watching the canal system heading towards a collapse would be nothing short of an economic crime'. The AP irrigation reform policy explicitly aimed to address the maintenance crisis. It promoted user involvement and ownership as a pre-condition for implementation of O&M activities.[15] The formation of WUAs in AP took place before the rehabilitation of irrigation structures.[16] The APFMIS Act declared that 'scientific and systematic development and maintenance of irrigation infrastructure is considered best possible through WUAs. These organizations have to be given an effective role in the management and maintenance of the irrigation system for effective and reliable supply and distribution of water'.

The reform programme emphasized rehabilitation of existing irrigation infrastructure and maintenance to address the problem of system degradation. To provide a momentum to reform policy

utilized and did not participate in decision-making. This further confirmed the nexus.

UTILIZATION OF FUNDS

During the 1998–9 period, the first funds for rehabilitation and maintenance under the reform programme were released. In April 1998, the government released a total grant of INR 1,064.7 million to WUAs, at the rate of INR 247 per hectare for the total localized command of approximately 4.4 million hectares. In 1999–2000, the focus of works was entirely on minor rehabilitation of irrigation structures. However, in the following year, the allocations for small rehabilitation works also decreased. According to the government policy of 2000–1, the WUAs were expected to meet the costs of regular maintenance works from internal resources and from the share of irrigation cess (charge) collection that they receive from the concerned revenue departments (RDs).

Although the government allotted uniform funds on the basis of command area to all WUAs, not all WUAs could utilize the available budget. Only a few WUAs formed under the MBC made full use of the minimum rehabilitation funds in 1998–9 (Table 3.1). Some WUAs could utilize only part of the funds, as they could not complete the required quantum of works.

The fund utilization pattern of WUAs shows that during the initial years, about 35 per cent of the funds were spent on earth works, clearing jungle weeds, and removal of silt from the canals. The remaining 65 per cent was spent on repairs of irrigation structures like sluice gates, culverts, and other small structures. This expenditure pattern raises the question why more budget was spent on irrigation structure rehabilitation than on maintenance works.[17] It also raises the question whether providing funds uniformly on the basis of command area is rational for WUAs. The field evidence suggests that it is not, as more funds tend to be needed for tail-end WUAs than for head-reach WUAs.

The WUA presidents, especially from tail-end areas, claimed that their resource needs were much more in terms of repairs, as the structures had been tampered with by the users in the past. Further, many users had changed their cropping pattern from wet to irrigated dry crops due to water scarcity. This has implications for

TABLE 3.1   Pattern of Fund Utilization by WUAs under MBC
(DC 14 and DC 15): Data from Individual WUAs
and Irrigation Subdivision Records

WUA	Command (in ha)	O&M (×Rs 1,000)	Minimum Rehabilitation (MR) (×Rs 1,000)		
DC14		1998-9	1998-9	1999-2000	2000-1
169	2,214	172,820	733,519	436,489	111,451
170	1,497	307,451	NC	290,473	76,039
171	1,631	79,548	NC	255,459	46,221
172a	1,491	84,691	NC	245,268	72,446
172b	1,199	24,000	NC	130,738	38,632
173	1,803	NC	NC	163,093	66,905
174	2,549	117,114	592,120	474,079	105,142
175	2,447	232,695	NC	451,865	74,446
176	2,309	87,169	140,506	433,711	101,980
Total	17,140	1.105,488		2.881,175	693,262
DC15		1998-9	1998-9	1999-2000	2000-1
177	2,214	232,695	NC	260,750	77,139
178	3,010	285,678	NC	411,052	146,826
179	1,651	105,768	NC	291,836	52,141
180	2,035	205,894	NC	364,334	48,623
181	1,951	186,400	NC	269,953	47,281
182	1,906	247,772	NC	301,460	75,030
183	2,610	284,713	NC	460,788	100,676
184	2,535	232,975	NC	472,660	114,442
185	2,457	207,382	41,048	473,211	101,100
Total	20,369	1.989,277		3.306,044	763,258

Source: Nikku (2006).

Note: NC = not carried out. In 1998–2001, INR 100 was USD 2.1–2.3. In both DC 14 and DC 15, the sequence of WUAs in the table goes from head-end to tail-end.

the revenue for the WUA, as the crop cess for irrigated dry crops is less than for 'wet' crops (water-demanding crops such as rice and sugarcane). Hence, the revenue cess (charge) for tail-end WUAs has decreased compared to head-reach WUAs, which represent more 'wet' cultivation.

on 'canal *gasti*' (canal guarding by private people) to bring water from the head reaches to their tail-end plots. They thought the situation would change with the introduction of WUAs in 1997. However, as Box 3.1 shows, tail-end farmers continued to be deprived of their legitimate share of water, and the practices of water distribution did not change after the reforms.

I conclude from such evidence that WUA leaders have been uninterested or unable to take over water distribution functions from the irrigation department, nor does the department want to lose control over distribution. In fact, in the absence of strong, functional WUAs and front-line workers with strong powers of control, WUA representatives, farmers and irrigation department staff still had to work together to ensure water distribution. These findings are similar to those of Oorthuizen (2003) who noted that joint action was still needed between front-line staff and farmers to distribute water in conditions of shortages or power struggles. Contrary to policy objectives, water distribution seems to have remained an arena in which different actors primarily protect their own interests at the expense of the legitimate rights of those users without power to influence the distribution.

JOINT AZMOISH AND IRRIGATION FEE COLLECTION

The then Ex-officio Secretary of the Irrigation and Command Area Development Department, Government of AP, J.R. Peter (2002) stated that 'if the WUAs function as the government anticipates, the revenue recovery will go up from 64 per cent to 95 per cent and the area under irrigation by more than 3 per cent a year'. Vaidyanathan (1999) argues for India as a whole that it is the poor service of irrigation that led to the poor payment of water charges. For example, water availability has never been a problem for the head-reach farmers. But these farmers too often do not pay the water cess, due to the weak collection mechanisms (see Reddy 2003).

The APFMIS Act empowers the state RD to collect irrigation revenues related to irrigation water supply by the irrigation department. To achieve better results, the so-called 'joint azmoish' (JA) of irrigated commands by both revenue and irrigation departments had become established in the state. JA is a joint supervision or survey of an irrigated command area in a hydraulic unit. As part of the irrigation reforms, it became a joint survey conducted by the representatives of the WUA and staff members of the departments of irrigation, revenue, and agriculture, to report on irrigated area and type of crop. The government has legitimized the participation of WUAs in JA through the APFMIS Act, and issued order number 155 on the estimation of irrigated acreage by the WUAs and submission of the estimation to the RD for cess collection.

The activities of each participating department are jointly coordinated and complementary to each other. The policy envisages that the concerned staff members and the WUA representatives together make a walk-through survey in the designated command of the WUA at the end of each kharif and rabi crop season.[18] Through the survey, the irrigated area and types of crops grown in the WUA command area are determined. Based on the survey results, the RD collects the water cess from the farmers in the command villages. The WUA is expected to facilitate the collection process. The share of cess collection is to be debited to the WUA bank account by the revenue officials.

To facilitate the process, the government fixed a particular month to conduct the JA: September for the kharif crop and February for the rabi crop. The policy also states that WUAs should be involved in collection of water cess and shall keep their share of revenue out of the total collected cess.[19] The rest will be deposited in the government treasury. Incentives for WUAs to perform better were also considered.

JOINT AZMOISH AND REVENUE COLLECTION IN PRACTICE

Contrary to these policy objectives, the involved departments (irrigation, agriculture, and revenue) contested this policy idea by not fully participating: bureaucrats involved continued their earlier practices and defended their interests by not participating in the process. There seems to be a common understanding between the officials; none of them have taken any initiative to improve the process. As a result, they continued to use different formats to report the command statistics and underreporting continued. Box 3.2 provides an example.

a manner as to make available the services of one or more luskar for each WUA. The government also proposed another rule regarding the method of payment, through an order issued stating: 'Payment of salary to luskars shall be made subject to the furnishing of a certificate by the concerned farmers' organization president that they had attended to their duties during the month.'[21]

The government order further stipulates that 'the concerned Chief Engineer shall complete the process of reallocation of the above personnel to the farmers' organizations and furnish details to the government through the engineer-in-chief by 30th of October 1998 positively'. In the case of resistance, the government gave them three choices: (*i*) a voluntary retirement scheme; (*ii*) work according to government instructions; and (*iii*) punishment of erring staff by the government.

According to a DC president, even before the government order requiring certification of a luskar's work, the WUA leaders demanded that the WUA would be provided with additional manpower to carry out its activities. He explained how they had shaped government orders:

Twenty-five WUA presidents passed a resolution in August 1997, immediately after they got elected to the office at Kadam, specifying that luskars should work under WUAs and meet with the managing committee members regularly. They also made a request to the deputy executive engineer of the area to release the salaries of the luskars based on their recommendation or a certificate issued for the respective month. However the luskars opposed the idea.[22]

The luskars opposed the government orders and WUA leaders' demands. 'We are accountable to the government but not to the WUAs' is a common statement given by WIs and luskars. According to the president of the Luskars Union at Kalluru division office, there were about 3,500 luskars working with the AP irrigation department by 2002. This Luskar union president stated that:

The government gave us three options to choose from in case we are not willing to work under WUAs. We see the government move as a blow to our self-respect. We are enjoying our status as government employees and receiving respect in society. We will lose our respect if we work under the WUAs. We are ready to lose our jobs but not our respect by working under the associations.[23]

The luskars and work inspectors represented by their unions filed a legal case in the state high court against the government. They have succeeded in getting a stay order from the court. As a result the government order of handing over the luskars and WIs to the WUA has not been implemented. The government has been unable to vacate the stay and implement the order.

According to the union representative, their lobby received support from their higher-level officers too. The engineers' association opposed the transfer of roles and responsibilities of luskars to WUAs. The Luskar union's president concluded that 'the officers supported us, because they do not want to lose us to WUA presidents. We have been used to do the personal work of the officers. So they cannot afford to lose us'.[24] The result was that the frontline bureaucrats could mobilize their resources and networks to defend their status and interests against the government reform policy.

* * *

The AP irrigation reform programme aimed to reform the state irrigation bureaucracy and to empower farmer organizations to improve services. It also expected attitudinal changes among the bureaucrats while implementing the PIM policy and programme. This chapter has focused on how policy, technology, and institutions interacted and produced results that are not in line with the policy objectives. Irrigation governance is not only determined by the irrigation technology, but also by the self-interests of the policy actors and broader institutional priorities. Local-level processes crucially shaped irrigation policy implementation, and hence the outcomes of policy reforms are very different from the original policy objectives.

At the local level, the additional powers that the lower-level engineering staff acquired through the APFMIS Act resulted in some cases in a special relationship between them and the WUA leaders in execution of works, but policy aims and outcomes became weakened by problems of budget releases and fee collection. The lowest level of the luskars saw the reforms as a threat to their jobs initially, but coped with the process by mediating it. They were successful in lobbying with the state and using legal means to resist transfer to the WUAs.

Act 1997 in this chapter have been sourced from http://www.apmitanks.in/templates/files/Others/AP%20FMIS%20Act.pdf (accessed December 2009).

[12] Following a popular upsurge in the TDP, Mr Naidu was unanimously elected as the Chief Minister of AP on 1 September 1995, dislodging Mr N.T. Rama Rao. Following the elections to the State Assembly wherein the TDP led by Mr Naidu emerged as a winner, he was sworn in on 11 October 1999 as the chief minister for the second term. The critics said that he did not have charm and a mass base like the former chief minister N.T. Rama Rao and hence wanted to create his own image and stronghold on the party by revamping the administration, for which he chose the path of reforms.

[13] To strengthen the irrigation reforms, many administrative reforms were introduced to increase the efficiency of the irrigation department (Nikku 2006).

[14] The highest level of government supervision comes from the Minister of Major and Medium Irrigation, and the Minister of Minor Irrigation. At agency level there is the Principal Secretary of I & CAD department, also with three to four secretaries, each of whom is assisted by three to four deputy or joint secretaries. All of them belong to the Indian Administrative Service. These bureaucrats exert much influence on policy-making and implementation.

[15] As a precondition, researchers and practitioners argued that the irrigation structures should be fully repaired prior to or after management transfer to the WUAs (Johnson *et al.* 1995; Meinzen-Dick 1997; Vermillion 1997).

[16] The popular IMT/PIM discourse in India and elsewhere stresses that the turnover of irrigation O&M works to WUAs will achieve quality of water distribution, low costs of irrigation to farmers and governments, and transparent and efficient water use (Mitra 1992; Sengupta 1991; Uphoff 1986).

[17] Maintenance work involves low-level expenditures with low allowances (100 INR per acre), while rehabilitation is a higher level of activity and expenditure is higher to rehabilitate the system. In practice the definitions are mainly important for public finance procedures under the programme, and are otherwise part of a continuum due to the poor condition of the systems (Oblitas and Peter 1999: 24).

[18] Kharif is the summer or monsoon crop, while rabi is the winter crop.

[19] The revenue in major irrigation systems is shared in a proportion of 50 : 25 : 10 : 10 : 5 per cent to the state (irrigation department), WUA, DC, PC, and Gram Panchayat, respectively.

[20] Lipsky (1980) showed how the working conditions of street-level bureaucrats influence policy implementation. He argues that the peculiar character of their work structures policy transformation. They are squeezed

between the demands and expectations of their superiors and those of their clients. For the Philippines, see Oorthuizen (2003).

[21] GoAP, 1998, GO Rt. No. 338, dated 12 October, I&CAD Department.

[22] Interview followed by personal communications with Y. Rama Mohana Reddy, President of Distributory Committee, Tungabhadra Project High Level Canal, Anantapur, during fieldwork conducted 2000.

[23] Personal communication Y. Rama Mohana Reddy.

[24] Statement of the president of the Luskar Union of Kalluru irrigation division. The interviews were held during fieldwork conducted during 2002.

4 IRRIGATION TECHNOLOGY AND IRRIGATION MANAGEMENT REFORM: THE CASE OF THE TERAI REGION IN NEPAL

Puspa Raj Khanal

Many countries have engaged in irrigation reform programmes called irrigation management transfer (IMT) or participatory irrigation management (PIM).[1] Vermillion and Sagardoy (1999) define IMT as the relocation of responsibility and authority for irrigation management from agencies to non-governmental organizations (NGOs) like water users' associations (WUAs). Such programmes aim to achieve better service provision through users' involvement. Other possible objectives are reduction in public expenditure and empowerment of farmer groups. The motivations for IMT are usually well described. However, the philosophy of programme design—in the choices of new institutional and technical structures promoted, processes for their joint analysis and adoption, and the role given to technology redesign—are rarely presented.

This chapter presents a study of these design issues for the Irrigation Management Transfer Programme (IMTP) of Nepal, based on research in three irrigation systems in the Terai of Nepal (Khanal 2003). It studies how the IMTP used action around technology to facilitate management changes and wider organizational transformations. In this, it looks at the critical role of participatory technology development (PTD) in supporting and building the new WUAs promoted as a new form of irrigation governance.

In Nepal, IMT to new local organizations in agency-managed irrigation systems (AMISs) began in the 1990s. The reasons for pursuing this reform were threefold. The first was to tackle the increasing dependency on the government for irrigation system development and management, while system performance remained relatively low. The second was the influence of and dependency on foreign donors. Donors favoured less government-sector and more private-sector involvement in development activities. Third, the reforms were inspired by the strong tradition of farmer-managed irrigation systems (FMISs). Thus, IMT was formulated around a decentralized and user-centred approach emphasizing participation and new organizational development.

IMT in Nepal involved both institutional reform and technical rehabilitation to facilitate this decentralization. However, the primary philosophy behind programme design was a 'modernization' approach to institutional innovation. The irrigation policies and programmes developed reflect both broader framings of modernization of society and also technological concepts seen as relevant for (more) modern water control. Modernization is viewed here as a process of technical and managerial upgrading of irrigation schemes to improve efficiency, productivity, and environmental sustainability of irrigation schemes and water delivery to farms.

The policies for IMT in Nepal recognize that regeneration of irrigation infrastructure is an essential prerequisite for local organizations to take up management responsibilities. However, the dynamics behind technological change to support institutional reform were less debated. This chapter aims to increase understanding how IMTP translated policies into on-the-ground changes in the evolution of WUAs, and how actions around irrigation technology related to these. It also hopes to improve understanding of how participatory approaches can be applied beyond just an instrumentalist perspective. It shows the relevance of a focus on technology design in establishing service-oriented water control—where user organizations develop technological and institutional changes that can provide water deliveries in ways appreciated by users—by supporting local ideas for new infrastructure and management.

The first section of this chapter presents the case study sites and the study methods. The next section introduces the context of the IMTP given the history of irrigation development in Nepal, and reviews its implementation framework, the institutional frameworks promoted, and efforts to promote new WUAs. The third section describes how the WUAs changed in terms of key management performance and water delivery. The fourth section looks at wider social outcomes and transformations with organizational evolution, and WUA evolution as state programme support ceased. The fifth section focuses on the role of PTD in these change processes to improve service-oriented water delivery consistent with WUA management capabilities. The concluding section of the chapter argues that future reforms must move beyond contemporary models of PIM and IMT, to bring a new realism in reforming technology and institutions together.

CASE STUDY SITES AND STUDY METHODS

The research yielded qualitative and quantitative data to illustrate three processes. The first was to recall and 'restructure' the processes and actions under IMTP, both as documented and as they actually took place. This was done through documents and interviews. The second was to study local management and its evolution through direct field observation (including some measurements of water flows and silt deposition), interviews with key actors, and secondary literature. The third was to understand the effects of change, primarily through asset surveys.[2] For two of these, I joined two independent surveys started then;[3] the third, Panchakanya, was done during my PhD research (Khanal 2003).

The reviewed schemes were selected from the first phase of IMTP policy implementation: Panchakanya Irrigation Scheme (PIS), Khageri Irrigation System (KIS) and Nepal West Gandak Irrigation System (NWGIS) (see Figure 4.1). The PIS and KIS

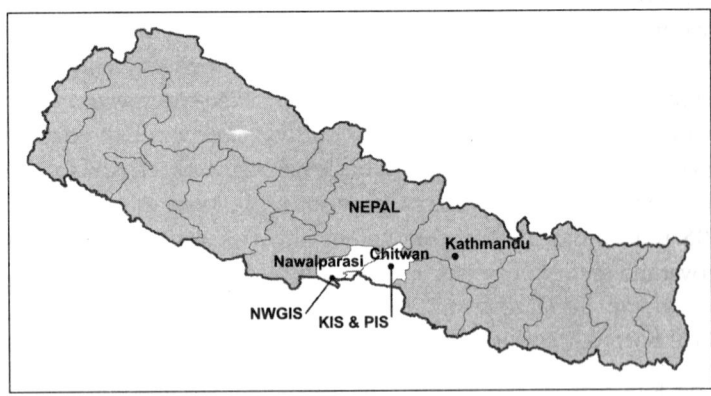

FIGURE 4.1   Location of the Irrigation Systems

Source: Author.

were under the remit of the Chitwan Irrigation Project (CIP). These sites were selected for different reasons. PIS and KIS had motivated and innovative farmers, simple water control structures, and a relatively water scarce situation, recognized as favourable for collective action. NWGIS was selected as scheme design with the potential to provide year-round irrigation, but also having complex water problems.

The Panchakanya Irrigation Scheme

The PIS was established over 200 years ago by the Tharu inhabitants of the Terai to provide supplementary irrigation to some 100 hectares. In 1974, and later in 1983–4, the government took control with rehabilitation and expansion. However, management takeover by the government was not achieved as planned, and the anticipated 600 hectares of irrigated area were never achieved due to poor construction. This made farmers interested in taking over. The agency rehabilitation had brought new headwork controls and lined branch canals, but the system had no tertiary unit development and associated water control structures.

The asset survey in the IMTP showed a remarkable array of structures, with all types of canal sections and construction materials possible. These reflected not only the different interventions and the absence of standard design procedures in the Department of Irrigation (DoI), but also struggles between farmers of certain areas, leaving a suboptimal layout in physical terms. The system was operated through rotation, with the main canal divided into three sections for this rotation among eight branch canals. The CIP was responsible for system management with government funding. Canal operations were carried out by water guards of CIP, and maintenance through private contractors. However, maintenance activities were rarely undertaken due to lack of funding. PIS was considered for complete handover to the WUA, with the government only providing technical support to the system.

The Khageri Irrigation System

The KIS was developed to provide supplementary irrigation for monsoon rice for 3,900 hectares. It has been water-scarce since finalization of its construction in 1967 that had started in 1960. KIS was established to support the livelihoods of new settlers and supply foodgrains to Kathmandu. The DoI had little experience then, and there were virtually no agro-ecological data of the area. Originally, 6,000 hectares had been planned, but the scale was reduced when there was a greater understanding of agro-ecology. Like PIS, KIS had a further rehabilitation in 1984–5, introducing canal lining and new control structures of cross regulators.

KIS is a typical medium-sized extensively developed surface system with a barrage, a 23-kilometre long main canal, eight branch canals and four minor canals. Performance was low but better than that of other systems because of its relatively simple water control technologies, functional canal networks, and silt-free water of the Khageri river. However, the variable flow at the source meant that the target command area was never met with. With IMT, the DoI expected improved system performance because of the relatively knowledgeable farmers and other favourable conditions mentioned above. In KIS, only the branch canals were to be handed over to the WUA, with responsibility of main canal and headworks remaining with the government, making it a joint management arrangement.

THE NEPAL WEST GANDAK IRRIGATION SYSTEM

The NWGIS was constructed under an agreement between India and Nepal to use the water of the Narayani river (popularly known as the Gandak river in India). Construction started in 1963, taking about seven years. The overall system now irrigates about one million hectares in India through two large canals serving areas of Uttar Pradesh (UP) and Bihar. Under the agreement, Nepal receives about 300 cusecs (cubic feet/second = 8.5 m^3/s) of water to irrigate about 8,700 hectares through an offtake 600 metres upstream of the barrage (the focus of this chapter), and also water to irrigate another 1,600 hectares by drawing water from the western canal serving areas in UP.

The NWGIS has year-round irrigation potential but complex water supply challenges like operation and maintenance (O&M) of canal control structures and heavy silt deposition, which raises maintenance costs. Constructed by the Indian government, when it was handed over in 1979 it had no canal networks at field level. The scheme came under a Command Area Development Project (CADP) providing USD 11.2 million for intensive development of irrigation facilities down to 7–12 hectare blocks. In the hierarchy, eight different types of canal network were developed, depending on irrigated area and canal discharge capacity. Drainage and flood control structures, and rural roads were also constructed.

This high level of services could not be sustained after finalization of CADP. Initially, the system had more than 100 supervisors and gate operators down to the tertiary level. It had been planned for continuous flow to all canal levels during the monsoon and up to the tertiary level in the winter season. Each turnout was provided with check structures employing manually operated gates. Overall system management responsibility lay with the DoI, which could not carry out the required O&M, primarily due to lack of funding. The DoI expected that, with the introduction of IMT, it would be able to increase the resources in the system and involve users in O&M activities to improve performance. Despite its size and complexity, it was planned to turnover the whole system to the WUA.

The NWGIS area was not newly settled like much of Chitwan district: there had been older FMIS taking water from small streams. However, there has been an influx of migrants from India and adjoining hill districts of Nepal. Indian migrants came to settle after famines in the 1930s and 1940s, while migrants from the hills started settling after the malaria eradication programme in the 1960s. So, most irrigator families were present before the construction of the NWGIS started in three different communities with different origins and caste orders.

Floods and silt-laden water create operational constraints, and water availability in the monsoon season is variable. The system deteriorated rapidly: by 1992 only about 5 per cent of the planned area received canal water. However, farmers and operators saw this deterioration as a result of faulty designs and budget constraints, especially in the lower-order canals. Farmers were never consulted about these control structures, so that within two years many were removed or made dysfunctional. The CADP created three levels of water users groups (WUGs), mainly after construction was over. A total of 132 WUGs and 11 federations of WUGs were formed, but by 1992 only a few remained.

IRRIGATION REFORM IN NEPAL AND ITS IMPLEMENTATION FRAMEWORK

This section discusses the evolution of the IMTP implementation framework, and its strengths and weaknesses as an enabling environment for institutionalizing the WUAs. Until the 1950s, irrigation development was largely a farmers' domain, most systems being developed and managed by farmers' groups. Many FMIS emerged from *birta* and *jagir* land tenure: land grants to private individuals. These owners, called *jemdar* or *jamindar*, had judicial and administrative powers over land use and could mobilize labour for canal construction (Regmi 1978). The main objective behind land grants and irrigation development was to increase revenue. Irrigation systems also developed under so-called *guthi* systems, endowments of land to support religious and charitable activities. Although developed and managed by local communities, they were linked with local forms of political control.

Birta and jagir types of land tenure were abolished in 1964 as part of wider agrarian reforms; irrigation development then gradually became a state affair. Construction of large and medium systems started especially after the mid-1960s under bilateral and multilateral funding, expanding the government's role in irrigation development and management. The irrigation bureaucracy also expanded: by 1988, the DoI had offices at all regional and district levels. It has remained one of the largest departments, focusing most of its attention on construction of large and medium irrigation schemes, with very little attention to system management.[4]

Despite considerable investments in infrastructure and a well-trained cadre of technicians for design, development, and operation, the performance of public sector irrigation schemes remained poor (see APROSC 1978). Government funding of O&M always remained lower than the requirements, resulting in deterioration of the structures and need for rehabilitation. The government first tried to address the problem through increased investment in rehabilitation under command area development programmes, but without sustainable results (as in NWGIS). This poor performance remained a key concern to government and donors; policy reforms, promoting participation, and local institution building under IMT were the next effort.

In 1992, the government initiated several policy reforms and legal changes to promote participatory management. These were part of the government's wider neo-liberal policies, curtailing the role of the state and promoting private sector involvement in tune with the ideas of major donors.[5] The major policies affecting public sector irrigation were the Irrigation Policy (1992) and the Water Resources Act (1992) (MoWR 1997), which set out legal provisions for building WUAs and handing over the public irrigation systems to them. IMTP originated from the 1992 Irrigation Policy. With this policy, management responsibility for AMIS could be transferred to WUAs (Figure 4.2).

The transfer programme aimed at complete turnover of the O&M responsibilities of small and medium schemes to WUAs. Joint management of larger schemes could take several forms, depending upon size and technical complexity of the systems. The most general form is that the irrigation agency operates and maintains infrastructure to a certain point of delivery, from which a local organization takes over responsibility for water delivery. However, joint management can also be achieved without partial turnover of the systems. In this situation, a shared responsibility is defined between the state and the users for O&M of part or whole of the system. The joint management domain in Figure 4.2 may be an intermediate stage to achieve full turnover or a final destination for the management of large-scale irrigation systems.

Under the Nepal policy reform, all systems under 2,000 hectares in the Terai and 500 hectares in the hills would be gradually transferred

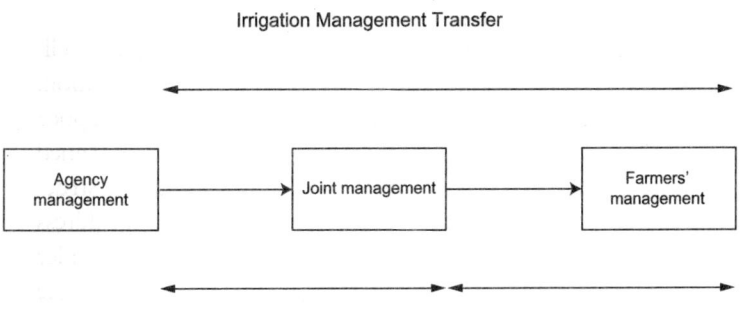

FIGURE 4.2 Framework of Irrigation Management
Transfer in Nepal

Source: Author.

to WUAs, whereas larger systems would be retained in the joint management domain. In a joint management arrangement, the government would handover all branch canals smaller than 2,000 hectares and retain other sections like the main canal and the headworks. However, the policy does not prevent handing over systems larger than mentioned above, if jointly decided by the DoI and the WUA. Under these guidelines, PIS was considered for turnover whereas KIS was planned for joint management. However, despite its scale, NWGIS was considered for turnover. The DoI considered that NWGIS had many assets (land, canals, forest, roads, tolls, and so on) which it could utilize to generate revenue for canal management purposes.

The IMT policy was transferred into programme action with the IMTP funded by the Asian Development Bank (ADB) and the United States Agency for International Development (USAID). In total, 11 irrigation projects were selected, with pilot intervention on three schemes: the Panchakanya, Khageri, and West Gandak systems, which are the focus of this chapter. The implementation framework of IMTP is presented in Figure 4.3. It consists of four stages: programme initiation and institutional development; action plan preparation; implementation; and the post-turnover phase (ADB 1995). The processes were developed on the basis of experiences from previous participatory intervention programmes in Nepal, such as the Irrigation Management Project (IMP), Irrigation Line of Credit (ILC), and Irrigation Sector Programme (ISP).[6] The development of WUAs was steered by the concerned irrigation project office looking after that system, usually with support from external professionals. The DoI had its own association organizers (AOs) in all district-level offices and a sociologist at local level. Farmer organizers were also recruited from among the farming community to act as an intermediary between agency and farmers especially in the participatory methodologies of irrigation system inventory.

Nevertheless, the reform process was initiated by the government: users were not consulted about their future role in water management. They were required to participate in a process whose agenda had already been set by the government. However, generally users did not object to this, except for some local concerns facing the irrigation systems which shaped negotiations over project acceptance, as will be shown ahead. There were several reasons for

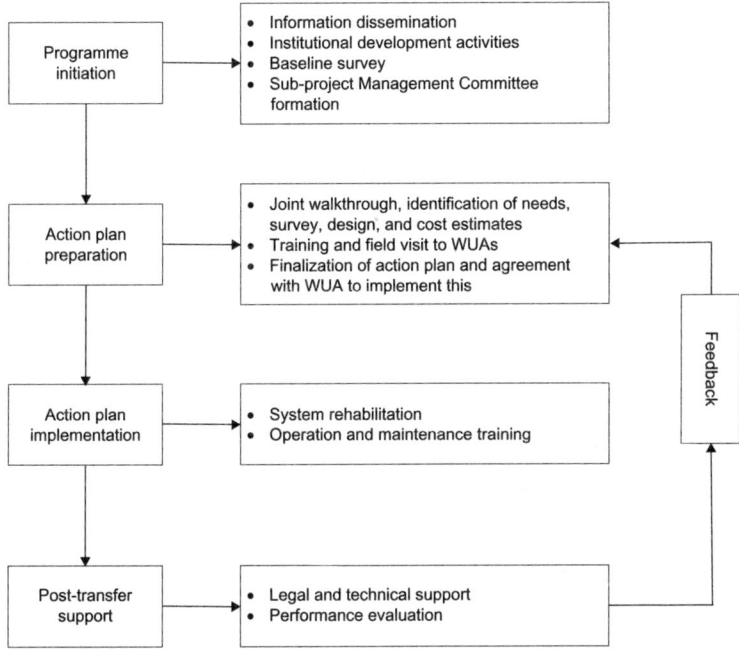

FIGURE 4.3   Project Framework for IMTP Implementation

Source: Author.

this. First, farmers had no alternative: they feared they might lose access to government funding if they did not agree. Second, IMT coincided with the wider process of political decentralization, as WUA development paralleled the start of multi-party democracy in Nepal. Farmers viewed WUAs as a platform to increase their economic and political power to negotiate with the government and other water management institutions. As the case studies will show, the WUAs have indeed played this political role. However, the WUAs' actions show quite different outcomes in terms of their success in changing management performance.

As shown in Figure 4.3, IMT mainly involves WUA formation and further support through technical rehabilitation following a participatory design approach. It also involves new management arrangements between the government and the WUA. This model of policy implementation for IMT is similar to those widely mentioned in irrigation literature (Geijer 1995; Groenfeldt 1999; Vermillion and Sagardoy 1999). It is influenced by the idea to design irrigation policies under which desired institutions successfully emerge; for example, those that ensure functional infrastructure, debate type, and size of organization, and allow user involvement at all stages and levels of project implementation.

The framework shows a top-down approach to designing and implementing policy reform, which Kloezen (2002) refers to as institutional engineering. Farmers were not involved in the design of this framework, nor informed about the policy reforms, which were induced by the government. However, it is not entirely dominated by instrumentalist perspectives. The framework allows actors to (re-) design, negotiate, and transform the guidelines in practice, in line with the local environment of the intervention programme.

The programme also directed all aspects of water control—organizational, technical, and socio-political—although its main focus was the organizational component. Depending upon system size and technical complexity, a multi-tier WUA would be formed, trained and given legal and technical support needed to carry out management activities. Rehabilitation is a technical intervention, but its objective is to facilitate management by providing better working conditions for farmers, making technology compatible with their management capabilities. The handing over of system management with new laws, regulations and conditions of operation brings a new socio-political environment to irrigation systems. However, the means to enable complementary evolution of these elements were not really defined. Rather they were just expected to develop through the different phases of IMTP implementation.

Despite the comprehensive brief of the reform, the design of its implementation had several practical limitations. While the programme design tried to locate conditions for collective action, rule systems in new laws were not always clear. First, it was unclear what forms of property (land, machines, project-owned forest) would be transferred to the WUAs and under what terms and conditions. Second, WUAs lacked government support in collecting the irrigation service fee (ISF). The laws failed to consider how to sustain the reform socially and financially beyond the launch phase. The process was also heavily biased towards technical improvement works to support organizational evolution, but failed to consider the important issue of choice of technology in water management. Besides, systems themselves are not always amenable to improvement in water delivery, given their wider agro-ecological environment. Specific examples follow in later sections for the three schemes studied.

The process in all three schemes was based on this implementation framework. Initially, DoI social organizers set up meetings with key farmers and local leaders to inform them. As explained, usually these people did not express objections. After several discussions, by-laws, rules, and regulations of the WUAs were crafted together with the farmers, leading to the election of the WUAs. The elected WUAs were then registered at the District Water Resources Committee as per the Water Resources Act (1992).

ACTION PLAN PREPARATION AND IMPLEMENTATION

After registration of WUAs, an action plan was developed for technical and institutional development works, following participatory tools and approaches including PTD. The plan also outlined roles and responsibilities of all parties involved (mainly the DoI and WUA) in implementation and the post-turnover phase. A memorandum of agreement (MoA) for IMT was then signed between the DoI and the WUA, to become effective after implementation of the action plan. The handing over of management to a WUA is recognized as a key event: it determines future success, gives a formal stature, and clarifies institutional rules.

There was a proposal to set up a sub-project management committee (SMC) for each system to design the action plan. Yet few people in the Chitwan office knew how to do this. Farmers recognized the need for an SMC for day-to-day work, but asked who should be chairman, and how many representatives there should be. In the end, they conceded the chairmanship to the project manager, but got what they wanted in having WUA executives in the SMC. These were needed before the participatory walk-throughs could begin. It was agreed that this would be done by key members of the WUA (chairman, secretary, member), an SMC member, an engineer, and an overseer from the project office.

However, after a walk-through, objectives and priorities have to be set, since there may be lists of demands beyond what is feasible or what a project can support. After an action plan was agreed upon, a memorandum of understanding (MoU) needed to be signed between the WUA and the project office on behalf of the government. The project environment for IMTP recognized these participatory procedures, but they also brought frustrations. Despite the availability of 64 project staff for the three systems, only one-third of them was permanent. With others on annual contracts, there was turnover and numbers also declined. Most staff involved were engineers, committed but with little experience of participatory processes. Training and capacity development were needed for project staff and WUAs, but these did not happen. This made initial development of contacts difficult.

In PIS, the formation of an action plan went smoothly. Users had several years of experience in O&M, and there was a common working background between DoI technicians and farmers, who knew each other from the Chitwan project. The system management had also done its best to get the scheme running well earlier. However, even in an ideal situation, many problems were overlooked; for instance, to carry out the survey when water is running, and not only with closed canals. Problems were noted without understanding their causes, and first walk-throughs were done without system operators. However, good relations meant that walk-throughs could be repeated and problem diagnosis made. Problems of seepage, silt deposition, and encroachment by farmers as water supply declined were discovered, all emphasizing complex problems for decision-making.

The WUA agreed that the key objectives were to cover the targeted command of 450 hectares, improve equity, reliability and delivery, and reduce O&M costs, with priorities set for the main canal and headworks (with no new control structures and branch level improvements left with farmers). The WUA was then left to make concrete proposals, but came back with a shopping list of demands. Even with a relatively good participatory framework, there were over-expectations to deal with, while people were also over-demanding. With the system being turned over to them they did not know when the next chance for any funding would arrive. However, eventually the action plan was agreed upon, with clear guidelines of work between WUA and SMC, and systems for payment of work, handling of disputes, and handing over. An MoU was signed in a formal ceremony in the presence of DoI and WUA officials.

The KIS had more complexities as farmers had not been directly involved in O&M. The WUA had been formed only in 1993, and recent elections had provided a WUA where 50 per cent of the representatives were new. Khageri was larger, so walk-throughs needed to be done at both the main system and branch canal levels. The WUA also had rather exaggerated expectations based on what they thought the project could offer rather than what they needed. Learning from PIS helped, and here a different way of working was tried with information collected by farmers' groups on branch canals. This, however, did not work as they had too little experience, so conventional walk-throughs were then followed. Once again objectives and priorities were set, this time focused on increasing water availability in the field especially at the tail-end areas. Although the canal conditions were better than in the PIS, the problem of extravagant demands at KIS was greater, including for extensive lining. It had received more visitors, and staff also seemed to emphasize construction projects. Finally, negotiation and local knowledge allowed a new solution to emerge around widening embankments with work that could also lessen seepage. Still negotiations were extensive and demanding, not only between the WUA and the project, but also within the project, and between the project and higher-level DoI and donor staff. However, ultimately, an action plan was drawn up and an MoU was signed. As the scheme was over 200 hectares only branch canals were transferred to the WUA. So, the signing of the agreement was done by different representatives for the main and branch canals, as approved in the general assembly (GA) of the WUA.

The joint action process in NWGIS also began with several rounds of walk-throughs, with committee members of branches asked to join. However, the sheer intensity of structures and the constraints of the system meant that very few activities for improvement were identified. There were no debates and arguments related to causes of problems. Once again, farmers and the WUA had limited experience to articulate such problems. Also, since water was abundant, farmers tended to take a narrow view of problems like siltation. Very few people were involved in the walk-through processes. Often committee chairpersons prepared reports without discussion with the committee. If group discussions were held, they were often dominated by influential persons. Thus, any agreed basis for joint action was limited. In fact, several demands related to disputes about structures and access between groups later emerged, that had not been noted in the walk-through and not itemized in the original action plan. An action plan was nevertheless drawn up, and an MoU signed. However, the MoU agreed for a turnover of the whole system. That this could happen for such a large system, whose WUA had so little experience and operated for only two years, raised questions. Evidence suggests that this situation was shaped by government aspirations for IMT, but also that a few WUA leaders were attracted to full transfer because of the resources besides irrigation fees that the system could provide, including forest resources and road taxes.

However, after the implementation of the action plan, the WUAs still confronted local challenges before signing for final handing over. These were mainly resolved through local negotiations and, in some cases, through higher authority intervention. In KIS, for example, the WUAs did not immediately accept IMT. They feared that their water supply regime would be severely affected by a settlement programme in the water source of the Khageri river initiated by the government. The WUA even challenged the government in the Supreme Court of Nepal on its move to settle a new village. The problem was settled after the prime minister himself intervened and assured the KIS farmers that their water sources would be augmented if any flow depletion caused by the settlement was noticed. In PIS, some farmers' groups still wanted the government to support them financially and would not agree to turnover at first.

WATER CONTROL AFTER MANAGEMENT TRANSFER

The IMTP intervention in the three systems was carried out from 1995–9, involving technical improvement and capacity enhancement of the WUAs. A study was made of satisfaction with the irrigation water supply situation two years after the transfer as part of the irrigation asset surveys in which the author was involved, whose results are summarized in Table 4.1.

A study on financial sustainability over the same period showed that PIS had a collection rate of ISF of more than 80 per cent, enough for O&M. However, the figures for Khageri and NWGIS

TABLE 4.1 Farmers' Perceptions of Water Supply and Operations after IMTP

Farmer Perceptions		PIS	KIS	NWGIS
Percentage judging supply to be adequate during	Monsoon	81	48	55
	Winter	85	8	30
	Spring	40	15	9
Percentage judging distribution to be fair	Between branches	85	66	13
	Along branches	75	60	18
Main constraints in different seasons	Monsoon	Water shortage	Water shortage	Operation
	Winter		Water shortage	Operation
	Spring	Water shortage	Water shortage	Operation
Percentage judging supply and operation of main system to be	Acceptable	86	83	21
	Poor	10	10	74
Percentage judging supply and operation of main system to be better compared to 5 years ago		72	10	74
Percentage judging supply and operation of main system to be the same compared to 5 years ago		22	68	21
Percentage judging supply and operation of main system to be worse compared to 5 years ago		3	15	74

Source: Adapted from Khanal (2003) and field survey conducted by HR Wallingford (2001).

stood at 62 and 22 per cent, respectively. System assets were found to be maintained in both KIS and PIS with more than 80 per cent of the farmers satisfied about the condition of the infrastructure. In NWGIS, with low fee collection, only 52 per cent of the farmers believed that infrastructure was properly maintained. These perceptions are consistent with the results of an asset survey conducted in 2001, which showed that the percentages of defective canal sections and associated structures were 12, 16, and 42 per cent respectively for PIS, KIS, and NWGIS.

On the institutional front, PIS and KIS transformed from a two-tier organization into a three-tier organization to respond to the new water distribution pattern, and especially to improve delivery in the lower-level canals. However, in NWGIS no institutional transformation could take place. The WUA became ever more engaged in party politics and failed to deliver irrigation services to farmers. This is also reflected in the survey on farmers' perceptions on WUAs shown in Table 4.1: in NWGIS only 20 per cent believed that the WUAs were doing their job and were accountable to farmers, whereas in PIS and KIS the figures were 72 per cent and 83 per cent, respectively. Likewise, only 19 per cent of NWGIS farmers became WUA members, whereas WUA membership in KIS and PIS were 56 and 82 per cent.

The above results confirm that farmers in the PIS saw an improvement in water availability, a related increase in irrigated area and a change in cropping pattern. Local people believed in their organization; the WUA was accountable to its members and financially capable to take up new management responsibilities. In the KIS, there were improvements in water delivery schedules, an increase in irrigated area, and a change in cropping pattern, but not at the scale seen in the PIS. The system fell short on financial viability, but farmers showed strong support for their organization. In NWGIS, the WUA was dysfunctional, losing its credibility and acceptability. Attempts to improve system performance through the new organization here proved disappointing, frustrating, and demoralizing for the local community.

Incidentally, the scale of change in these systems has run parallel with their service area. The PIS (600 hectares) has good outcomes in terms of service delivery, while NWGIS (8,700 hectares) has experienced problems of management incompetence. However, these variable outcomes cannot be simply explained in terms of their areas. They must also be related to the wider challenges of regulation and control (both social and technical) for irrigation and how these are addressed during the design.

ORGANIZATIONAL EVOLUTION AND MANAGEMENT TRANSFORMATION

Development of local organizations is the first and also a key activity in IMT. WUA development is largely shaped by existing social conditions, resulting in different outcomes in different systems, despite similar design principles and IMTP approaches. The acceptance of a WUA as legitimate by users depends on the quality of service delivery and accountability. Organizational design thus not only involves the structural design of the organization, but also needs to consider the social and political relations in the irrigation systems and the technical demands and inequities of water distribution.

As already mentioned, it was not clear what IMT would entail, even though allowed by law. There were no guidelines on types of documents to be prepared and rights and responsibilities to be transferred. It was unclear how to proceed in situations of full or partial transfer. This was the situation even at the end of the IMTP. These issues had to be discussed between relevant parties in detailed negotiations. Finally, the condition set was that the status of the transfer was the 'right to use' infrastructure only. WUAs could not pledge or sell infrastructure and were also prohibited from damaging or changing it. However, changes could be made in the structures and canal network for expansion. Further, the WUA could not transfer the system to another person or organization. The transfer would also include the property that belonged to the system—resources like forests and roads that had previously belonged to the DoI. For systems under joint management like Khageri, only the resources of the branch canals were to be transferred to the concerned WUA.

Support for future development was more difficult to agree upon, as the government wanted to avoid any immediate new commitments after these improvements. Ultimately the WUAs agreed to a five-year period before they could seek any government help for development and expansion. Another issue was future technical services from the DoI. It was agreed that a WUA requiring technical advice on repair and maintenance could request the concerned Irrigation Office, which shall provide technical advice.

The WUAs considered post-turnover support even more important than ownership issues. This was necessary for boosting confidence in managing the system in the future, with clarification of support needed for emergencies, development, and expansion. Criteria were needed to decide what was beyond the capacity of farmers to repair after a calamity, requiring government support. The PIS farmers defined such criteria on the basis of their past expectations.

PANCHAKANYA

The WUA in PIS evolved first as a two-tier organization, a management committee (MC) at system level and branch committees (BCs) at branch level, with a GA as a general policymaking body for the WUA. It has since made several changes in structure, most notably to allow outlet committees to emerge below BCs to help water management, and has taken initiatives for more women to sit on committees. The WUA has established new principles of water distribution, as farmers increasingly show interest in growing early paddy. Implementation is possible thanks to new control structures introduced with IMTP and their understanding of the system; farmers said that water availability had doubled.

ISF collection started in 1995. These were immediately increased after handing over. In 1997 another service fee, the maintenance fee, was introduced separately from the ISF and used only to clean and maintain the main canal. This fee (NR 300) replaced the earlier call for three labour contributions. It was introduced to improve work quality: labourers are now hired from among member farmers unless not available. Since this innovation, farmers also collect money for cleaning branch canals to replace earlier labour contributions. The WUA can now claim that 450 hectares are under irrigation for monsoon paddy, its key objective. However, the share and membership of the WUA have been registered for only 350 hectares, and fees were paid for only 259 hectares in 2000–1 (due to farmers arguing they do not use canal water), but at the time of study, there was a 100 per cent fee payment for winter and spring crops. Estimates suggest that collection rates covered the costs for O&M; if the fee increases, KIS might even pay for some rehabilitation. A serious process of dialogue of change between WUA and agency, but especially WUA and farmers, was worked through for a viable and accountable management system, and service-oriented water control.

In Panchakanya, traditions of collective action shaped the negotiations and the process of group formation. An organization was established and recognized by the farmers who were familiar with collective management. Farmers designed the WUA in accordance with their existing management practices. The technical change brought about by the project could work as a catalyst for new management, bringing better water availability and new production options that helped achieve financial self-sufficiency and acceptable management practices. The WUA thus acquired the necessary skills, generated resources, and maintained accountability in service delivery. One main cause of the rapid management change in PIS was that the project intervention increased the command area under early paddy more than six fold (from 25 to 150 hectares) and contributed to doubling the command area through increased water availability. Farmers were thus willing to pay the full costs of canal O&M.

KHAGERI

The KIS began with a similar management structure, but did not see the same degree of changes and evolution as in PIS, as there was not so much change in water availability to drive changes. This has continued to cause concerns and struggles. The O&M cost of the main canal remains with the government. ISF is only for one crop, monsoon rice, except where early rice is grown, and this fee has not been increased. Costs were low given the DoI support for the main canal and the simple technology of the system overall. However, it might also be hard to raise fees as long as service is not as adequate as farmers want. Nevertheless, farmers were satisfied with the way the WUA was operating (HR Wallingford 2001). It was also gradually changing to a three-tier structure by adding outlet committees. The WUA has gradually shifted toward implementation of strong rotational practices. Both are directed to help farmers below the outlets.

After the formation of the WUA, it took over control of canal operation from the government. The WUA did this because otherwise it had virtually no job, and because it wanted full control to avoid conflicts with the agency about water distribution. It decided to make operational decisions itself, although the government gate operators would implement these. It decided on principles to deal with the uncertainty of water delivery at the intake. Gradually, as this operational plan became embedded, more responsibilities for allocation were taken up inside BCs, where outlet committees also emerged. This helped to offset the retirement of most government gate operators. This explains the findings in Table 4.1: while water was still considered scarce, two-thirds of the farmers considered distribution to be fair, and thought water supply had not worsened since turnover. Once again, efforts in the crucial stage of joint planning and PTD made a difference in raising understanding of realistic options for management transformations.

In Khageri, head-end and tail-end farmers struggled within their own WUA development; IMTP gave them new spaces to claim and negotiate their water rights and representation. This early conflict gave farmers an opportunity to learn about their system and find solutions satisfying both parties. The tail-end farmers got equal representation in the organization, something the head-end farmers had initially objected to. Likewise, the head-end farmers continued to enjoy their prior right over spring water, which tail-end farmers wanted renegotiated before the management transfer. Because of their relatively higher educational level and political consciousness, Khageri farmers were in a position to craft their WUA as demanded by their context.

The technical intervention in Khageri was in accordance with the farmers' service delivery pattern, and the WUA managed to maintain water delivery as farmers required. The WUA here continued to develop its economic and political networks with other local actors and institutions, strengthening its legitimacy and power.

Nepal West Gandak

In NWGIS, the WUA became dominated by party politics, and by its third election in 1998 realized its initial structure was ineffective. People only aspired to be elected to the upper tiers, and often even MC meetings were cancelled as inquorate. The WUA members continued to lack knowledge on O&M and the WUA could not escape from wider politics. A new structure emerged from a GA meeting. A board of directors, executive committee, and MC for everyday management were each selected from the members of the committee below it. Thus, the manager of the MC now became the most important person provided he had support from other key committees in the WUA. While it is easy to blame politics for the poor working of this system, in fact poor project structure, lack of accountability, and policy gaps of the government need some recognition. The project was said to be fully handed over because of strong local demand, although Khanal (2003) found no evidence for this. In 2003, only 29 per cent of the farmers were members of the WUA. As a fully turned over system, O&M had to be borne by the WUA. However, the study did not find any documents regarding requirements for O&M. There had been neither increase in ISF collection nor progress in mobilizing funds from other resources of the government. This reflects lack of proper mechanisms, conflict and corruption in the WUA, which has fuelled other problems in maintenance and water supply.

With the handing over of the system, a new water distribution pattern was developed by the consultancy firm involved, with the main canal divided into four divisions, each with a fixed allocation. According to the WUA, the operational plan was developed only for a few selected canals and was unclear to them. In fact no new management or funds could be established as part of a post-turnover supporting regime, because the system ran under 'no management'. A member of the five-person team to operate the main canal quit after a few months when they were not paid, and the WUA remained dependent on post-turnover support funds from the government. The scheme deteriorated rapidly. Field studies in 1999 showed that whoever wanted water had to organize in groups to come to the main canal to divert water to their branch. Except for a gate operator at the intake of the main canal, there were no workers to look after water distribution. This explains the low levels of satisfaction about water delivery shown in Table 4.1. Despite being a focus under CADP and IMTP, NWGIS has never achieved a basic standard of water delivery. Its complex infrastructure and problems suggest it should have remained an agency-managed system.

In NWGIS, a few farmers were able to capture the reform process, while the others were silent because of the local patterns of social and political dependence. The WUA became used as a platform to monitor the strength of the political parties. One of the reasons for a high level of (party) political influence on system management has been its large scale and local importance. The area of the NWGIS overlaps with two parliamentary constituencies. It is a rural area, where opportunities for economic and political activity outside irrigated agriculture hardly exist. So the WUA was the first and only organization where political parties could enter to increase their influence. Elsewhere in the Chitwan valley, other spaces of political and economic representation were already more developed.

IMT did not bring any visible change in NWGIS. The WUA lacked skills and resources to operate and maintain such a large complex canal network. The technical improvement work largely failed to recognize the requirements of use of this technology and to communicate these to the WUA. It focused only on remoulding the old structures and design. The participatory design process was limited to discussions with the local elites, without any investment in system learning in the technology change process. NWGIS farmers had never been involved in canal O&M, and thus had no experience with system constraints. This knowledge gap was a key limiting factor in the design process, as the farmers failed to interact meaningfully with the designers. Another limitation was that participation was mostly confined to the top level of the WUA, and these key representatives decided the main changes and modifications for the whole system. Through its failure in delivering the necessary irrigation services to the farmers, the WUA lost its credibility at the local level, weakening its leadership.

The variable outcome of the WUAs should also be studied in relation to the organizational structure taken up. The WUA of NWGIS was structured in a unitary model, giving opportunity to the local elites to capture it, exactly as Freeman *et al.* (1989) had warned. The WUA had been developed on the basis of hydraulic boundaries. It had a highly unequal representation: 68 per cent of the MC was reserved for farmers cultivating 19 per cent of the land only, whereas farmers cultivating 58 per cent of the land controlled only 11 per cent of the membership. The NWGIS has 35 offtakes, 3 for the main and large-size branch canals, 8 are medium-size offtakes, and 24 are small outlets. Since each offtake is allocated to one member in the MC, the three large branches, which cover more than 59 per cent of the area, had only 3 members, whereas the 24 outlets which roughly cover 19 per cent of the area, had 24 members. The chairman, vice-chairman, and secretary of the WUA had to be elected from these 35 MC members only, and not from the GA of the WUA, which has 172 members, equally representing the various geographical blocks.

However, both Khageri and Panchakanya avoided this elite control by considering geographical area of the schemes, and not limiting organizational design to hydraulic boundaries. The resulting WUAs were a mixture of units with a unitary and a federated character, avoiding power concentration with a few individuals. The key members of the WUA (chairman, vice-chairman, and secretary) were elected from the general assemblies, not from within the MC, as in NWGIS. This shows that the organizational design, though usually based on hydraulic boundaries, must also consider the geographical or political boundaries within the network characteristics of the particular systems.

Type, size, and membership of an organization have been given considerable attention in the literature on WUA design (Patil and Lele 1995). This research shows that this depends on the management pattern adopted for the WUA and the physical layout of the canal systems. The WUAs presented in this chapter were designed as multi-tiered associations with varying sizes in the lower unit of the WUA structure. However, the size and structure of these WUAs changed under the management transformation processes because of the new operational requirements. In Panchakanya, the WUA changed itself from a two-tier to a three-tier organization, with groups formed even down to 5-hectare irrigation blocks. Likewise, Khageri transformed from a two-tier to a three-tier organization, with a lower unit of 20 hectares for group formation. Both WUAs were operating in a participatory model, with users directly involved in canal O&M. This necessitated the formation of WUA sub-committees at a much lower level, due to the increased O&M requirements.

The NWGIS WUA was also designed to operate in a participatory model, employing the users in direct O&M. Owing to the large area and need to mobilize many farmers, it could not function as designed. It thus tried to change its organizational structure, adding a separate 'management committee' to look after the management tasks by hiring the required technicians. This effort to change from a participatory to a management model failed due to lack of financial resources to pay for the technician team.

WUAs cannot be designed just by following a set of routine activities. The above examples show how existing social conditions determine the outcome of organizational design. Likewise, the irrigation system and its wider environment (physical, technical, and social) largely determine the organizational structure and its evolution.

IRRIGATION TECHNOLOGY AND THE REFORM PROCESS

IMTP emphasized participatory design and construction of technology to create better working conditions for farmers. The underlying assumption for adopting PTD has been that it creates a feeling of ownership among the users and produces technology compatible to farmers' management. The evidence from the case studies shows that PTD is not only a means to develop the technology that people want. It also builds a more stable project environment, which is essential to facilitate participatory change processes and to sustain local water management. Likewise, PTD should be about how to build accountability of the WUA for future governance and management.

Another argument for PTD in the context of IMT is that it is a way of establishing service-oriented water control. The new service-oriented design must be based on future operational plans of the users. This requires bringing together users, canal operators, and designers. The design should proceed only after the WUAs are in a position to formulate a performance-oriented action plan based on their experience, as in KIS and PIS. In both systems users had prior experience in O&M, and were clear what delivery pattern suited them in accordance with water availability. Users did not propose any radical changes in the system, but aimed to achieve improved water delivery through incremental changes. Rather than freezing the infrastructure into an inflexible distribution pattern, this allowed for different future operational strategies. PTD, therefore, requires that the irrigation agency involved adapts an iterative and locally interactive process of design and invests in initial training of the WUA. It may also need to accept departure from the design standards to meet farmers' requirements and allow sufficient time for consultation and design processes.

In both Khageri and Panchakanya, technical change was guided by emerging organizational capacity and management objectives of the WUA. Users in both systems preferred to have different kinds of rotational delivery patterns depending on water availability across the seasons.[7] They chose to continue with the manually adjustable gated water distribution technology, but also relocated some gates and modified their technical characteristics to suit the new water delivery requirements. The resulting redesign remained in accordance with the locally preferred rotational delivery system. The design also took the institutional capacity of the WUA into account. Both systems were extensively developed systems with fewer water control structures and canal networks. The design tried to maximize both technical and social control in ways relevant to farmers and the WUA, keeping the number of control structures to a minimum. As a result, the overall technology was simple to operate and required fewer operators and less cost to operate and maintain, while in accordance with farmers' delivery objectives.

The West Gandak system had complex canal networks with many water control structures requiring O&M, placed during the earlier command area development programme. It also suffered from heavy silt deposition in the main canal, requiring expensive repair and cleaning. Though efforts were made to reduce siltation by construction of a silt ejector, it did not function due to its poor location. So, system operation continued to be constrained by the large number of check structures and the silt-laden water. The WUA had neither the expertise nor the resources to operate and maintain this complex canal network, and lost its credibility when unable to deliver water adequately.

The study also identified several major constraints in the current approach to PTD in large and medium irrigation systems. Care is needed in the way criteria and preferences are incorporated in the design process. Users may construct their ideas and priorities on the basis of project opportunities rather than of what is technically required, and may have too high expectations of what new infrastructure can improve.

Lack of learning by key actors about the system environment is another problem. As irrigation systems are socio-technical systems, design should begin by considering both human and physical dimensions, and how these are linked. The strength of the participatory design approach depends on what users and designers know about the system, its opportunities, and constraints. Only then can an interactive design process be maintained. In both Khageri and Panchakanya, farmers were familiar with their systems' opportunities and constraints, and able to interact with the engineers in the design processes. In West Gandak, lack of farmers' knowledge of their system limited the outcome of PTD.

PTD also has a political dimension in that users have to contribute to the process. In the cases mentioned, users' contribution was set at 26 per cent, which WUAs vehemently opposed as they had not been consulted. Ultimately, the farmer contribution did not exceed 10 per cent. Another political dimension is the involvement of different actors at different institutional levels in setting criteria for interventions, which can cause contradictions, uncertainties, and conflicts. In IMTP, for example, important decisions, like the level of user contribution and the project framework, had already been decided by the central authorities in Kathmandu and the ADB officials funding the project. Other government authorities were not in a position to negotiate user contributions nor to allow the project team any flexibility in implementation, which is essential in a participatory process. User contributions required were only negotiated after concerted WUA action.

There is ongoing debate on whether system improvement should be done before or after the management transfer. There is no single answer to this. This study has shown that a key issue for establishing service-oriented water control is an interactive design process in a project framework, linking different actors together in a learning process that allows flexibility in technical, institutional, and financial norms in WUA and irrigation system development.

IRRIGATION MANAGEMENT REFORM AND THE MATCHING OF IRRIGATION TECHNOLOGY AND INSTITUTIONS

Almost two decades have passed since irrigation management reform became the key focus of irrigation intervention in Nepal. The effort followed a 'modernization approach' through both technological and institutional change to support the policy reform for improved agriculture productivity. This chapter shows that the outcome of the reform has been mixed. The envisaged institutional reform and decentralization of irrigation management functions have not always been realized.

The study has shown that including only technical rehabilitation components in the reform process is not enough to support new local organization. A broader understanding of irrigation infrastructure, its pattern of service delivery, and requirements for use is needed. Technology design influences organizational capacity for operating and maintaining the system. Technological change to support institutional reform thus should consider the institutional capacity of the WUA, style of the local management organization, and management objectives in terms of water delivery preferences, operating requirements, and cropping choices. These are dynamic and differ from system to system. Hence, there can be no universal rules assuming that a particular technology is suitable for all situations. As noted by Burt and Styles (1999), many management problems could be minimized or eliminated with proper designs and operational patterns.

The cases of Khageri and Panchakanya show that technical change can provide an incentive for local organization to take up management responsibility. Here, the main issue has not been inclusion of technical change in itself, but rather its design process, guided by the management objective and institutional capacity of the local organization. Farmers continue using gated structures that allow them to implement water distribution in accordance with their service requirements, with particular concerns for ease of operation, workable institutions, and water availability.

The variable outcomes documented here also suggest that project support alone does not provide incentives for local organizations to change. Rather, it depends on the approach of the project, and on how it translates the opportunities and constraints of the system environment into practical design to bring better service. Critical to a participatory approach is the presence of designers able to become familiar with the users, the system, and its environment, and to reflect on its design. Their instrumental objectives will not be met if they proceed without adequate knowledge of these. This will lead to failure of new infrastructure, water delivery, and institutions, as happened in the NWGIS system.

Limitations in the application of PTD have come from a too narrow focus and understanding of the participatory development process in the current development discourse. Future design approaches should be based on, but move beyond, these contemporary models of technology change, and move forward towards establishing service-oriented water control, in which technology change will proceed with the organizational capacity and management objective of the local organization. Designing for service-oriented water control would also overcome the limitations of PTD to date.

Even when project support proceeds with system learning—the key to participatory development—this forms only a base for future management. Sustaining water management beyond a project launch depends on wider water control dimensions: socio-political, organizational, and technical. These are dynamic, and continuously bring new challenges in the management continuum. Local organizations need to adapt to their environment to sustain water management, to build up necessary skills, generate resources, and develop accountability between the key actors.

Future design of irrigation reforms should build on, but move beyond, blueprint models of PIM and PTD to bring a new realism into reforming technology and institutions together. One approach can be the transforming concept of irrigation modernization, applied with a better understanding of local performance priorities, possibilities, and constraints. This can help local organizations design service-oriented management approaches in accordance with the resource base of their irrigation systems.

Notes

[1] Also state disengagement, management devolution, privatization, turnover, or handover. See Vermillion and Sagardoy (1999).

[2] An asset survey in irrigation documents the amount and condition of physical infrastructure and equipment for system O&M.

[3] The author joined the HR Wallingford team to conduct the asset surveys in KIS and NWGIS, part of a comparative study on transferred/non-transferred systems in India and Nepal (HR Wallingford 2001).

[4] For example, the Sunsari Morang (66,000 hectares), Narayani (29,000 hectares), Kankai (7,000 hectares), and Bagmati (60,000 hectares) systems.

[5] The changes were also in response to policies of the ADB and the World Bank, which favoured more market-oriented economies.

[6] The IMP was initiated in the early 1980s with USAID support to study and formulate participatory irrigation policies in Nepal. ILC and ISP were funded by the World Bank and the ADB, targeted at rehabilitating FMIS following PTD processes on a pilot basis. ILC and ISP were expanded to large-scale national programmes, with funding from both agencies, but confined to FMIS only. Lessons learnt from these programmes provided a foundation for the design of IMTP.

[7] The PIS had two delivery systems operating in different periods, depending on water availability. One is continuous: all offtake gates are open and each branch is allowed to take water as per its capacity. The other is sectional rotation, used with lower water supplies: water is rotated between three canal sections with time allocation. KIS had a similar pattern, but one more arrangement besides the above two: weekly rotation, in which the whole command area is divided into two parts and water rotated among them on a weekly basis.

5 IRRIGATION TECHNOLOGY, AGRO-ECOLOGY, AND WATER RIGHTS IN THE MID-HILLS OF NEPAL

Umesh Nath Parajuli

In Nepal, irrigation development is central to policies of food security and poverty alleviation. Irrigation takes place in a wide range of agro-ecological zones across the country. About 70 per cent of Nepalese irrigation systems are farmer-managed irrigation systems (FMIS), in which traditions of self-governance and strong community participation are important and common features. Developed over time through local ingenuity and skills, these FMIS are also responding dynamically to changing societal and environmental contexts.

This chapter focuses on FMIS in the hill areas of Nepal, and how social, technical, and ecological conditions interact to shape irrigation technology and its use. It describes the diversity of types of hill irrigation systems, their choices of technology and how these shape management requirements, and shows the materialization of water rights. In particular, it discusses the presence and use of the indigenous technology of the proportioning weir. The use of this structure, especially in mountain environments in Asia, has been widely documented (Ambler 1990; Dani and Siddiqi 1989; Leach 1980; Parajuli 1999; Yabes 1990), but the diverse factors shaping the presence of this technology are poorly understood and documented.

Through case studies, this chapter develops an explanatory framework on why particular irrigation structures have come into being in FMIS in different landscapes and operational conditions of water flow in the hills of Nepal. It also analyses the relationship of these structures with the development of water rights to the irrigation water they deliver, and the institutional rules and roles present for system management that reinforce rights and operations. Demonstration of these social dynamics integrating technology design, operation, and use is possible through use of interdisciplinary frameworks of analysis. Technology can be considered as a mediation between nature and society (Benton 1992), which can be studied by looking at particular conditions of agro-ecology, culture, and society as they influence technology.

After a short theoretical framing of the research, the chapter introduces two classifications developed for the water distribution systems found in FMIS from field research. The first is a classification of types of water division structures: subsequently, this chapter concentrates on the occurrence and use of the proportioning weir, which is a fixed control type of structure. The second classification evolves from analysis of agro-ecology and social conditions, to differentiate hill irrigation systems with reference to their physical setting and location in the landscape, and their options for water control. These principles link agro-ecology with social and cultural choices to shape selection of irrigation infrastructure, where technology can have a symbolic as well as a functional role. The chapter then illustrates these principles and classifications through case studies of the operational realities of three FMIS: *Sankhar*, a slope-hill irrigation system of some 37 hectares with proportioning weirs; *Julphe*, a river valley irrigation system of 200 hectares with proportioning weirs present from cultural choice as well as functionality; and *Patne*, a slope-hill irrigation system of some 21 hectares operating with an ad hoc type of division structure, where proportional weirs were not in use. The chapter is based on findings from four[1] detailed case studies and a rapid rural appraisal of 20 systems conducted in the mid-hills of Nepal in 1996–7 (Parajuli 1999).

FRAMEWORKS FOR ANALYSING IRRIGATION INFRASTRUCTURE

This chapter uses socio-technical and agro-ecological frameworks to unravel the motivations of infrastructure designs. The first framework enables study of the social conditions of and requirements for use of technology, its social construction (in knowledge, artifacts used, and social dynamics of water delivery), and its social effects (see the Introduction to this volume). Gutierrez (2005) further developed this framework to study how irrigation systems are appropriate to users, in technical and productive terms. She sees designs as operationally appropriate when they achieve equity in water access and are manageable by local users and organizations (Gutierrez 2005). She shows that this operational–organizational appropriateness is present when:

- water rights and rules of work are clear and socially accepted;
- there is an organization able to assume water distribution;
- there is adequate competency to distribute the water as agreed and consistent with norms of equity and fairness;
- there is ability to maintain infrastructure.

These social requirements of use can be traced back into the infrastructure, and are discussed further in the short case studies.

Agro-ecology[2] frameworks can show how producers use and gain control over the landscape and habitat they cultivate, the social functions of design in a given ecosystem, and also the cultural forces and ideas shaping techniques of resource capture and cultivation. Irrigation systems are made by human workmanship for a working purpose, within limitations set by local knowledge and resources. Thus, besides technical considerations, socio-cultural and agro-ecological factors shape the design of FMIS.

WATER DISTRIBUTION IN FARMER-MANAGED IRRIGATION SYSTEMS

Nepalese FMIS show two types of water distribution on the basis of system operation: downstream control and upstream control. These can be further sub-classified, as shown in Figure 5.1, in the type of distribution system possible and whether distribution can be made proportional or adjustable.

Downstream control refers to a situation of being 'demand-driven'. Here, an individual farmer can draw a required amount of irrigation water from the distribution system as and when needed. This is possible, and used, where supply exceeds demand. Upstream control refers to a situation of being 'supply driven': water delivery is related to its availability irrespective of demand. This type is used when demand exceeds supply, requiring rationing. Thus, an upstream control system imposes restrictions on unlimited delivery of irrigation

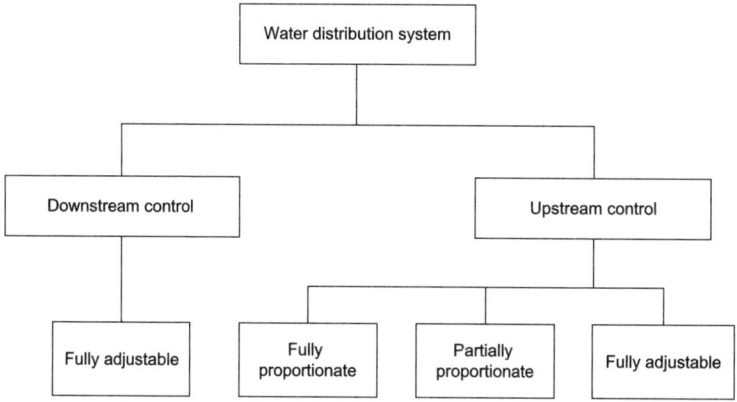

FIGURE 5.1   Types of Water Distribution Systems Used in Many FMIS in Nepal

Source: Parajuli (1999).

water and requires synchronization of demand and limited supply. Upstream control can be further classified into three types of water distribution: fully proportionate, partially proportionate, and fully adjustable. As Figure 5.1 shows, a fully adjustable type of distribution system is also used under downstream control, and is the only type possible in this situation.

A distribution system is said to be fully proportionate if all individual farmers receive water continuously on a fixed proportional basis, irrespective of their demand. This type of distribution system is equipped with a device having fixed openings, which can either be an orifice or a weir. In the hills of Nepal, mostly fixed proportioning weirs are used for this purpose. As shown later, they are usually made of wood where available, are set for a season by a meeting of farmers, and need no operation (although they may be checked to be operating consistent with allowances). In a partially proportionate system, continuous flow is delivered on a proportional basis to certain sections of the system. Within these areas, water is intermittently delivered to farmers on a proportional time share basis, using an open–close type of water division structure.[3]

A distribution system is said to be fully adjustable when it delivers irrigation water to farmers on demand (often based on the combined criteria of available supply, crop need, and land type). Here, flows can be adjusted frequently to match the dynamics of water demand and supply. In FMIS in Nepal, water division structures of this type of distribution system are simple. They consist of an open cut (turnout) in the canal bank where flows are adjusted on an ad hoc basis, either by changing the dimensions of an open cut (for example, by adding mud and stones) or by changing the operating hydraulic head.[4] Thus, this structure type is termed here as an ad hoc adjustment structure. Not all these water distribution methods are viable everywhere, and choice between them is largely shaped by the local context.

THE PROPORTIONING WEIR

A proportioning weir, the main focus of this chapter, is a simple control structure, placed across the direction of flow (Figure 5.2). This structure divides the water in a canal into two or more parts, each of which corresponds to the water share allocated to a farmer or farmers' group served by the branching canal. Widths of notches in a proportioning weir represent farmers' water shares (water rights) passing through them. The correctness of flows passing through notches of a proportioning weir is judged by measuring and comparing their widths. In the past, a notch width used to be designed, measured, and compared with reference to the thumb width of a socially recognized person. However, in weirs fabricated recently, farmers have started using an equivalent

FIGURE 5.2   Indigenous Proportioning Weir

Source: Parajuli (1999).

dimension, at about 0.75 inches (2 centimetres) for a thumb width. The basic objective of proportional distribution is to proportionally distribute irrigation water to each branching canal. This means that the quantities of water flowing in the parent and branch canals are equally affected by any variations in the level of water in the parent canal.

In Nepal, most proportioning weirs are made of timber and have several rectangular notches of uniform depth cut into them. To ensure that the flow of water is proportional to the width of the notches, the sills of all notches are kept at the same level; only the notch width is varied. Once water distribution by proportioning weir starts, the irrigation system becomes rigid. Unless there are agreed alterations in notches, farmers have no opportunities to take extra water (except for socially negotiated temporary flow changes or stealing). The system runs automatically with no need to open, close, or adjust the flows. Thus it eliminates the need for an operator.

For proportioning weirs, equity in water distribution is judged by measuring and comparing the widths of the notches. Except for one rule—no farmer is allowed to block other farmers' notches—the irrigation system runs without further regulations after the weir is set. Water distribution by proportioning weirs has the following distinct characteristics: there is visible distribution with reduced guesswork, and it is simple, understandable, and measurable by a farmer. Hence, flows can be easily compared with allocations to other farmers. These characteristics reduce water-related disputes. Also, when weirs are nested together, it reduces management activities by reducing the number of control points.

CONDITIONS DETERMINING THE CHOICE OF WATER DIVISION STRUCTURES

Water distribution in these hill FMIS employs mainly three types of water division structures, and choice is largely dictated by agro-ecological conditions and socio-cultural conditions. Table 5.1 presents these conditions:

AGRO-ECOLOGICAL CONDITIONS

Physiographic location and categories of irrigation systems
Based on its physiographic location, irrigated land in the hills can be categorized into three types: foothill terraces, slope-hill terraces, and

TABLE 5.1   Conditions Determining Choice of Water Division
Structures

Agro-ecological Conditions	Socio-cultural Conditions
• Physiographic location of irrigated area	• Knowledge and practices
• Availability of water	• Labour resources and water rights
• Altitude (construction material)	• Operational modality
• Topography	
• Shape of the irrigated area	
• Crops	

Source: Parajuli (1999).

river valley terraces. Accordingly, irrigation systems irrigating foothill, slope-hill, and river valley terraces are classified as foothill, slope-hill, and river valley irrigation systems (Parajuli 1999).

*Foothill terraces (*khola khet*) and foothill irrigation systems*

These terraces are located along the banks of a small stream (khola) in the foothills in the gorge, and are known locally as khola khet. Such valleys have rough, narrow V-shaped cross sections with moderate slope in their lower sections (in the foothills) and steep slopes in their upper sections. In general, the khola khet is steeper than the slope hill terraces (tar) described ahead. The khola khet usually consists of a long strip of land parallel to the source river, often heavily dissected by gullies or cliffs into smaller patches. Therefore, the foothill irrigation systems irrigating khola khet are usually small in size. By virtue of their location in the gorge, many khola khet areas are shaded during parts of the day. Farmers live up in the hills, known locally as the upper villages, and use khola khet mainly to cultivate paddy, which cannot be grown profitably closer to the upper villages due to water shortage or low temperature.

The source rivers for both foothill and slope hill irrigation systems are relatively small in terms of flow carrying capacity. Thus, the intake is usually a simple temporary diversion weir constructed across the river. Although these need frequent maintenance, this is not considered a big task. In most of these systems, irrigation begins a short distance after the main canal branches off from the intake. Due to the elongated shape of the irrigated area, many of these systems have only one order[5] of canal, supplying water directly to a series of field channels aligned down the hill through their respective turnouts.

By virtue of their location along the river bank, foothill irrigation systems exercise prior rights[6] to water over the slope-hill irrigation systems. As a result, water supply to these systems is relatively high. The combined effect of high water supply and steep topography makes the controlled entry of flow into terraces more important than the distribution of available water. As a result, the ad hoc adjustment type of division structure (with a characteristic of downstream control) is more viable. Because of the much shorter idle length of the main canal and relatively easy intake, their maintenance is relatively simple. Consequently, these systems lack well-defined organizational and institutional arrangements, as these are not required for system functioning, although clear social roles may be accorded to people given operational responsibilities.

Slope-hill terraces (tar) and slope-hill irrigation systems

These are terraces located on hill slopes much above the large or medium rivers flowing through the bottom of a deeply incised valley. Locally they are known as tar, in which series of levelled and bunded terraces are constructed and managed by the farmers, especially for the cultivation of monsoon paddy. These terraces are the most striking features of irrigated agriculture in the hills. Unlike in the khola khet, houses are located near the tar and many farmers depend on it for their livelihoods. Hence agricultural activities in the tar are very intensive, with a very high cropping intensity under irrigated conditions. Given the great differences in level between these tar and the large rivers flowing in the deeply incised valley floor, this river water cannot be conveyed for irrigation. Rather, water from the tributaries of these large rivers, often located in a different or side valley, are tapped and conveyed to irrigate these tar.

The main difference in irrigation infrastructure between the foothill and slope-hill systems is the idle length of the main canals, which for slope-hill systems is usually much longer and passes through difficult terrain.[7] Therefore, concerns for the stability of the main canal, high conveyance losses, and high resource mobilization for maintenance are common characteristics of slope-hill irrigation systems, which make water a valuable commodity. Thus, these systems tend to have an elaborate network of distribution canals with relevant water division structures (with upstream control) supported by well-defined organizational and institutional arrangements. Most of these systems are managed by specific irrigation organizations, which have established rules and regulations to operate the system, although these may not be formally documented. Social values and norms developed over many years are the basis of system operation. The examples in this chapter are the Sankhar system, which has proportional weirs, and the Patne irrigation system, which has ad hoc water division structures.

River valley terraces and river valley irrigation systems

Like the khola khet, land in this category is located in the same valley as the source river. However, unlike the khola khet, river valleys allow a relatively large irrigated area, have no effect of shading by mountain slopes, and are almost flat. As in the tar, houses are located close to or within the irrigated area. Most farmers depend solely on this land for their livelihood. Hence, agricultural activities in the river valley are intensive and the cropping intensity is high.

The source rivers irrigating river valleys are relatively wide, have relatively flat gradients, and discharge substantial amounts of water with frequent flash floods, especially during the monsoon. Thus, building and rebuilding of temporary diversion weirs, often washed away by flash floods, is the principal challenge in system management. This challenge makes water a valuable commodity. Hence, in many systems, water allocation is tied in with the original farmers' investments in developing the system, and the ability to mobilize labour resources for maintenance. Thus, these systems tend to have elaborate networks of distribution canals with water division structures (upstream control) which allow water distribution proportional to original investments and regular operation and maintenance (O&M). Unlike slope-hill systems, these systems usually have earthen main canals. Because of the relatively easy terrain conveying the flows is not a problem. The Julphe irrigation system discussed below is an example. Table 5.2 summarizes the water control characteristics of different hill irrigation systems.

TABLE 5.2   Characteristics of Foothill, Slope-hill, and River Valley Irrigation Systems

Type of System	Size of Irrigated Area	Levels in Distribution System	Critical Component of Physical System	Available Flow per Unit Area	Control Structure	Types of Water Division Structure Mostly Used	Organizational and Institutional Support
Foothill	Small	One level	–	Relatively high	Downstream control	Ad hoc adjustment	Low
Slope-hill	Small–medium	Two–three levels	Conveyance canal	Relatively low	Upstream control	Mixed (depending on the local context)	Moderate–high
River valley	Large	Two–four levels	Intake	Relatively low	Upstream control	Mostly a form of proportional device	Moderate–high

Source: Author.

Availability of water

Availability of water influences the choice of water division structure, especially whether a fully or partially proportionate system is more appropriate. Figure 5.3 shows the relationship between the water division system and the flow available per unit of irrigated area. From this relationship, the available flow can be classified into upper, medium, and lower ranges, established from the maximum and normal levels of flow used in design.

The maximum requirement level is the level at which controlled entry of flow into fields to prevent damage is the prime consideration, rather than the distribution of available water. The range of flow greater than the maximum requirement level is the upper range. The normal requirement of flow will vary in relation to crop choice and allowances for operational factors (like canal filling and seepages, and cultural practices in allocation and application). It is difficult to estimate exactly, especially in the hills, but a general guideline for a normal system requirement for rice may be around 2–2.5 litres per hectare per second.[8] The medium range of flow falls between the normal and the maximum requirement.

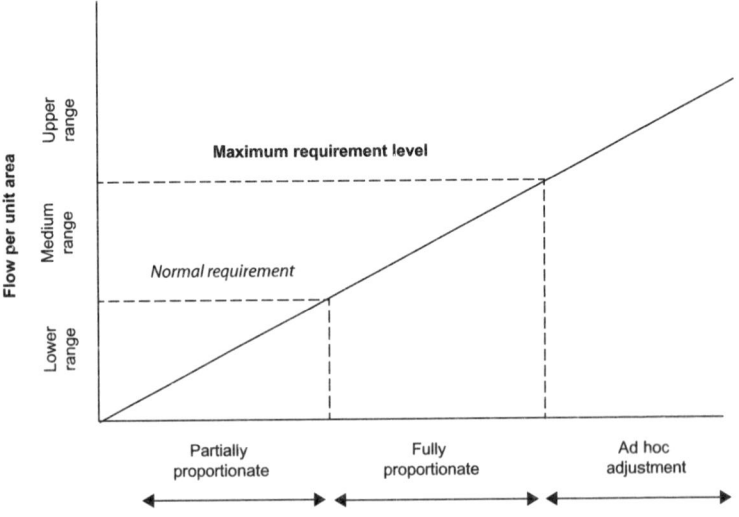

FIGURE 5.3   Flow Per Unit Area in Fully and Partially
                Proportionate Systems

Source: Author.

The lower range of water availability, below normal requirements, implies water scarcity. This requires a distribution technology that brings high water use efficiency. A partially proportionate system is more viable here, although requiring relatively high management inputs for open–close type of water distribution at its lower end. A fully proportionate system, which requires division of flows down to the individual farmer's share, is not practical, especially when the available flow is small. Thus, with the lower range of supply, farmers use a partially proportionate system to balance water scarcity, which maximizes required management inputs.[9] This situation will be seen in the Julphe system case study, where division structures differ across the diverse branches of the canal system, and which has a well-developed water management organization.

Systems with a medium range of flow allow a fully proportionate system.[10] Farmers receive a continuous flow to manage irrigation activities independently in their field. Although a fully proportionate system has a lower irrigation efficiency, the advantage of more water is also enhanced by minimizing management inputs.

Systems with an upper range of flow allow neither fully nor partially proportionate systems. In such a system, flow into terraces needs to be checked to protect the crops and terraces for stability. Thus a fully adjustable system (with a characteristic of downstream control), which can check the entry of flow, is more viable.[11]

A system may receive all three ranges of flow depending on the season. During monsoon, for example, the available flow may lie in the upper range while during winter it may decrease to the lower range. The use of division structure varies accordingly. Farmers are capable of adjusting operations to this differing water availability, as the case studies will show.

Altitude, topography, system shape, and crop diversity

In the case study areas, altitude was insufficient to affect crop choices: these were more affected by topography and insolation. However, in Nepal altitude also influences construction materials. The hardwood sal tree (*shorea robusta*) is used to construct a wooden weir because of its higher resistance to rotting in water. Its growth is limited above an altitude of about 1,000 metres. Hence, irrigation systems with wooden proportioning weirs are often found below this height.

Topography has diverse influences on irrigation: slope rarely restricts crop choice but is an issue for land stability. Irrigation takes place at hill slopes up to about 45 degrees for cultivating paddy or wheat under terracing, but at steeper slopes stability of terraces becomes a major concern. In many FMIS in steep terrain, large water flows are usually conveyed only through natural gullies, and are a concern in setting the maximum flow allowed for an irrigation system and choice of distribution structures. In a fully proportionate system the canal network density remains high, so all individual farmers receive water through their personal field channel from the respective turnout of a proportioning weir. Further, all branching canals (including field channels) should also be able to convey their share of peak flow.[12] As steep terrain limits such developments, a fully proportionate system is not viable in a steep topography.[13]

The shape of the irrigated area also determines the viability of distributing water through proportioning weirs. As already explained, where the irrigated area is a long strip of land, the distribution system is usually a single order of canal network: the parent canal directly supplies water to individual fields through a series of division structures. This makes the relative difference in the flow size between the parent and the offtake canals too high.[14] Consequently, the viability of distributing water by proportioning weirs decreases. In such a situation, the ad hoc adjustment type of structures is mostly used, as seen in the Patne system case described in this chapter.

A fully proportionate system is best suited to mono-demand occurring simultaneously throughout the irrigated area, especially for paddy cultivation. A partially proportionate system, on the other hand, is even suitable for diversified cropping.

SOCIO-CULTURAL CONDITIONS

Knowledge and practices

Knowledge and practices in system design and management are significant in influencing choice of structures. Usually, farmers in localities with similar agro-ecological conditions use a similar type of water division structure and have comparable management practices. For example, in the Kali Gandaki river valley, from the southern part of Parbat district down to Tanahu district, proportioning weirs are very popular. It is difficult to find a slope-hill irrigation system without proportioning weirs here unless the agro-ecological conditions mitigate against them. Elsewhere, other types of structures are used, even though agro-ecological conditions allow for the use of proportioning weirs.

Labour resources and water rights

Most FMIS are labour-intensive; O&M are highly dependent on the ability to mobilize labour. Therefore, in many FMIS mobilization of labour resources and water rights are interrelated. Variations in farmers' labour investment during system construction and maintenance create differential water rights among farmers in different parts of the system. The more one invests, the more rights one obtains over water and irrigation infrastructure ('hydraulic property'; see Chapter 1).

Rights are attributes of persons against other persons, and they are actualized with the help of certain rules, institutional arrangements, and physical devices. An individual may have a right to water, but unless this is defined through physical devices and rules (how much water is one authorized to get, from where, when, and how?) it is not possible to actualize these rights. Thus, the types of rights over water existing locally have direct influence on the choice of water division structure.

Operational modality

Operational modality refers to the requirements for use of infrastructure, shaped by local social preferences. Design of FMIS is also guided by operational objectives. Key operational objectives found in FMIS are equity, simplicity, and flexibility. If equity and simplicity are the prime considerations, a fixed proportional water division structure or a combination of fixed proportional and open–close types are mostly used. With fixed proportional structures, equity in distribution is real and can be quantified and judged by measuring and comparing the notches of a proportioning weir. It is also one of the simplest devices.

Unlike fixed proportional structures, open–close structures require a functional organization able to coordinate its members, enforce distribution rules, and resolve disputes. With diversified demand, negotiations for shifting and borrowing water time between farmers may occur. This necessitates frequent changes in rotational schedule, making it complex. Thus, the open–close division structure option is less simple and equitable, but more flexible compared to a fixed proportional one. If flexibility is the prime consideration, the ad hoc adjustment type is the most suitable. In ad hoc adjustment water division, equity can be judged neither with respect to the flow nor to the time for which water is received, but rather on the basis of the actual field conditions. Hence, measuring equity is subjective.

CASE STUDY 1: THE SANKHAR IRRIGATION SYSTEM

The Sankhar irrigation system in the hills of Syangja district is a slope-hill system with proportioning weirs, irrigating about 37 hectares of land belonging to 112 farmers. The Keladi Khola, a perennial stream and tributary of the Kali Gandaki river, is its main source of water. The irrigated area (khet) is situated on the left bank of the Kali Gandaki river in a tar locally known as *phant*, whose altitude varies between 343 and 367 metres. It slopes from north to south with an average slope of about 2–3 per cent. Its relatively low slope gives wider terraces. Paddy is the principal crop grown during the monsoon season, followed by wheat in the winter season. In the spring season, spring paddy is cultivated in about 15 per cent of land, and spring maize in the rest of the area.

The system has four levels of canals: the main, branch, distributary, and field channel. The main canal is 1.35 kilometres long. The system has two branch canals known as Thado Kulo and Terso Kulo. The Terso Kulo is aligned along the contour from the uppermost terrace, while the Thado Kulo is aligned down the hill. The system has 12 distributary canals and a large number of field channels. Any canal belonging to an individual farmer is termed here as a field channel. The average density of canals down to the field inlet (terminal points of proportioning weirs) is about 182 metres per hectare compared with the Food and Agriculture Organization (FAO)-recommended value of 50 metres per hectare for good water control (Ambler 1989). Aside from this density of distribution canals, the most striking feature of this system are the wooden proportioning weirs installed at all canal bifurcation points to distribute water continuously to all users, especially during cultivation of the monsoon paddy (Figure 5.4).

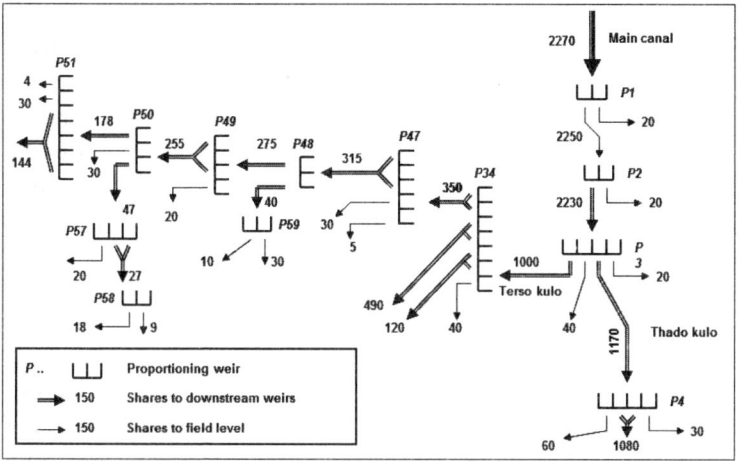

FIGURE 5.4   Schematic Layout of Proportioning Weirs Installed
at the Canal Bifurcation Points of a Section of the
Sankhar Irrigation System

Source: Adapted from Parajuli (1999).

SYSTEM OPERATION: WATER ALLOCATION AND DISTRIBUTION

In this system, water is allocated on the basis of the number of shares
held by each farmer irrespective of his irrigated area. The total water
flow available is considered to be 2,270 shares, known as *mato muri*,[15]
which is distributed among 112 farmers according to their respective
shares. Water shares are not attached to land. Farmers can transfer
their water shares from one place to another. They can even practice
water trading, temporarily or permanently.

Water distribution starts with the start of paddy transplantation during the early monsoon in mid-July, when diversified water
demands exist. These include a water demand for land preparation,
paddy transplantation, and to maintain standing water after paddy
transplantation. As not all farmers transplant paddy at the same time,
water is distributed as needed by them. However, as transplantation
proceeds, competition for water increases with demand. Consequently,
farmers install proportioning weirs at the canal bifurcation points to
divide water in accordance with their shares. By the time each farmer

completes paddy transplantation in at least one of his uppermost terraces, all proportioning weirs are installed, and the system functions as fully proportionate.

During the monsoon season, the canal base flow varies between 70 and 100 litres per second (average 85 litres per second; 2.29 litres per second per hectare). With such a flow in the main canal, the stream size available to an individual farmer is too small, especially when continuous flow is delivered to all farmers.[16] Although such a small stream size is enough to maintain the standing water in the paddy field, it is not enough to prepare land for paddy transplantation. For this reason, in this system paddy transplantation (in the peak period) is performed only with the rain, which substantially increases the canal base flow. Farmers store their share of such peak flow in their upper terraces, which have already been transplanted. The stored water is then discharged into the lower terraces to manage land preparation at times suitable to them.

During the winter season, water supply is greater than demand. As a result, farmers do not have any specified rules for water distribution. Farmers can irrigate their fields as and when required. During the spring season (mid-March to June), flow available at the canal varies between 15 and 25 litres per second (average of about 0.54 litres per second per hectare). With such a small flow, continuous flow to all farmers is impossible. So, at this time, only the main proportioning weir installed at the bifurcation of the two branch canals is set to function to supply water continuously to both branch canals. Within a branch canal, water is distributed to farmers through an open–close distribution system based on time shares. So during the spring season the Sankhar system is partially proportionate.

The time unit per share within the branch canal is decided according to the agricultural activities to be performed. For example, from mid-March to mid-April (during paddy transplantation), nine minutes is allocated per share of water. A farmer having 20 shares (or mato muri) receives water for 180 minutes. From mid-April to June, however, a shorter time of three minutes per share is provided.

The system has a single-tiered irrigation organization led by a chairman known as *akhtiyar*, an influential person elected through general consensus, who continues to work as long as he enjoys the confidence of farmers. The remaining farmers are general members; major decisions are taken in the general assembly (GA). The main function of the akhtiyar is to mobilize resources, especially labour, and to fix the start and closing dates of regular system maintenance. Maintenance involves desilting of the canals and tunnel, mud-lining of porous areas of the canal and strengthening of canal banks. In 1996, some 315 man-days were mobilized over 5 days for the work. The basis of maintenance is the water share—a farmer with 30 units of water shares needs to provide one labourer a day; and a person with 15 units provides a labourer every alternate day—and there are fines for non-reporting. During spring, water acquisition is the responsibility of farmers taking water, but during the monsoon it is a collective effort for which the akhtiyar is responsible; every day four farmers are deputed to watch and repair the canal and intake. The fixed proportioning weirs work automatically so few institutional roles are required at this time, and a main overall objective is resource mobilization for maintenance. However, for the periods when timed rotation is needed, a task force is present—an informal organization headed by experienced and socially recognized farmers. They oversee rotation around a prepared schedule, although this has frequent changes requested by farmers. Because many farmers with different water shares at different times of the year are involved, management of timed rotation of water needs careful attention. This causes high administrative costs because it demands time and resources. However, this cost is taken up among farmers, and a high level of individual supervision of distribution is possible because water distribution tasks are assigned to users.

CASE STUDY 2: THE JULPHE IRRIGATION SYSTEM

Julphe irrigation system is a river valley irrigation system with proportioning weirs, located in Nawalparasi district, in the foothills of the Mahabharat mountain range, in the river valley known as the Siwaliks or Doon valley. This area has a sub-tropical climate. This system serves about 200 hectares of land belonging to 340 farmers residing in the villages of Julphe, Kolia, and Basantpur. The irrigation system in each of these villages operates independently as a sub-system of the larger system, the Julphe irrigation system. The irrigated area constitutes flat land which slopes from North to South at an average slope of 1.06 per cent. The altitude of the area varies between 200 and 240 metres. Paddy is the principal crop grown during the monsoon season, followed by wheat or potato in the winter, and rain-fed maize in the spring season.

The Girwari river, the main source of water for the system, drains out of the gorge of the Mahabharata mountains into a wide river valley just a few hundred metres upstream of the intake, and has formed several unstable river channels at the intake site. A temporary diversion structure is built across these unstable river channels to divert water to the main canal; maintaining this temporary diversion is a key challenge. The system has a four-tier hierarchy of canals: the main canal (2.1 kilometres), sub-main canals, secondary canals, and field channels. The two sub-main canals below the first proportioning weir are the primary canals for the sub-systems. A canal at the lower end of the distribution system supplying plots is known as the field channel.

Julphe irrigation system was first developed by local farmers around 1910–20 to irrigate about 14 hectares of land belonging to Julphe village, which later increased to about 32 hectares by the 1950s. Locally, this original land area is known as *numbari* land. Since the 1950s, many people migrated from the nearby hills to settle here by encroaching on the forest. By the 1970s, the population had increased considerably. Later, due to increasing population pressure, farmers from two nearby villages, Kolia and Basantpur, and new settlers of Julphe village collectively mobilized labour resources and widened the existing canal from three to five *haath*,[17] and extended it to their areas. By doing this they acquired rights over the water. Since then, the water shares that a farmer or group of farmers own became the basis for both water allocation and mobilization of labour and resources to maintain the temporary intake and the canals. Farmers no longer consider their water shares as attached to land, but as—tradable—private property.

To divide the water according to the agreed shares, wooden proportioning weirs were introduced. The first one, placed at Girwari, divides the incoming water into two parts, one of which goes to Julphe and the other to Kolia and Basantpur. The shares of Kolia and Basantpur are further divided by the proportioning weir installed at Hattigaunda (Figure 5.5). Each of the sub-systems is further divided into blocks and sub-blocks to distribute water as per agreed shares.

The command area of the Julphe sub-system is first divided into two blocks. The first block, consisting of the numbari area, is

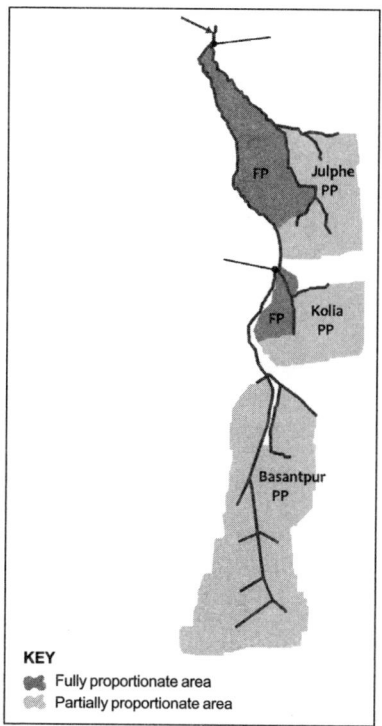

FIGURE 5.5   Physical and Operational Characteristics of the Julphe
Distribution System

Source: Parajuli (1999).
Note: FP refers to fully proportionate system area, and PP refers to partially
proportionate system area.

designed with a fully proportionate layout served by 32 proportioning weirs. The second block is divided into seven sections with the characteristics of a partially proportionate layout, each of which served by the respective turnout of the proportioning weir. The Kolia sub-system has a similar layout of its distribution network—fully proportionate at the head-end, and partially proportionate at the tail. In contrast, the entire area of the Basantpur sub-system is designed with a partially proportionate layout. Table 5.3 presents the physical and operational characteristics of the distribution system.

TABLE 5.3   Physical and Operational Characteristics of the Distribution System

| Sub-systems | Physical Characteristics | | | | Operational Characteristics | | |
| | Type of Layout | No. of PW | Served by One Turnout of PW | | Water Share per Land Unit (kattha/ha) | Flow Rate (l/s/ha) | |
			Average Area (ha)	Average No. of Farmers		During PT	After PT
Julphe subsystem:	FP	32	0.83	1.62	16.79	11.58	8.39
Numbari area	PP	7	5.77	16.5	5.62	3.87	2.81
New settlement area							
Kolia subsystem:	FP	4	0.53	1	9.25	3.88	3.05
Head reach area	PP	2	8.22	18	4.86	2.04	1.60
Remaining area							
Basantpur subsystem	PP	8	14.0	29.5	6.62	2.32	1.72

Source: Parajuli (1999).

Note: FP refers to fully proportionate systems; PP refers to partially proportionate systems; PT refers to paddy transplantation; and PW refers to proportioning weir.

System Operation: Water Allocation and Distribution

As noted earlier, water is first allocated to the three sub-systems on the basis of water shares they own. The total incoming water available at the head end of the command area is considered to have 1,595.45 units of water shares, locally known as *kattha*. Of these, 750.45 units belong to the Julphe, 745 units to the Basantpur, and the remaining 100 units to the Kolia sub-system. Within each sub-system, water is allocated to farmers on the basis of the number of shares held by them. Whenever the main canal remains 'on', all sub-systems receive continuous flow corresponding to their water shares. However, the mode of water distribution within sub-system varies among them, as explained below.

Water distribution begins with the start of paddy transplantation in early July. In this period, neither the Julphe nor the Kolia sub-system uses proportioning weirs to distribute water. Rather, paddy transplantation takes place from head to tail of the sub-systems. This is because, although both the Julphe and Kolia sub-systems have a fully proportionate layout in their head reaches, in each case distributing water continuously on a share basis would result in simultaneous small flows to all turnouts.[18] Such small flows are enough to maintain standing water in the paddy fields after transplantation but insufficient to transplant paddy.

So in these sub-systems water (for transplantation) is distributed to farmers as per their requirements irrespective of water shares and irrigated areas. Mutual consensus is the basis of water distribution. In such a situation, with a low level of transparency, it is difficult to maintain equity. To cope with this situation, ad hoc committees are formed, especially for paddy transplantation, to manage water distribution with close monitoring at farm level.

However, after transplantation all proportioning weirs are installed in their locations. The numbari area of Julphe and the head-end area of Kolia sub-system start functioning as fully proportionate, and the other areas as partially proportionate distribution systems. With the start of water distribution by proportioning weirs, the irrigation system becomes rigid, and equity in water distribution is judged by measuring and comparing the width of notches. The other water distribution characteristics of proportioning weirs are similar to those explained earlier for the Sankhar irrigation system.

In contrast, all the proportioning weirs of the Basantpur sub-system are installed prior to paddy transplantation. It functions as a partially proportionate system; all of its seven sections receive continuous flow. Within the section, the flow available is distributed among all farmers based on a time-share method. Equity in water distribution is judged in terms of time and not in terms of quantity of water received, even if the flow is erratic.

In all three sub-systems, overall irrigation management activities are performed independently by the sub-system Water Users' Committees (WUCs). At the sub-system level, where proportioning weirs distribute water, involvement of WUCs in water distribution is nominal. However, in areas with partially proportional layout, each section has an independent water users' organization known as a *tol samiti* (section committee) which remain very active during the paddy cultivation season. The section committee performs all the day-to-day water distribution management tasks independently, including management of timed rotation within the section.

CASE STUDY 3: THE PATNE IRRIGATION SYSTEM

The Patne irrigation system is a slope-hill irrigation system with ad hoc adjustment structures, located in the hills of the Tanahu district, which irrigates about 21 hectares of land belonging to 56 farmers. Bilmade Khola, a seasonal stream, is its main source, which is supplemented by a natural perennial spring located close to the main canal. During the monsoon, the system receives a flow of 80–100 litres per second (about 4.3 litres per second per hectare), which decreases to about 10 litres per second (about 0.47 litres per second per hectare) during the spring season. Paddy is the main crop grown during the monsoon season.

The irrigation system is situated on the right bank of the Madi river (about 400 metres altitude) in raised alluvial terraces (tar), which slope from west to east at an average slope of 2–3 per cent. The irrigated area constitutes a long strip of land parallel to the main canal, which is aligned from the uppermost terrace of the irrigated area. The total length of the main canal is about 1,650 metres; its average width and depth in its idle length are about 1.2 and 0.5 metres, respectively. Most landholdings lie perpendicular to the main canal: they are long and narrow strips with their upper ends connected to the main canal. A number of 54 turnouts,[19] corresponding to each landholding, are placed in the main canal. The turnouts consist simply of an open cut in an earthen canal bank. The shape of a turnout is fixed with turf and stones placed on either side of the cut.

The ad hoc adjustment type of water division structure is used here as others are not viable. An open–close type of division structure is viable only at the lower end of the distribution system, where flow is relatively small and can be rotated among farmers without further dividing it. As the main canal directly supplies water to individual farmers' holdings through a series of turnouts, the higher relative difference in flow size between the parent and the offtaking canal restricts the viability of dividing water through fixed proportioning weirs. Further, the water allocation principle adopted in this irrigation system does not allow for the use of fixed proportional type of distribution structure.

System Operation: Water Allocation and Distribution

The principle of water allocation in this system relates the amount of water provided to the combination of size of a farmer's holding, type of terrace, and crop stage. Water distribution starts with the start of paddy transplantation. As all farmers do not transplant paddy simultaneously, water is distributed on a rotational basis according to the needs of the farmers.

After the completion of paddy transplantation all farmers gather at a place, known locally as *saameli*, to clean the canal and fix flows in all turnouts. On that day, a few experienced farmers fix flows in all turnouts by setting the turnout size, in the presence of the system representative (*paani jimmawal*).[20] Thereafter all farmers receive continuous flow. Once the turnout size is fixed, only the paani jimmawal or his representative can adjust its flow. No other farmer is allowed to increase or decrease the flow by changing the size of individual turnouts. The reason for this is two-fold. First, during periods of scarcity, all turnouts should receive their authorized shares of water. Second, during heavy rains when the main canal gets swollen, water should also be dispersed equally among all fields to minimize the chances of terrace collapse. Every day, three of the farmers (excluding the chairman and his secretary) perform a routine check of all turnouts and of the water levels in every terrace. If necessary, they adjust the flows through turnouts, with permission from the chairman. Every 19th day, all farmers organize a meeting (saameli) to evaluate the water distribution arrangement of the previous 18 days.

In any run-of-the-river system, water supply in the parent canal fluctuates considerably. For equitable distribution, such fluctuations need to be uniformly distributed among all farmers. This means that flows through these turnouts must be adjusted frequently, requiring human involvement. It is, however, difficult and cumbersome to keep on adjusting the flow through the turnouts to synchronize the frequently changing supply level with the demands. For this reason, the equity of water distribution is not judged on the basis of the flow through it, but on the basis of actual field conditions. This is explained below.

In the hills, during paddy cultivation, water flows continuously terrace to terrace, from the uppermost to the lowermost terrace. Farmers always check the flow through their lowermost terraces and compare it with other farmers to ascertain if equal amounts of water are being received or not. If the last terrace of one farmer is dry, he expects that the last terraces of other farmers with similar holdings should also be dry—the same also for flowing water. If a farmer does not find these conditions similar, he requests the paani jimmawal, to adjust the flow in his turnout. The distribution rules do not permit him to adjust the flow by himself. Depending on the situation, the paani jimmawal can issue an order to increase the flow.

* * *

The technologies found in the FMIS of Nepal merit sustained attention for the way their designs deal with the agro-ecological and socio-cultural conditions of the area in which they are found. This chapter has examined these relationships with irrigation infrastructure, especially for water division structures in hill areas and the presence of the proportioning weir. It has shown the importance of knowledge present to build such systems in hill environments, and of social concerns in how structures materialize water rights and management options. Critical factors include the quantity and availability of water and choices in the operation of systems, the shaping of rights to water, and the choice of operational modality between simplicity, equity, and flexibility.

The choice of water division structure is also shaped by the requirements for use of infrastructure, which are in turn shaped by the operational objectives. If equity and simplicity are the prime considerations, a fixed proportional type of water division structure is mostly used. If flexibility is the prime consideration, the ad hoc adjustment type of division structure is best suited. Local irrigators have shown remarkable understanding of their local environments, and social commitment to collective action in the design choices of technologies for hill irrigation systems.

As concerns of food security and economic transformation continue to drive investments to support FMIS, it remains vital to understand how technologies and institutions still have elements of choice, but also are shaped by each other and the environment they operate within. Such knowledge can prevent the imposition of blue-print institutional models and technical designs. Recognition of these relationships between agro-ecological systems, hydraulics, and local social preferences for the use of infrastructure can guide policies for improved use of water resources and livelihoods of local communities.

Notes

[1] Three systems are documented here. The fourth case, Bachcha Irrigation System (BIS), is not documented for reasons of space. References to some of its characteristics are found in the text and in Notes 11 and 13 further ahead.

[2] Agro-ecology is used here as a concept integrating study of the biophysical, technical, and socio-economic dimensions of agriculture, recognizing that the term has changing usage over recent decades. It is used here particularly for the increased attention it provides to broader landscape and habitat dimensions that shape production options, use of materials, and resource management strategies. This use encompasses older work on irrigation from human ecology perspectives (Mabry 1996), which focused on interaction of people with their environment, and the recent concept of socio-ecologies now in use in some water management studies in South Asia (Shah and Raju 2001).

[3] The flow in 'open–close' types of division structures is fixed and there is no need to split the flow into two or more parts. When closed, the flow is zero; when opened the water flow is total. The open–close type of water division structure is mostly used at the lower end of a distribution system where individual farmers can easily handle the flow.

[4] Operating hydraulic head is the difference in water levels between the parent and the branching canals at the point of bifurcation.

[5] Canal order or level is related to the number of divisions of water flow in a system. In a system with only one level or order of canal, the entire flow of the main canal is rotated among users. In a canal with two levels, a main canal feeds branch canals to supply different tertiary units.

[6] A legal system of prior rights protects a water right that is historically older from changes created by new water abstractions. In Nepal, FMIS organizations regularly take legal action if their water flows are affected by new developments upstream.

[7] Based on a study of 41 irrigation systems in the hills, Ostrom *et al.* (1992) noted that the average length of the main canal is 2,159 metres (76 metres per hectare and 34 metres per household).

[8] Normal system water requirements depend on many factors. In the hills, its actual estimation is difficult. For this reason, many authors have recommended values varying between 1.5 and 3.0 litres per second per hectare for the cultivation of monsoon paddy (Jacob 1995: 101; MacDonald and Hunting Technical Services 1982: 30). Thus, as a general guideline, normal system water requirements may be around 2–2.5 litres per second per hectare.

[9] Chow (1960) notes that rotational water distribution can potentially save 20 to 50 per cent of water in comparison to continuous distribution.

[10] Examples are the head-end of Kolia and Julphe sub-systems of Julphe irrigation system, which have water allowances of 3.05 and 8.39 litres per second per hectare, respectively.

[11] The khola khet areas of the BIS are examples of this (see Parajuli 1999).

[12] Being a fixed control structure, a proportioning weir distributes both the low and the peak flows in the same proportion.

[13] BIS is an example of this. Though it could have a fully proportionate system from the perspective of water availability, its topography did not allow for this (see Parajuli 1999).

[14] An example is the Patne Irrigation System.

[15] Mato muri is a traditional measure of land. In hill irrigation systems the water share is also sometimes named mato muri: one mato muri of water is the share of water allocated to one mato muri of land.

[16] The 85 litres per second of water, when divided into 2,270 shares (mato muri), one mato muri gets about 0.0374 litres per second. In this system, about 80 per cent of the farmers have the average water share of 20 mato muri. This means a flow of 0.75 litres per second is available to an individual farmer.

[17] Haath is a measure of distance. One haath, which is equal to the distance from the elbow to the tips of the fingers, is approximately equal to 0.45 metres.

[18] Average land per turnout in the numbari area of the Julphe sub-system is about 0.83 hectares, and average flow during paddy transplantation is about 11.58 litres per second per hectare. Thus, the average stream size per turnout available in the numbari area is about 9.6 litres per second. This stream size is considered inadequate for preparing land for paddy transplantation.

[19] Thirty in the main canal, and the remaining ones in three small branch canals.

[20] The Patne Irrigation System has a single-tiered irrigation organization with two executive members, the chairman and the secretary. Locally, the chairman is known as paani jimmawal and the secretary as *sachiv*. The posts of both the paani jimmawal and the sachiv are hereditary.

6 GROUNDWATER IRRIGATION DEVELOPMENT, CONJUNCTIVE USE, AND THE EVOLUTION OF WATER USE COMPLEXES IN THE NEPALESE TERAI

Suman Rimal Gautam

Interventions in water resources development for irrigation in the Nepalese Terai—the largely flat southern part of Nepal bordering India—have developed around various water sources in this water-abundant region. The Irrigation Policy of the Government of Nepal (2003) stresses the need to promote conjunctive use of groundwater and surface water for irrigation systems. The policy document states that the 'feasibility for conjunctive use of surface and groundwater' should be the basis for selection of projects, and that the 'concerned stakeholders' would be 'coordinated in the process of project selection' (Irrigation Policy 2003: 4).[1]

In practice, the target-oriented nature of intervention programmes in both groundwater and surface irrigation has always ignored the fact that various water resources were already in use and collectively managed in the area. Subsequent irrigation sector reforms[2] after 1992, promoting turnover of systems to water users' associations (WUAs), have also ignored these realities.

Implementation of shallow tube well (STW) and deep tube well (DTW) projects has been carried out through different models of intervention. Yet, even though groundwater is seen as critical for rural development, little knowledge exists on how these different technology choices transformed the social relations around water management and production in areas with multiple sources of water. Nor is there much information on how farmers make their choices of technologies and institutions for irrigation and how these affect the technological and organizational performance of surface and groundwater interventions when they exist side-by-side. The study presented here is all the more relevant given the focus on groundwater for irrigation development and the repeated calls for conjunctive water management.

This chapter discusses the institutional and production changes taking place around groundwater technology for irrigation in the Tinau river basin in the western Terai district of Rupandehi, where multiple sources of water exist (see Figure 6.1). The chapter is based on in-depth

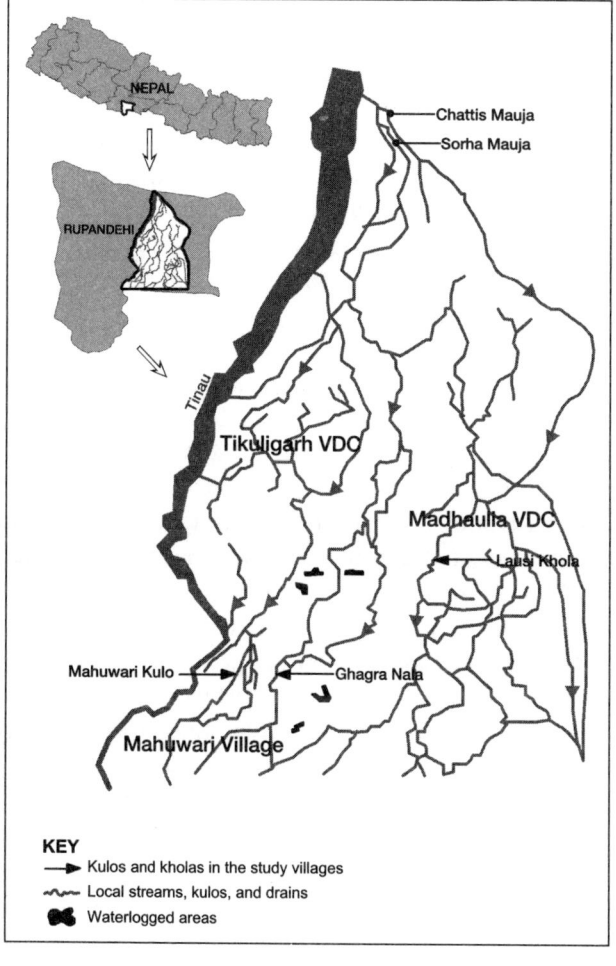

FIGURE 6.1 The Research Area

Source: Author.

case studies in three village development committees (VDCs) with a long history in surface irrigation management prior to intervention in groundwater. The research was undertaken in 2001–2 and revisited in 2004. Two of the sites, Tikuligarh and Madhaulia, were part of the Bhairahawa Lumbini Groundwater Irrigation Project (BLGWIP) which promoted DTWs. The third site, Mahuwari village in Hatti Bangain, lies outside the project area. The BLGWIP installed DTWs in Rupandehi in three phases of implementation from 1975 to 1999.

A socio-technical approach (Mollinga 2003) was adopted to study water resources management and governance. Diverse technological intervention programmes in irrigation were studied alongside the existing forms of water use within the broader water resources systems in which they function. The evolving irrigation practices were understood as water complexes with physical–technical, organizational, and normative–legal dimensions of water control, developed in larger agro-ecological and socio-political contexts. The heterogeneous nature of interactions between different sources of water, technologies, and people, and the way they mediate water supply is likened to a 'development arena' (Jorgensen and Sorensen 1999). Here, the actions and behaviour of different actors are embedded in the larger context in which they operate (McCay 2002; Mollinga 2003; Mosse 1997b; Roth 2003, 2006).

Intervention processes in irrigation also interact with, and are embedded in, existing power relations at the local level. They 'interact with already contested domains of power and meaning' (Li 1996, cited in Mosse 1997b: 499). Understanding of power and control relationships helps in analysing how different actors choose, strategize, reject, manipulate, and adjust ways to irrigate in order to improve their production. They make use of various technical, organizational, and normative–legal options to negotiate their water rights. The way they behave has to be examined by knowing their backgrounds, the social entities they represent and help reproduce, their histories, values, and significance (McCay 2002). Studying power and control relationships helps to analyse the capacity of farmers to use or transform processes of intervention, work out different relations in (of) production, and come up with innovative solutions for irrigation. Diversity of rule-making was examined from the perspective of legal pluralism (Bruns and Meinzen-Dick 2000; Roth et al. 2005).

This chapter is divided into six sections. The next section provides a description of the study area, showing the overall processes of water resources development in the Terai, and groundwater irrigation in particular. The following three sections present the case studies of the research. The final section summarizes the conclusions drawn from these case studies, and a short discussion of the implications.

THE RESEARCH AREA

Planned development for irrigation in Nepal started only after the end of the Rana regime in 1950. Land reform measures were introduced in the 1960s. Prior to this, numerous streams flowing through the Terai had been harnessed for irrigation by farmers' initiatives. From the time of unification in 1768 to 1951, the agrarian economy of the country was characterized by a feudal system. Rulers assigned tracts of land to subjects under various tenurial conditions. The majority of the surface irrigation systems (*kulos*), later called farmer-managed irrigation systems (FMIS), emerged from the land tenure system prevailing during that period.

Like other parts of the Terai, Rupandehi district has always been the focus of irrigation development by the government. This narrow strip of land covering around 14 per cent of the total area of the country, is the only 'real cultivable area', thanks to its relatively flat topography. Even though planned intervention in groundwater irrigation had started in the 1970s, the main government priority until the 1990s was the expansion of surface water sources. However, large-scale projects developing deep groundwater for irrigation like the BLGWIP did receive equal priority with its counterparts in surface irrigation; unlike shallow groundwater, which was left more to the private sector, albeit with subsidies.

It was only after 1995 that the national plans and policies made investments in groundwater irrigation a priority. STW irrigation remained in the domain of the Agricultural Development Bank of Nepal (ADBN) until the late 1990s. The bank promoted this technology through loans and technical support until 1999. After this period, under the influence of the neoliberal policies adopted by the government, subsidies on STWs were completely removed. Presently, both the government and non-governmental actors are involved in

STW installation across the Terai. The focus on groundwater was justified on the grounds that it provided year-round supply of water, necessary to overcome low flows in canal irrigation, and its capability to reach lower-income farmers and provide them with cash-intensive farming opportunities.

Intervention processes in both groundwater and surface irrigation in the Terai are still ongoing through various approaches. Government agencies, together with non-governmental organizations (NGOs) and bilateral projects, have also been involved in improving irrigation services through adoption of 'demand-based participatory approaches' (see Khanal, this volume). Yet most agency programmes work in isolation from each other and implement their own models for intervention. Problems and constraints are identified and dealt with individually. The fact that other technologies and sources of water coexist with those introduced is not denied. However, the lack of coordination between intervening agents and agencies helps aggravate the problems of a 'piecemeal' approach to (water resources) development.

Like other districts, Rupandehi is home to numerous FMIS.[3] In addition, it harbours the largest FMIS that are still functioning in Nepal, constructed nearly two centuries ago. The Sorha Mauja Irrigation System (SMIS) and the Chattis Mauja Irrigation System (CMIS) function as separate systems, but they have a joint committee at the point where they share the water from the Tinau river. Intervention in DTWs through the BLGWIP predominates development of collective groundwater-based irrigation systems. The first case study site, Tikuligarh VDC (TVDC) is part of the Sorha Mauja network, while part of Madhaulia VDC (MVDC) lies in the Chattis Mauja system. Mahuwari, the third site lies outside both the BLGWIP and the CMIS–SMIS systems, although in the past it did have a kulo that has now disappeared.

The Tinau basin is underlain by multiple aquifers of different depths. Hence groundwater was identified as a source of irrigation water to augment agricultural production options. The BLGWIP intended to open-up parts of Rupandehi district for cultivation by supplying year-round deep groundwater in the tail-end of existing kulo systems by tapping into these aquifer resources. A unique hydro-geological feature of the area is the presence of subsurface springs (*jharan*) that flow on their own, but also continue and increase as natural flow after rains fall and supply recharge in the upstream catchment, and as drainage flows after irrigation begins. Free-flowing artesian flow is found in several parts of the district.

The original inhabitants of this area are called the Tharus. After the eradication of malaria in the 1950s, Rupandehi has seen a high influx of migrants from various parts of Nepal, predominantly from the adjoining western hills. There is a substantial population of Nepalis called Burmelis.[4] Originally, these people had left the western mountains to settle in Burma. In the southern parts bordering India, as in Mahuwari village, the dominant ethnic group is of Bhojpuri origin.[5]

The BLGWIP installed 182 DTWs in three phases between 1976 and 1999, covering a net area of 13,185 hectares. BLGWIP is quite unique in the history of agency-initiated irrigation development in Nepal, as it is not only the biggest groundwater irrigation project, but also the most expensive irrigation project implemented so far. The cost per hectare amounted to USD 5,174. It also included a 92-kilometre rural road and electricity transmission development programme. These have substantially transformed the rural landscape in the district and also encouraged migration from adjoining mountain districts. The project started with intensive state involvement typical of the 1970s trend in irrigation development worldwide. Until 1989, design was done without any farmer participation (Olin 1994). Later on, the project incorporated a 'participatory approach' characteristic of irrigation reform policies in the 1980s and 1990s. This included the incorporation of the 'demand-based participatory approach' in project design for the DTWs installed in the third phase of implementation. All DTWs that had been installed before 1992 were transferred to a water users' group (WUG).

There have been changes in the design of the DTWs as well. The older tube wells in the project area were equipped with high capacity motors of 75 kilowatts and an open flow conveyance system. In the third phase, they were equipped with lower capacity motors of 35 kilowatts and an underground pipe distribution system. In all cases, each DTW unit was designed to operate as a separate irrigation unit. The average design command area was around 120 hectares. The transfer process was a highly politicized affair, and it took the project nearly two years to really hand over the tube wells to the farmers. The farmers were not willing to take over the entire responsibility of maintenance and operation of these DTWs, which entailed substantial sums of money as a demand charge (flat rate of electricity), besides the usual operating charges.

The study also shows how farmers incorporated groundwater before and after handover into their systems of water use. An intricate relationship had emerged between different kulos, jharans, DTWs, and STWs, such that farmers made use of a combination of water sources to fulfil their irrigation requirements. The choices they made with respect to these sources gave rise to unique complexes of water and related institutions. This is examined with respect to changes in demography, as the population changed from homogenous to ethnically diverse, as well as changes in agriculture, irrigation, energy pricing policies, and socio-political conditions in the country.

Research on the water use complexes formed by accessing different sources shows how farmers inside a VDC were grouped together through common sources of water. Such a complex formed the dynamic hydrological boundary that they had defined. Taking the village or VDC as the focus of study made it possible to locate these water complexes within physical, social, and administrative spaces, and as part of the larger water resources system.

INSTITUTIONAL CHANGES IN IRRIGATION IN TIKULIGARH VDC

This case study shows the process of transformation of water use practices in TVDC after intervention in groundwater irrigation, and the emergence of different water use complexes around its water resources. It examines how the introduction of DTW technology interacted with the historically grown relationships around various sources of water, differences in power structure that existed in the villages before intervention in groundwater irrigation, agrarian structure, and shifts in power from one group of farmers to another.

Eight DTWs were installed in two different phases of implementation, in different wards of TVDC.[6] Four DTWs, installed in the mid-1980s, were 75 kilowatts capacity tube wells with an open flow conveyance system. The other four DTWs had a 35 kilowatts capacity with an underground PVC pipe system for distributing water, installed in 1999. The mode of implementation of the older tube wells was more focused on infrastructure, as was typical of irrigation development in that period. Participatory concepts were added to project design in the course of the 1980s. These DTWs were then handed over to water users' groups in 1992. The new DTWs, like the one in Durganagar, were installed through farmer participation. The three case studies in TVDC—in the villages of Tikuligarh, Supauli, and Durgarnagar—highlight the story of three DTWs that are historically, hydrologically, socially, and politically connected to each other, even though they were installed to function as separate units. The study shows that, though interventions followed similar 'models' and approaches, the outcome is different in Tikuligarh and Supauli villages.

The SMIS was the main source of irrigation for people living in TVDC, prior to the advent of BLGWIP. One of the oldest *maujas*[7] of SMIS, Tikuligarh village, after which the entire VDC was named, had been the stronghold of the local Tharu landlord or Chaudhary. Supauli village lies south of Tikuligarh, and is the tail-end mauja of SMIS. Durganagar village lies south of Supauli and was never part of the SMIS. Besides the DTWs, there is a prominent presence of old kulos, jharan flow, and shallow groundwater for irrigation. The BLGWIP project area overlapped with seven maujas of the Sorha Mauja network. However, four maujas inside this VDC retained their water rights with SMIS, even after the project had provided free groundwater. The distribution systems of the DTW projects were installed without clear reference to how they overlaid the existing network of kulos and jharans, and this happened in diverse ways and with diverse impacts. Some DTW distribution systems cut off existing maujas that shared a common kulo. Elsewhere, parts of the DTW conveyance joined communities in such a way that people who belonged to different maujas were now sharing a common DTW.

Official government records denote the entire area as 'deep tube well irrigated'. Fieldwork showed that, in reality, farmers in this area were organized around different sources of water and technologies in unique patterns of organization that are very dynamic. Table 6.1 shows how farmers were originally grouped around different sources of water and how these patterns of organization changed through the years. This process can be understood by examining the timeline of intervention processes, understood at three points in time: prior to BLGWIP, when DTWs were installed (1982 and 1999), and after 1992, when the older DTWs were handed over to WUGs. All villages were either part of SMIS or used jharan flow for irrigation.

TABLE 6.1 Process of Transformation of Irrigation in Tikuligarh VDC

Ward No. and No. of DTWs	Year DTW Installed	DTW Location (Ward)	Water Use Prior to BLGWIP	Water Use Complex after BLGWIP	After Handover of DTWs (1992)	Complexes in 2004
1 (1)	1982	1	Surface	Deep groundwater and surface		Deep groundwater (2003)*
2 (1)	1982	2	Surface	Deep groundwater and surface		Deep groundwater and surface
3 (1)	1999	3	Surface	Deep groundwater and surface		Deep groundwater and surface
4 (1)	1999		Surface	Deep groundwater and surface		Deep groundwater and surface
	1999	5	Jharan	Deep groundwater and jharan		Deep groundwater and jharan
5 (1)	1982	5	Member of surface irrigation using jharan	Deep groundwater and jharan		Deep groundwater and jharan

6 (1)	1982	Jharan	Deep groundwater and jharan	Shallow groundwater and jharan	Shallow groundwater and jharan
7 (3)	1982	Surface	Deep groundwater and surface	Surface (largely)	Surface (largely)
	1982	Surface	Deep groundwater	Deep groundwater and surface	Deep groundwater and surface
				Deep groundwater and shallow groundwater	Deep groundwater and shallow groundwater (largely)*
7	1999	Surface	Shallow groundwater		Deep groundwater
8 (2)	1982	Surface	Deep groundwater and surface		Deep groundwater and surface

Cont'd

TABLE 6.1 *Cont'd*

Ward No. and No. of DTWs	Year DTW Installed	DTW Location (Ward)	Water Use Prior to BLGWIP	Water Use Complex after BLGWIP	After Handover of DTWs (1992)	Complexes in 2004
9 (1)	1982	1	Surface	Deep groundwater and surface	Deep groundwater	Deep groundwater
	1982	Madhaulia VDC	Jharan	Deep groundwater and jharan		Deep groundwater and jharan
	1982	8	Surface	Deep groundwater and surface		Deep groundwater and surface

Source: Field survey conducted in 2001–4.

Note: * denotes year in which changes occurred.

The villages where DTWs were installed in 1982 immediately used groundwater but tried to retain water rights in SMIS. After 1992, some villages continued conjunctive use, while others changed to groundwater use only. The WUGs in charge of managing the older generation of DTWs faced similar constraints throughout the project area, after handover. These problems were largely due to reluctance of farmers to pay a demand charge,[8] a resurgence of interest in surface sources, and unwillingness to contribute money for DTW repair and maintenance. Reduction in irrigated area meant that the rates each irrigator had to pay would increase. Most farmers were willing to pay for what they used (that is, the operating costs for tube wells), but were reluctant to pay the demand charge. Though these issues were common to Tikuligarh and Supauli, they took a different turn in each area. The cases show how different actors strategize differently to gain control over different sources of water within their villages.

DISINTEGRATION OF THE DEEP TUBE WELL WUG IN TIKULIGARH VILLAGE

In 2004, the DTW in Tikuligarh was on the verge of collapse. The few people struggling to retain it were facing insurmountable problems, as the already deteriorating social relations in the village with respect to surface irrigation had spilled over into DTW management.

The older kulos of SMIS had been constructed by the Chaudhary's family through their subjects. They collected revenues and held absolute power over them. As migration increased, hill migrants started settling in the upstream parts of the kulo. This had a bearing on social relations, both in the village and in irrigation. As kulos expanded, hill migrants came to dominate the irrigation committees (kulo samiti). Older water users had now become tail-enders, and were not happy with the way the migrants carried out canal cleaning operations (kulahi). The landholding pattern became more inegalitarian—with a large group of smallholder owner-cultivators, and a small group of landowners with larger landholdings. These were mostly ex-servicemen, with monetary income from pensions as well as farming.

Two incidents in this village have had a bearing on how the DTW was managed after handover. The first was the case of the Chaudhary who filed a complaint against the irrigation committee. The former had filed a complaint to the zonal commissioner because he did not get his share of water. He lost the case against the kulo samiti, which alleged that he had not contributed labour for kulo cleaning. Another incident, in 1966, led to a scuffle between the Chaudhary and some migrants. A number of families had settled there after clearing the forest. An individual from a leftist group among the new settlers was involved in an incident in which the Chaudhary was attacked when holding a SMIS meeting in his yard. The Chaudhary had filed a case against 46 villagers, thus creating a rift between groups in the community. Meanwhile, other irrigators too had been cheating on their share of contribution for kulahi.[9] They wanted the landlord to be brought to justice before they participated in canal cleaning. These actions weakened the kulo management, as it was not able to cope with the fines the village had to pay to the main SMIS water users' committee. Fines are paid when maujas are unable to maintain their labour obligations. When the BLGWIP started building a distribution system for the DTWs, many farmers let the project build permanent canals over existing kulos, and entirely stopped the call for kulahi. The village discontinued its relation with SMIS after the project installed a DTW in Tikuligarh.

The ADBN had started providing subsidies to group and individual STW installations across the Terai. Farmers in this village installed STWs through this programme. They refused to pay the demand charge for the DTW for their entire irrigated area and, instead, claimed that they used STWs to irrigate the larger portion of their plots. In 2004, the WUG collected money for only 35 out of 187 hectares, which was an 81 per cent reduction. This was a loss for the DTW committee, but not entirely for farmers as the large volume of water obtained from the DTW in a short time was still sufficient to fulfil their water needs for both paddy and wheat. By 2004, the DTW committee changed for the fifth time, and none of the committee members were willing to stay beyond their term. The only persons forced to take leadership were the local politicians and village leaders. It was not easy to enforce regulations, as they too were involved in absconding from payments to the DTW fund. It was important to keep the DTW running, as a large section of the population still relied on it.

CONJUNCTIVE USE IN SUPAULI

While farmers in Tikuligarh were facing problems in managing their DTW, Supauli village was successfully managing two DTWs and the kulo together. People in Supauli agreed to contribute land for a second DTW in the VDC, after the Chaudhary had not allowed the project to install one in his village. The Chaudhary did not want the earthen kulos destroyed and land fragmented by the lined DTW canals. The kulos were part of his heritage, and he had a special interest in preserving it. The migrants in Supauli, eager to be rooted in their newly adopted society and to build networks with the political and bureaucratic authorities, came forward to contribute land for a second DTW. This village also maintained its surface water rights with the SMIS throughout the project implementation period. The only difference was that, instead of the usual eight shares of their labour obligations, they had settled for two with SMIS. As in Tikuligarh, farmers here were reluctant to pay the demand charge. In 2004, the WUG was drawing only 50 per cent of the demand charge from two DTWs. In spite of the constraints it faced with respect to DTW management, this is the only village in the study area that was able to make joint rules regarding the use of both sources of water for irrigation.

The DTW canals had overlain large parts of the earthen kulo in Supauli. Farmers, irrigating from lined canals, had stopped contributing to the kulahi for SMIS even before the tube wells had been handed over to WUGs. However, several people figured prominently in leading the village in securing water for agriculture in Supauli. Their personal strategies, political ambitions, and strong kinship relations have led them to continue to manage both groundwater and kulo, and enforce joint rules. The inegalitarian nature of landholdings in the village, together with the advantages that the combined layout of the water distribution of groundwater and kulo collectively presented, enabled the leaders to enforce these regulations on joint use.

The main actors in this process were two migrants from the western mountains who made their entry into local politics. The first, Tek Bahadur,[10] came here as a young army retiree at the age of twenty-nine, but could make his debut in politics only in the 1990s, after the start of multi-party democracy. The second, Hari Bahadur, was elected ward chairman in the first democratic elections in 1991 from the United Marxist–Leninist Party (UML). In 1995, Tek was elected ward member from this party. As elected members they had to oversee all developmental activities in the ward, including irrigation. Hari was in charge of the southern DTW, Tek Bahadur of the kulo and the northern DTW. After management transfer of DTWs, they realized that it was necessary to exert some form of control over both groundwater and surface water. They called a village meeting in 1993, and put forth their conditions for water use, obliging farmers interested in surface water use to pay the demand charge for DTW. This rule was put in effect after a majority had endorsed it. This did not hamper water distribution, as groundwater is used mostly in May and June and sometimes in the following months, when the monsoon is late. Groundwater is used again in the winter months if the village does not get surface water in time or is scarce. The two DTWs in this village, like all other DTWs installed by the project, were designed as separate units. However, both DTWs in Supauli were linked by the SMIS kulo so that, if there is a crisis in the south, the users are confident that they can distribute water from the northern DTW.

The leaders in this village were able to enforce rules for joint use for several reasons. First and foremost, the boundaries of DTWs, kulo, village, and ward coincide here. Rules could not be made through the platform of the DTW committee alone. The DTW boundaries were not synonymous with the village boundaries, but kulo boundaries and village boundaries coincided. Historically the kulo has bound social life in the northern and southern sides of the village together. Besides, for irrigation water delivery, the kulo committee is responsible for organizing religious and social activities, fee and fine collection, and other infrastructural needs like road cleaning operations. It was necessary for the whole village to be together to formalize the rules and regulations for water management.

Enforcing rules for joint use was part of the process of gaining control over both surface and groundwater resources. It was a big challenge for those responsible for this to keep the DTWs in running condition, to devise rules and regulations, and enforce them. The DTW committee is one platform amongst many used by aspiring local political leaders. At the time of the research in 2004, management had changed only once in the southern tube well since it had been handed over, and twice in the northern tube well. Both positions were taken up by new migrants, eager to establish their network in society. One of them had left his job as an agricultural technician in the government and came back to live in the village and build a political base there. The other person had retired from the Indian army and was new to farming as well. Both, like their predecessors, belonged to the same ethnic group and were involved in the formation of a new political party, the Rashtriya Samata Party, in 2001. The former technician had connections with the bureaucracy, which was regarded by the community as a powerful tool. In addition, both were in constant discussions with the older chairmen of the DTW.

DEEP GROUNDWATER AND JHARAN IN DURGANAGAR

Durganagar received a DTW 17 years after Tikuligarh and Supauli. Before 1999, it irrigated from jharan and STWs. The village had always wanted a DTW of its own, as the farmers had observed the convenience it provided their neighbours. However, they were also aware of the plight these DTWs brought upon fellow villagers after handover. Initially, Durganagar used drain water from Tikuligarh. It lost this source of water when Tikuligarh discontinued its surface water rights, so Durganagar farmers then used the only remaining sources: jharan water and STWs. A jharan-in-charge managed the cleaning operations and management of jharan. Attendance at canal cleaning is mandatory for all farmers, and absentees are fined. Labour obligations are proportional to landownership. Maintaining the jharan is crucial for Durganagar, as it is a rich source of water in the paddy transplanting period and helps drain off excess flood water in the monsoon.

This village was very active in developing relationships with project authorities to get a DTW. It put in a request for the project in 1988, but the DTW was installed only in 1999, as political changes from 1989 to 1992 had affected all development works. In accordance with the participatory concept for irrigation development, water users contributed 25 per cent of the cost for layout of the distribution system. It was collected in proportion to the land owned by each farmer. Four people, including the jharan-in-charge and the VDC chairman, were very active in this process. The project could easily convince the farmers of the benefits of the DTW, as these leaders were involved. Changes in design and capacity of the tube wells in the third phase also meant lower demand charges and costs of operation. By 2004, Durganagar had become a role model in groundwater management, surpassing not just the older generation of tube wells, but also its counterpart villages with tube wells of similar design. This has to do with the entrepreneurial qualities of the community and the leadership of the village, where vegetable cultivation has come up as a lucrative cash crop.

Experiences of its neighbours had made the village wise. The farmers took care not to obstruct jharan canal cleaning operations. Pipes were laid parallel to the kulos instead of letting them cross these canals, so that the floating debris from upstream did not get trapped in the canal. They convinced farmers in neighbouring villages to share a loop from the four-loop distribution system of the DTW, as three were enough for the village. Most farmers involved had either sold their pump sets or stopped using the STWs by 2001. The people here paid the lowest demand charge compared to other tube wells because of a lower motor capacity of 37.5 kilowatts. When all the loops were in use for irrigation, the rate per hour further decreased to NR 40 per hour. The ease with which the PVC pipes could be dismantled and assembled again helped the farmers to make adjustments after a year of operation.

The WUG for managing the DTW has evolved in a very positive way in Durganagar. There was increased interest in leadership positions in the WUG. The first executive committee consisted of seven members, including all four people who led the efforts for a DTW, other members of the community, and a pump operator. The village later decided to increase the number of members from seven to eleven because the community was interested to 'get a representation across all castes, ethnic lines, and gender'. The second committee included the loop leaders, people from different castes, ethnic groups, and a woman. Most former members were deeply involved in party politics and needed time to prepare for upcoming elections. The former VDC leader, who had been chairman in the first DTW committee, opted to stay in the capacity of a member only. One reason for the politicians not to leave the DTW 'connection' is the success of the tube well. By 2003, people from various agencies visited Durganagar to observe this first-hand. The WUG executive conducted frequent meetings between themselves and all members in the village, and set up rules to set fees for demand charge, operating charges, and other fines.

The people who are important in management of the DTW and the jharan in this village are also the main actors in the VDC. They have also been making significant changes in TVDC as a whole. This is because they worked both as residents of this village and in a network of people who were deeply involved in decision-making about water resources management in the VDC. The executive committee of the WUG consists of seven members and a pump operator, but not all members are equally involved in the activities. It mostly becomes the responsibility of one or two individuals within the committee (for example, chairperson, secretary, treasurer, or pump operator) or those involved in the day-to-day working of the tube well.

The democratic elections of 1991 put the newly elected VDC chairman under immense pressure to support tube wells in his jurisdiction. Large parts of TVDC needed deep groundwater. Expenditure on DTWs does not formally qualify as a candidate for the 'development budget'. Moreover, the cost of the demand charge per year amounted to around NR 24,000. The annual budget allotted per VDC had to be shared equally by its nine wards. The VDC chairman, who came from Durganagar, appointed the DTW chairman from Supauli to investigate the issues around DTWs. He was supported by a network of people who belonged to the same political party. These included all four leaders from Durganagar, Supauli, and other wards in TVDC. There was a general consensus to set up a trust fund to support the 12 DTWs in this VDC. Money came from the gravel industry that had emerged as a lucrative business on the banks of the Tinau river, and from road tax on the trucks and tractors that came in to fetch the gravel. This increased the budget by over NR 7.000,000 in 2000–1 (1 USD was NR 69 in 2000). Each DTW was allotted NR 25,000 as an initial fund. The WUG kept this as a fixed deposit in the commercial bank to use the interest. After 2000, the VDC gave NR 12,000 to every tube well. Such support is unique for TVDC in the whole BLGWIP area. The TVDC chairman was popular among most people in his jurisdiction, except amongst those involved in 'party politics'.

CONTROL OVER MULTIPLE WATER SOURCES IN BIHULI, MADHAULIA VDC

Bihuli has access to two different sources of surface water, the CMIS and a local stream called Lausi Khola. Along with Bihuli, 10 other maujas upstream had discontinued their surface water rights with CMIS. In Bihuli, the farmers stopped irrigating from both sources of surface water immediately after the DTW had been installed. However, seven years after abandoning their customary rights to surface sources, their interest in them was revived when they had to pay for groundwater. They then started optimizing use of deep groundwater.

The actions irrigators took to secure water rights in Bihuli can be understood against a timeline from 1990 to 2001 (Figure 6.2). They drastically reduced groundwater use following handover of the tube well. In the summer of 1993, the project had locked the pumps in its attempt to put pressure on the farmers to take over the tube wells. That year paddy was rain-fed. The curve denotes the extent

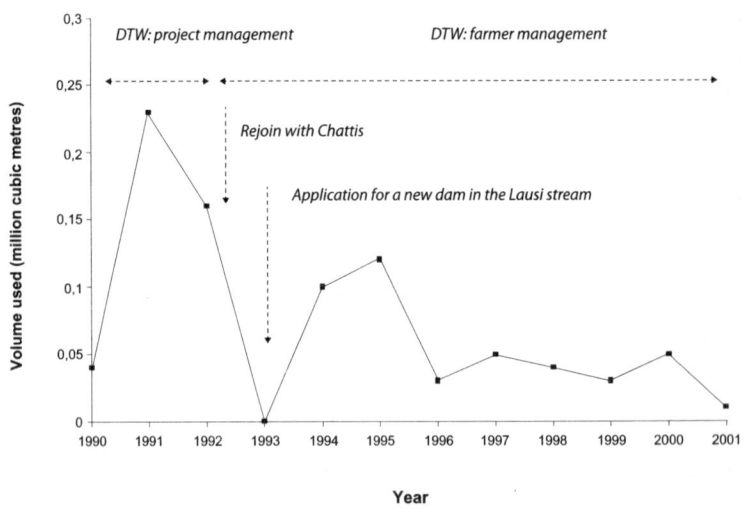

FIGURE 6.2   Actions Taken in Bihuli to Control Different Sources of Water

Source: Fieldwork conducted in 2002–4.
Note: DTW water use records 1990–1 and BLGWIP 2002.

of extraction from the aquifer through the DTW in Bihuli between 1990 and 2001.

The CMIS constitution enforces strict regulations on water rights. Entitlements to water and jharan that come up when upstream farmers irrigate are tied to the ability of a village to maintain labour obligations to the kulo system. The irrigation system, however, has provisions that allow villages to renew water rights, change outlet locations, or even discontinue rights. Any village interested in renewing water rights must make a request to the CMIS main committee and prove its commitment to labour obligations. This has to be endorsed by the majority in the general assembly (GA). Farmers in Bihuli took this rule as a reference to justify claiming back their water rights.

Two men from the village belonging to the UML took up the responsibility for organizing this effort. One was a Tharu, who was also the democratically elected chairman of the VDC and a landowner in this village. The other was a migrant and aspiring politician. A village meeting was set up. After consensus had been reached, the request was put forth to the CMIS GA.[11] In 1992, a mass meeting of Chattis Mauja[12] endorsed this request, allowing Bihuli to be an official member of the irrigation system again. Instead of the two *kulara* shares that they had been taking earlier, Bihulians opted for only one share this second time. With an increasing number of people moving away from agriculture, the village realized that it was not in a situation to harness the labour required for two kularas. Rejoining or leaving CMIS entails payment of certain fees to the kulo system. In 1989, the fee set for rejoining was NR 10,000 per kulara, while the fee for leaving had been set at NR 12,000. Money was collected from all households in proportion to area irrigated. Other villages in the upstream area also renewed their water rights with CMIS.

Several villages in MVDC irrigated from Lausi Khola. This is a perennial stream from a subsurface spring north of Bihuli. It flows right through the VDC on to the urban area of Bhairahawa and then into India. The first intake is that of Bihuli village; the second is a brushwood dam, from where the farmers in ward number two and three had historically been taking water from the Lausi. The third is a permanent dam constructed by the District Irrigation Office (DIO), through its FMIS rehabilitation programme. Another village,

Gangobalia, north of Madhaulia, also irrigated from Lausi. Water in the stream was divided by means of a wooden proportioning device called a *saacho*, and each village was assigned a share of 75 *haath*[13] of water (see Parajuli, this volume). After transfer of DTWs, this source of water, not taken seriously by implementers, became the focus of contestation inside the VDC.

Bihuli filed a formal request for a dam with the DIO in 1993. This request had not been met till 2004. An inter-village conflict had brewed up in the VDC because farmers in other wards had protested against the location of the dam that Bihuli requested. They feared that this would affect the water flow into their kulos. Since then, the people in the village tried different ways to strengthen their claims on the water. First, they approached farmers in neighbouring Babani to make a joint application. They later abandoned this strategy, and joined with another village in Gangobalia and one in adjoining Karaiya. Changing alliances was critical for Bihuli. They planned to change the position of the intake to a location higher upstream. They had Gangobalia's support, as this move would improve water supply to this village.

Changing the intake to a location higher up meant that the Bihulians would have to construct a new canal. Instead, they planned using the Gangobalia kulo on the border of the two villages. They would then transport it through a new short link-canal to the existing kulo into Bihuli. The WUG submitted NR 10,000, 25 per cent of the estimated total cost, as part of the 'farmer participation component' for the new irrigation system. The decision was still pending in 2004. A reason for this delay was the opposition from other villages, from none other than a former chairman of the VDC. As a lawmaker for the entire VDC he felt that it was not fair to take decisions that favoured one group over the other. The main factor behind the dam initiative was a small politician in Bihuli, of the same political party as the chairman. The more powerful leaders in Bihuli were not directly involved in this endeavour. There was much dissatisfaction between this politician and the 'bigger' leaders belonging to his own party. The alliance with Gangobalia made their case stronger, as decision-making now shifted to another VDC. In addition, Gangobalia possessed stronger water rights, being an older user of Lausi. This would make it difficult for opposing wards in Madhaulia to stall the process. The farmers in Bihuli, therefore, used different normative repertoires to legitimize their claims on Lausi. The government intervention programme offered them an opportunity (as it did for some villages downstream) to strengthen their water rights. They shopped around for the best way to strengthen their water claim by entering into or dropping alliances. Choice of Gangobalia, for instance, provided them a basis of stronger water rights but also another forum—the VDC—to strengthen their case.

DTWs presented similar problems here as for WUGs in other villages. The committee was active in working out strategies internally, at the village level, to support groundwater, and had also started networking with WUGs in other VDCs to garner support from outside. The committee made it mandatory for all farmers to pay fees, whether they used groundwater or not. This was especially true of farmers in low-lying areas. This strategy helped set up an initial fund for DTW management. In the subsequent year, the rate was reduced and fees were collected from groundwater users only. In 2004 only 24 per cent of the designed DTW command area was functional. The committee set regulations that allowed it to gather funds for the DTW. DTW canal cleaning was set up as a regular activity before the paddy and wheat season. These regulations also included issues like: selling of wood from trees around the pump house, charges to fill fishponds, determining a minimum time for which a person can request groundwater; and rules for distributing free-flowing artesian water from the DTW. Where a deep aquifer gave free-flowing artesian water, the project had made provisions that allowed this water to flow out into the canals.

In 2002, the government dismantled all local-level government bodies, thus incapacitating local politicians in the VDC. Due to the absence of the ward as a government unit, in Bihuli certain responsibilities had been shifted to the kulo committee. This committee, normally responsible for irrigation and certain pujas[14] performed by Tharus, was now overseeing the development budget of the ward. In addition, it had started adding fines collected in the village to the DTW funds.

Tharus make up the majority of the population in Bihuli, but power relations by the time BLGWIP embarked on handover of DTWs had shifted towards the hill migrants. Shifts from the DTW to conjunctive use were made possible through actors in this village who figured prominently in decision-making on water resources. One of them was the chairman of the DTW, named Sharma. He had held that post since handover of the DTW. He had taken early retirement from his job as government officer and come to live in the village. A large landowner with 14 hectares of land, he also oversaw management of land for an absentee landlord of the village. He was fully involved in village-level decision-making and had risen to be a fully-fledged politician and a prospective member of parliament for the UML party. He was the main person involved in negotiation with the CMIS.

Rules for renewal did exist in the constitution of CMIS. However, these still had to be worked out, interpreted, and negotiated. He used his contacts and strong political networks to negotiate with the executive committee of the CMIS. After his success in this process, he also became the chairman of the revived kulo in Bihuli village. After serving two terms as kulo chairman, he decided to quit the job. He then rose to the ranks of treasurer of the Chattis Mauja main committee. The executive committee of the irrigation system consists of a chairman, a vice-chairman, joint-secretary, and nine regional members. These 12 members nominate three other members from within the command area. One person is chosen as a treasurer. This village, which had stepped out of the CMIS for many years, is now linked up with mainstream irrigation development of the district through the different networks that such actors have carved out for themselves. Besides being the treasurer of the Chattis, Sharma was also treasurer of the joint committee of Sorha and Chattis Mauja, and chairman of the Rupandehi District National Irrigation Committee— the Rupandehi component of the National Federation of Irrigation Water Users of Nepal (NFIWUAN). In this role he was also involved in bringing groundwater issues under the attention of policymakers. He had networked with leaders from other VDCs to build a stronger coalition to advocate for DTWs.

Sharma was successful in the negotiations, and he did get the recognition and support for leading them. However, as the above case study also shows, a crucially important factor that makes it possible for actors to find negotiated solutions to water resources issues is the flexible nature of the irrigation system and its capacity to allow for uncertainties and changes.

PROFITS FROM AGRICULTURE AND
STW IRRIGATION IN MAHUWARI

Possibilities of profitable agriculture have shaped farmer choices for shallow groundwater in Mahuwari. Mahuwari lies in the downstream part of the Tinau river basin before the rivers and streams flow into India. This region, rich in shallow aquifers, had previously been irrigated by surface kulos, each with their own system of management.

Like the previous villages, Mahuwari was inhabited by Tharus. However, its present ethnic composition is made up of Bhojpuri-speaking people of North Indian origin. Only two Tharu households live in present-day Mahuwari. Most present-day farmers in Mahuwari have no knowledge of the kulo that once existed here. The village has completely stopped irrigating from surface sources and converted to shallow groundwater. The factors shaping the choice of farmers for shallow groundwater, how they try to gain control over it, the strategies developed, and the types of relationships that emerge around it are examined further.

Contrary to the abovementioned villages, this village did not have experience in state-sponsored irrigation development up to 2004. The only state initiative in irrigation had been the Participatory District Development Programme (PDDP) under the Local Governance Programme. Through this programme, a group STW had been installed in the village. Farmers shared the pump set within a group that belonged to a larger savings group. Land tenure is highly inegalitarian compared to the other villages. 85 per cent of the total village area is owned by just three persons who live in Kathmandu. A total of 82 households were living here in 2004; out of these, 15 were landless. The average landholding for 67 landowning households is very small: 0.17 hectares.

Mahuwari consists of two settlements: one with the same name and another called Kutti Tola. Collectively they are referred to here as Mahuwari. The village is located just 11 kilometres south of TVDC and Hatti Bangain VDC. It is connected by a gravel road to the Bhairahawa–Lumbini highway. The social and cultural links of the villagers with the villages in the neighbouring Indian state of Uttar Pradesh (UP) are very strong, as the border with India is around five kilometres away. Migration into the village took place in different phases, starting from the 1960s. Only a few households came from other villages in Nepal, while most of the women who married into the community came from India. Child marriage[15] is prevalent amongst the lower castes and illiteracy is high, with only 10 men who have completed high school.

After the Mahendra sugar mill had opened in Rupandehi in 1963 sugarcane was the main cash crop for this village until 1996. The factory had been a source of employment for many people in the area until it closed down in that year. Farmers were then compelled to switch over to a paddy–wheat–mustard cycle as they could no longer market their produce. They also grew vegetables as a cash crop. In 1998, the District Agriculture Office (DAO) introduced banana cultivation. A few enterprising farmers immediately ventured into banana cultivation. Some farmers, who could still muster transportation facilities for sugarcane to a factory in a neighbouring district, continued growing sugarcane. However, the rising insecurity in the country as a result of the insurgency hampered easy movement for the farmers. Hence, from 2003 the last of the villagers stopped sugarcane cultivation and diversified their regular paddy–wheat–mustard crop cycle with vegetables and bananas. Banana cultivation became a lucrative enterprise, as the market for the produce was good. Marketing was done individually by the farmers, and through wholesale dealers who came to buy the produce in the field.

The switch in Mahuwari from sugarcane to vegetables and then to bananas has been possible thanks to the fact that landless farmers and smallholders were provided with access to land and STW irrigation. Absentee landlords hired locals to act as managers (*sirbaar*), who took over the entire responsibility from choice of crops in sharecropping arrangements. For sugarcane they used to hire wage labourers; profit was shared between manager and landowner. Sharecropping (*adhiya*) replaced wage labour after the switch to paddy–wheat. Profit-sharing between the landowner and the farmers was done on a 50–50 basis. The start of banana cultivation led to the emergence of another new arrangement locally called *hunda*. This is a form of contract farming, in which the cultivators pay a fixed sum of money to landowners per year. Banana cultivation provided an opportunity for landless migrants to make a living and for smallholders to increase the size of landholding. Some farmers came in to take a contract from neighbouring villages in India through their kin who had been to the villages earlier.

A total of 30 households had taken up contract farming from absentee landlords through the sirbaar. In 2004 the contract rate was fixed at NR 17,700 per hectare for banana cultivation and NR 15,700 for those farmers who grow paddy, wheat, and a third crop. Farmers tried to combine contract farming with sharecropping because of the risks involved in contract farming. Adverse weather like hailstones sometimes damaged the crop and reduced the profit. In 2004, more than 30 per cent of the village land was under banana cultivation; neighbouring villages had also started banana cultivation.

Farmers tried to maximize profit in banana cultivation by reducing costs of inputs. They purchased pump sets at a lower price either individually or as savings groups. The number of tube wells rose remarkably. STW drilling has become a profitable enterprise for local drillers. Their knowledge and skills on drilling combines traditional methods with know-how gained from the experience of drillers who previously worked for the ADBN. Since drilling a well is much cheaper than buying a pump set, farmers have drilled wells extensively in many plots. The reason for this is that land fragmentation is very high. The traditional system of land inheritance divides land in portions amongst the sons, irrespective of where the plots are located. In 2004, there were 37 pump sets and twice the number of tube wells installed in the village. To cope with land fragmentation, farmers developed various relationships to access groundwater. Common practices were renting pump sets, lending and borrowing tube wells, and buying, selling, or exchanging water. Prior to the boost in tube wells, a charge of NR 10 per hour was set for using a tube well. Later farmers felt it was a *bhyavahaar* (social norm) to let neighbours use it free of charge. Sometimes farmers had the same neighbour (for example, in the case of brothers who had divided inherited land) at more than one plot location.

Profitability of agriculture was a major force driving Mahuwari farmers towards individual control over water through STW irrigation. Besides Mahuwari, five other former maujas have converted to STW irrigation. Floods had destroyed part of the kulo in Mahuwari, but other villages too had left kulahi. Farmers in these villages had been cash-cropping for many years and tried to work out farming strategies to incorporate other cash crops. This made them turn towards groundwater irrigation. Farmers in this village were not interested in reviving the kulo because they had already realized the importance of having total control over water. The case study shows how prospects of profitable farming induced the evolution of different farming strategies. This was also made possible by the availability of STWs, which further induced the movement away from kulo irrigation.

This chapter has also examined the different arrangements that farmers have made for irrigation. The arrangements that emerged in areas irrigated by shallow groundwater are influenced by specific characteristics of land tenure like land size and land fragmentation. The STW technology both facilitates as well as constrains these arrangements because it can be used as two components: a fixed tube well and a movable pump set. The farmers work out different ways within these constraints. A variety of relationships has evolved between farmers even around this very 'individualistic' technology. They try to work out irrigation practices that do not create a hassle and are compatible with local norms and values (bhyavahaar). The aquifers in Mahuwari are recharged by the Tinau river and the Ghagra Nala, and the farmers have not experienced reduction in discharge from the tube wells, so their diverse ways of trading water and sharing tube wells and pumps are unproblematic. The emergence of the local 'drilling industry' has also allowed tube well drilling at reasonable costs.

Most farmers use their own networks for accessing STWs. As a consequence of the individualized nature of the technology, their strategies are also individualized. Farmers try to find ways by which they can have more control over irrigation technology and groundwater. Farmers try to reduce costs of the technology by different means. Only a few farmers who own land in the village made use of a government agency to obtain a tube well.

In contrast with the situation in other VDCs, this transformation in irrigation has proceeded with minimal government involvement. This does not mean that it is the preference of the farmers. There is active involvement if an intervention option caters to their needs, as shown by the case of the emergence of the STW group initiated through external intervention.

* * *

Interventions in irrigation are implemented under certain assumptions about both existing conditions and practices, and the technological, institutional and other changes they will bring about. BLGWIP, implemented in the Terai of southern Nepal, was initially intended to supply year-round groundwater as the main source of irrigation water. The in-depth case studies on developments around various water use complexes presented here show that such assumptions do not hold out against the actual developments around irrigation technology, institutions, and farmer practices observed in the research.

The STW installations, implemented through different community programmes to cater to a specific group of smallholders, eventually became individually owned. This smaller technology, as the cases of Tikuligarh and Mahuwari show, had the power to transform existing relations around larger technologies like the DTW or surface canals. In Tikuligarh, STWs served as a weapon for farmers to bargain against paying fees and the demand charges for DTW operation, while in the southern parts of the river basin, several villages moved away from utilizing their surface water rights.

The handover process for DTWs followed a policy prescription for IMT. This was a crucial factor leading to the emergence of a multitude of institutions around various water sources. Instead of creating sustainable WUGs for DTWs, it led to a situation in which these WUGs started to 'piggy-back' on other institutions for support. Irrigation management re-involved itself with issues of security and insecurity in surface irrigation. The sustenance of DTWs, supposed to have been realized through returns from agriculture, now depended on the skills and capacities of local leaders to find investments for support.

Farmers are constantly involved in creating or maintaining legitimate access to, and organizing different combinations of water sources. They do this by making use of different technical, organizational, and normative–legal options in negotiating their water rights. They choose, accept, adjust, or reject technologies or institutions, while working out arrangements for irrigation that are most effective for local production options. They have recognized the heterogeneous nature of interactions between intervention processes around different water sources. In this context different technological innovations linked to a variety of organizational and institutional arrangements had to be worked out, and different strategies formulated and networks created.

Negotiations and interactions took place within and between a variety of irrigation (socio-technical) domains at various levels: a DTW as a unit; several DTWs together; between DTWs and surface sources; between DTWs and STWs; and between several STWs. In addition, these negotiations took place in villages, wards, VDCs, and surface irrigation networks, and even across the border, as in Mahuwari. Institutional changes take place as farmers strategize to gain control over water. The characteristics of the water use complexes that the farmers ultimately choose are shaped by the performance of technology, social networks, and opportunities provided by different sources of water at different points in time under specific agro-ecological conditions. The strategic behaviour of farmers and the way they define rights and responsibilities pertaining to various water sources are embedded in larger structures and processes. Understanding groundwater use in the study sites was to understand the shaping and transformation of groundwater technology within the water cycle in specific agro-ecological, socio-political, and cultural contexts.

The findings from the case studies provide better insight into strategies worked out by farmers for systematic conjunctive management. The diverse and often interconnected sources were approached by intervention programmes as separate and distinct sources to be developed. The farmers were the ones who first appreciated their potential for conjunctive use, which has made irrigation both more flexible and less costly.

The cases clearly reveal that there is a need to look beyond the stated objectives of water development interventions, to focus on their interactions with local water conditions and social processes. Only once resource use and technological performance are well understood, can future options for their transformation be devised. Such socio-technical analyses, usually not taken into account in policymaking and project design, have the power to modify a group's own water use practices and those of other irrigation systems in the vicinity. Farmers' choices for crops depend on inputs and market conditions, as well as on the availability of labour. Low levels of deep groundwater use are not just related to difficulties to form WUGs but also to costs and choices between alternatives. The study clearly shows the advantage of understanding complexes of water use and the social, political, and institutional spaces that they function in.

Notes

[1] See http://www.doi.gov.np/acts/irrigation_policy.pdf (accessed 5 September 2005).

[2] In line with neo-liberal policies initiated after the establishment of democracy in 1990.

[3] Gyawali and Dixit (1999) give an overview of water use institutions in the Tinau river basin.

[4] Nepalese who migrated to Burma (Myanmar) and returned to Nepal later. They are either family members of the British Gurkha troops taken there in World War II or of those who migrated independently in search of a better life.

[5] Ethnic group of north Indian origin.

[6] Like all VDCs, Tikuligarh is divided into nine administrative wards.

[7] A mauja here denotes a village irrigated by one common village canal of a traditional surface irrigation system.

[8] A flat rate of electricity, to be paid irrespective of whether the DTWs are used or not.

[9] Call for labour contribution by kulos, usually in proportion to land owned.

[10] Names have been changed to ensure privacy.

[11] Made up of the main committee as well as the kulo chairman of each mauja or their representatives.

[12] This refers to the meeting of the members of the irrigation system. Each kulara is entitled to four representatives.

[13] Haath literally means 'one hand'. This was the local measurement of water indicating the length from the elbows to the tip of the second finger.

[14] Pujas refer to prayers and offerings.

[15] Children marry at the age of 8 or 9. Girls stay with their parents until they reach puberty.

7 SOCIAL DIFFERENTIATION AND THE POLITICS OF ACCESS TO GROUNDWATER IN NORTH GUJARAT

Anjal Prakash

From the second half of the twentieth century, agriculture in India has gone through enormous changes with the introduction of Green Revolution technology.[1] This demanded more control over irrigation, which the surface water deliveries of large canal systems were unable to provide. Groundwater irrigation was seen as a feasible alternative to bureaucracy-controlled canal systems. In addition, groundwater irrigation could cover areas that were historically not within reach of surface irrigation schemes. These advantages of groundwater over surface irrigation led to a sharp increase in its use. However, as increasingly large areas are irrigated by groundwater, over-development and depletion of aquifer systems has become common.

The state of Gujarat in western India is no exception. Groundwater supported more than 77 per cent of its irrigation water requirements in the year 2003–4. Increased groundwater use and a rising pollution level in surface water bodies have resulted in water scarcity, leading to more groundwater exploitation. This turned many regions of Gujarat from water abundant into water scarce areas in just four decades. With this increased groundwater use, a spurt in water markets was reported from the early 1980s.[2] Dense markets developed in the alluvial central and northern regions of Gujarat, which were suitable for sinking deep tube wells (DTWs).[3]

This growth in water markets led to debates over their nature and functioning, often staying at the level of political rhetoric. Some academics have advocated dense and competitive groundwater markets on the ground of their assumed efficiency and accessibility to the resource, but without taking into account the nuances of unequal social relationships and natural and historical conditions that shape and determine groundwater access and use (discussed ahead in this chapter). Apart from a few recent studies, the water market debate lacked detailed methodological and empirical inputs for understanding how groundwater irrigation and markets function locally.

This is even more true for Gujarat: its model of groundwater markets has been widely discussed, but with little sociological enquiry on the process of groundwater development, paying little attention to the question of who gains and who loses. Most detailed accounts of social processes in tube well irrigation come from Tamil Nadu (Janakarajan 1994, 1997, 1999; Meinzen-Dick 1989). Apart from Bhatia's (1992) monograph on groundwater irrigation in north Gujarat, there was no detailed account of groundwater-led agrarian change in Gujarat when this study was conceptualized. Dubash's (2002) case study of two groundwater-dependent villages was published soon after this study was initiated.

Allocation of a scarce resource like groundwater has been looked at mainly from a narrow productivity and efficiency perspective, while the questions of individualized appropriation of a common resource and its impact on an unequal society have hardly been studied. Further, the nature of relationships between agrarian structure and technological change in the context of agrarian transformation has hardly been dealt with. There is an oversimplification of the problem, connecting groundwater irrigation with poverty alleviation without critically exploring other possible linkages.[4]

The intricate relationship between groundwater irrigation and development has been widely debated. Many believe that groundwater irrigation has a positive impact on productivity and generates year-round employment thanks to intensification of agriculture, and that therefore it is a weapon against rural poverty (Chambers 1986, 1988; Narayanamoorthy 2001; Roy and Shah 2003; Shah 1993; Vaidyanathan 1996, 1999). Irrigation is considered a powerful factor for food security, protection against drought, increased employment opportunities, and stable income, offering opportunities for multiple cropping and crop diversification. It is argued that access to irrigation leads to the adoption of new technology and higher cropping intensity, generating higher productivity and greater returns from farming. Besides this, it opens up new on- and off-farm employment opportunities, improving the income level of farming households (Hussain and Biltonen 2001).

However, unlike canal irrigation, groundwater development mostly took place in the private arena. In many locations, access to groundwater is chiefly determined by local water markets. Widespread tube well irrigation and groundwater markets have been reported from various parts of India, mainly from Gujarat (Bhatia 1992; Dubash 2002; Kolavalli and Chicoine 1987; Shah and Raju 1987; Shah, M. 1985; Shah, T. 1993), Tamil Nadu (Janakarajan 1994, 1997), Uttar Pradesh (UP) and Bihar (Kishore 2004; Pant 2003, 2004; Pant and Rai 1985; Shah and Ballabh 1997; Wood 1999), Karnataka (Chandrakanth *et al.* 1998), Andhra Pradesh (AP) (Satyasai and Vishwanathan 1997; Shah 1986), and Punjab (Jairath 1985). Most academic work revolves around groundwater development (Dhawan 1982, 1988, 1993), groundwater markets (Saleth 1998; Shah 1993) and management of groundwater resources in reaction to large-scale depletion (Moench 1994, 1999). The contribution and use of groundwater in irrigation development has been widely accepted.

However, questions have been raised on the nature and functioning of groundwater markets and market-based management prescriptions based on this analysis. The most prominent among these is Tushaar Shah's influential work on groundwater markets. Based on neo-classical economic analysis, Shah argues for dense and competitive markets on the ground of efficiency and equity of water use. Shah claims that groundwater markets are natural oligopolies and advocates policies that create a situation where oligopolists behave as if they operate under competitive conditions, making the resource available to the larger population through market mechanisms (Shah 1993). His policy prescription is to have flat-rate electricity pricing based on contracted load of the tube well motors, which will create incentives for farmers to access and sell surplus water.

Shah's analysis has been criticized by scholars for his policy recommendation of charging flat-rate electricity. It is argued that flat-rate electricity charges have led to widespread overexploitation of groundwater by resourceful farmers, with deleterious implications for small and marginal farmers, who are excluded due to their inability to chase the water table. Further, the issue of dominant caste or class appropriation of an economically and environmentally precious resource was not part of Shah's analysis (Bhatia 1992; Moench 1994;

Palmer-Jones 1994).[5] Much of the critique of Shah's work is available elsewhere; hence, I will not go into it here.[6]

This chapter presents an irrigation ethnography focusing on groundwater development from the perspective of agrarian change in groundwater-dependent societies. The reason to take this approach is the lack of detailed description of the internal characteristics of groundwater irrigation institutions in the literature. Irrigation ethnography has been a long tradition in canal irrigation systems (Attwood 1985, 1987; Chakravarti 2001; Gorter 1989; Mollinga 2003; Omvedt 1993; Ramamurthy 1995; Sengupta 1980; Wade 1988). It was influential in showing irrigation as a politically critical resource in societies characterized by inequitable distribution of power and resources. These studies showed how irrigation played a role in transforming social and production relations, and have had different implications for different classes and actors. Thus, it contributes to policy formulation for irrigation reform. However, apart from a few recent studies (as discussed above), this has been largely missing in most analyses of groundwater irrigation. Especially Gujarat's model of groundwater markets has been widely discussed without much sociological enquiry focusing on who gains and who loses in these developments.

This chapter fills this gap by focusing on the politics of ground-water markets and their interrelation with social differentiation and class–caste relations. It is based on anthropological fieldwork in 2002 and 2003 through an intensive village-based case study situated in the Mehsana district of north Gujarat. This district is famous for its widely developed groundwater markets and depletion of aquifers due to excessive pumping. The chapter shows how social relationships shape and determine access to and use of groundwater in the context of specific agro-ecological conditions, prevailing social relations of production, ineffective regulation to check groundwater exploitation, and inequality in access to and control over resources. In response to groundwater depletion and increased irrigation costs, new forms of sharecropping with interlinked land, water, and labour market transactions have evolved in the area. Farmers with large landhold-ings adjust to the declining profit from agriculture by appropriating surplus through elaborate local share contracts, on the one hand, and long-distance international migration, on the other. As the profit margin gets tighter, the exploitative class structure reasserts itself and the burden of resource depletion is transferred to those further down in the socio-economic hierarchy.

To gain insight into these concerns, the study follows a primarily qualitative research approach. It actually emanates from the criticism raised against the numerous quantitative studies done in north Gujarat, which fail to capture the social nuances and exploitative relationships in groundwater use. Thus, my choice for a primarily qualitative approach is based on a well-established methodological tradition in the social sciences. Scientifically speaking, this study also critically reacts to the choice of methodologies that create generalizations, models, and blueprint images. It stands in the critical interpretive tradition in the social sciences.

Further, the chapter continues with a description of the case study village, its agrarian relations, access to water, and the emergence of water markets. Next, the relationship between groundwater irrigation and social differentiation is explored in detail. In the subsequent section, the focus will shift to the political economy of groundwater irrigation in the study village. The chapter ends with a short conclusion.

THE STUDY VILLAGE, AGRARIAN RELATIONS, AND ACCESS TO WATER

Sangpura[7] is located in one of the talukas of Mehsana district in the northern part of Gujarat (Figure 7.1), registered as 'overexploited' and as a 'dark zone' in its use of the groundwater resources of the deep alluvial aquifer it overlies.[8] In 2001, Sangpura was inhabited by 628 households. Two major caste groups, the agriculturalist Patel and Thakore, made up two-thirds of the village population. The other major caste groups were Prajapati (potter caste, 70 households), Parmar (cobbler caste, 60 households), Vaghari (vegetable vendor caste, 25 households), and Darbar (former feudal lords, 12 households). The village economy was dependent on agriculture and animal husbandry. All water needs of Sangpura were met through groundwater, a resource that has shown increasing signs of over-development in recent years.

LANDOWNERSHIP

In an agrarian economy, land is the most precious resource in agricultural production. It is considered a basic resource around which

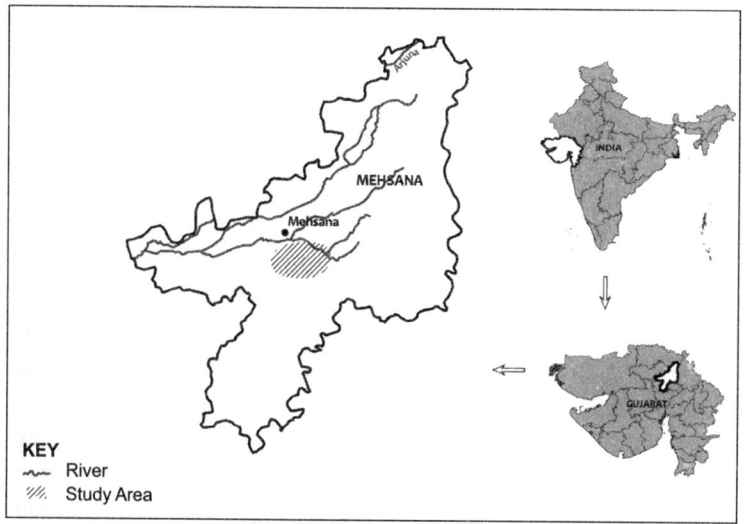

FIGURE 7.1   Gujarat and the Research Location

Source: Author.

other resources, such as water and other agricultural inputs, can be mobilized. In 2002, the Patels of Sangpura constituted around 32 per cent of the total households, but possessed 54 per cent of the village's agricultural land. The Thakores, who constituted approximately 36 per cent of the total households, had only 21 per cent of the land. The Prajapatis, who comprised 12 per cent of the village households, controlled 14 per cent of the village land. Other castes, who constitute around 20 per cent of the households, own 20 per cent of the land. Almost 50 per cent of the households were landless; 11 per cent of these were below the poverty line.[9] Thus, land distribution was inequitable. It closely followed the caste lines, with the Patels being the dominant landholding caste in the village.

ACCESS TO IRRIGATION

Access to groundwater partially determines the productive capacity of land and the ability to generate agricultural surplus. This is a crucial factor in the groundwater-dependent economy of Sangpura. The village has a history of well irrigation that dates back to before

1960. When dug wells started drying up, tube wells were installed for accessing water. However, tube well technology was expensive; this made it difficult for people to own tube wells individually. In response to this an institutional innovation in the form of shared tube wells emerged. In Sangpura, there were 34 functioning private tube wells in 2002; only 4 of these were individually owned.

An examination of the percentage of tube well shares owned by families showed that Patels, 32 per cent of the village households, had 65 per cent of the total number of tube well shares. Thakores, around 36 per cent of the village households, owned only 15 per cent of the tube well shares. The majority of tube wells were in the 45–60 horsepower class (41 per cent), while the others were almost equally divided between the 30–45 horsepower and 60–75 horsepower classes. Ownership of tube well shares closely followed the landholding and irrigation patterns, where 66 per cent of the total land under irrigation (through tube well shareholders) belonged to Patels, followed by 15 per cent belonging to Prajapatis and only 10 per cent to Thakores.

WATER BUYERS

Until the late 1990s, water buyers constituted a large group in the village. Since 2001, there has been a drastic reduction in the supply of electricity, and therefore the number of water buyers has also decreased. This is mainly because the total volume of water pumped from a particular tube well has, first of all, to fulfil the need of the shareholders. Only then can the surplus be sold. Though selling water is a profitable business, it is not done at the expense of the water requirements of the owners.

An analysis of water buyers in the village reveals that the total number of buyers decreased from 173 in 1999–2000 to 104 in 2001–2. The total area of land irrigated by 'bought water' also decreased from 124.5 hectares to a mere 48.76 hectares in 2001–2. Thakores, the largest water buyer group in the village, were the hardest hit as the total area of land irrigated through water bought from tube well cooperatives decreased from 45.41 hectares to 20.13 hectares in 2001–2. The process was detrimental to the buyers, as they were mainly from the lower strata of society, and could not afford to invest in tube wells. These marginal farmers, who formed the water buyers' group, later converted into sharecroppers.

Sharecropping started to pick up in the late 1990s. When the water markets began to shrink, the buyers were slowly pushed out of irrigated agriculture as they could only cultivate a monsoon crop in their field. As most buyers did not have skills other than agriculture, they started taking up sharecropping arrangements with medium and large farmers, who owned land and water rights in the area. The sharecropping arrangements influenced the crop share not only for the component of land, but also for labour and water. In this type of arrangement two-thirds of the crop production went to the owners, while sharecroppers got only one-third of the produce.

This situation favoured the land and water rights-holding upper class, which could appropriate a surplus. The large and medium farmers also received remittance money from relatives in the United States of America (USA). Hence they were economically relatively independent of agriculture. Thanks to this, their bargaining power in the sharecropping arrangements increased, while the erstwhile 'water buyers' had no alternative but to fall back upon such arrangements.

GROUNDWATER IRRIGATION AND SOCIAL DIFFERENTIATION

In Sangpura, the relationship between groundwater irrigation and social differentiation starts with the basic inequality in the ownership of land. The inequality in the distribution of groundwater resources builds on this, and therefore early advantages lie with landed farmers. The analysis shows that the distribution of the two most productive resources—land and groundwater—is skewed towards the higher classes. Further, the class structure closely parallels the caste structure in the village. The dominant Patels are the principal land and groundwater share-owning group in the village.

The organization of groundwater irrigation is based on access through shareholding in the tube wells and selling water to non-shareholders. In Sangpura, around one-third of the village households do not have any access to irrigation. These people belong to classes and castes lower down the social hierarchy. Access to irrigation water is dominated by the Patels, who belong to the higher caste as well as class. They have shares in 29 out of 34 tube wells in the village, amounting to 65 per cent of the total shareholdings. Thakores, the largest water buyer group, have only 15 per cent of the total number of tube well shares.

However, the situation for the Thakores was different in the 1960s, when irrigation used to be carried out through dug wells. In those days Thakores owned 19 per cent of the total number of village water shares around these dug wells, which were mostly owned individually. This shows that, while the total area under irrigation has increased since the introduction of tube wells, the accessibility for some groups of people has decreased. The lower caste groups neither own shares in the tube wells nor buy water. Many of them are near-landless cultivators who only cultivate rain-fed crops in the monsoon season. These groups, along with the landless families from the Thakore caste, currently form the wage labour force in the village.

HISTORICAL EVOLUTION OF SHARECROPPING

The system of sharecropping[10] in Sangpura dates back more than a century. However, it was not a very significant production relation until the rise of irrigated agriculture in the early 1960s. During that time irrigation was carried out in the winter season (October to March), and only if there was inadequate rainfall. Water was lifted through *kos*,[11] and the production under sharecropping used to be divided in three parts: the *bhagiyo*'s (tenant)[12] share for labour, the *dhani*'s (landlord)[13] share for the land, and the owner of the draught animal's share for lifting water.

However, this system was not very popular initially and was enforced slowly in the late 1960s and early 1970s, when the cropping system included irrigated winter crops. Earlier, the winter crops had been grown with water available from residual soil moisture; irrigation was not necessary. In this system, the shares were divided equally between the tenant and the landlord, as there was no irrigation cost component. With the changes in water table, the winter crop started to be irrigated using the open wells. From this time onwards, the one-third system of sharecropping was introduced.

The system of lifting water using kos was followed by well mechanization. As the water level went down, it fell beyond the capacity of draught animals to lift water. With well mechanization the area under crops increased significantly. The well yield almost doubled, making water available in surplus. This water was sold to people who did not have a well, and a water market started to develop. The water charge was based on crop sharing. One-third of the production went for the share of water. With the availability of water, the gross area under irrigation increased significantly, as summer crops were introduced together with the existing monsoon and winter crops. The changes in agricultural technology under the Green Revolution required more control over external inputs. This was different from rain-fed agriculture, where the external inputs used were minimal and required minimal care from the farmer. Slowly a large part of the land came under irrigation. The sharecropping system evolved as a response because the owners could not cultivate all their land. This situation existed until the full-fledged introduction of tube wells in the early 1980s.

During the pre-tube-well phase, the sharecropping system divided production again into three components for land, labour, and water. However, the expenses needed to access water increased under the mechanized well system, which used electricity or a diesel engine to lift water. Costs were higher compared to the use of draught animals. In mechanized wells, the machines for lifting water always belonged to the well owners, who were also landowners. Unlike draught animals, the machines for lifting water are immovable. Draught animals were used for carrying out multiple tasks and could be hired if they were not owned. In contrast, the electric or diesel motors had to be fixed to the well. Therefore, it was the well owner who invested in these machines. In this context, the landlord could claim two-thirds of the produce while the tenant received one-third. The well and the motor to access water always belonged to the landlord. This was the case in the early stage of the Green Revolution. The use of external inputs such as fertilizers, pesticides and seeds was still minimal, as the land was fertile enough to provide good harvests. Therefore, the net benefit was primarily thanks to the low expenses incurred for external inputs.

During the 1980–90 period, there was a slump in the incidence of sharecropping contracts. This period was marked by innovations in tube well irrigation. Because of round-the-clock electricity supply much more water was available than a decade before. The water markets became more extensive, covering a much larger area than before. It was also an opportunity for small and marginal farmers to access water. Hence being an owner-cultivator became a much more lucrative option than sharecropping.

Apart from a few drought years, the period is referred to as the 'golden years of agriculture' in which land productivity was at its maximum. As compared to the present day, the farmers recollect that the land was very fertile and had the capacity to retain soil and moisture levels that resulted in good production with less irrigation. For the tube-well-owning farmers it was a vending boom, and water markets expanded. The returns from the sale of water were higher than those from sharecropping. The good harvests created less willingness than before to share the produce with sharecroppers. Agricultural operations were carried out with household labour and hired labourers. For marginal farmers without a tube well, water was made available. Together with fertile land conditions and family labour, the returns were higher than those received through sharecropping arrangements.

However, this situation lasted only for a decade; in the late 1990s things started to change again. By this time the tube-well-owning farmers had generated enough surplus from the booming agriculture and invested in other sectors. The water level started to decline, as did the availability of electricity, making groundwater much more expensive. By this time, some of the tube wells had become old and their maintenance costs increased.

All this reduced the well yield, which resulted in a shrinking of the command area. Once again sharecropping contracts became preferable to both tenants and landlords for two reasons. As already explained, the shrinking of water markets pushed out the dependent water buyers and water was made available first to the shareholders. However, the number of irrigations also increased, together with input costs for fertilizers and pesticides needed to keep up the level of productivity as soil fertility and water-holding capacity declined. This made agriculture more expensive than in the past. The increased costs of agricultural production reduced the margin between investment and profit. The landlords wanted to share this burden. Alongside, many people who had been excluded from the water markets were available now, ready to take up sharecropping for survival, regardless of the terms of contract. This led to an elaborate and more institutionalized form of sharecropping, in which the contractual arrangements carefully followed the input share in agriculture to allocate the increased costs. Another cause of the rise of sharecropping was the increasing land division and fragmentation in the village, which reduced per capita land available for farmers. This was especially the case for marginal farmers. Their descendants have doubled in number but not their size of landholdings. More people have to survive on the same piece of land. Dairy production and employment opportunities outside agriculture supplemented the income of large and medium farmers. The marginal farmers, who had partially moved out of agriculture, were hit the hardest. Some families with small tracts of land on two different sides of the village preferred to have one sharecropped, as they could not manage both.

Land, Labour, and Water Interlinkages in Sharecropping

As discussed earlier, the incidence of sharecropping was closely related with the rise and fall of water markets. In periods with a sharp decline in water markets, the incidence of sharecropping increased strongly. The most popular form of sharecropping was the one-third share of gross production for each of the components of land, labour, and water. In most cases, water and land belonged to the landlord, while the tenant provided labour. In 2001–3, around 80 per cent of the irrigated area in the village was sharecropped. Actually, under the sharecropping system water is sold to tenants in return for a one-third crop share. This brings in a new linkage between land, labour, and water markets. In order to understand this, the economics and the rationale of the tenants and the landlords behind this institutional arrangement are illustrated.

To understand the economics of sharecropping,[14] information was collected from farmers for five major crops cultivated under the one-third share system. The information showed that the net gain for the labour share is highest in castor, a cash crop, while it is lowest in pearl millet, a food crop mostly used for household consumption. For a large part of the village population, especially the poor, pearl millet is the staple diet. Wheat supplements pearl millet for the middle and upper strata of the population. From the landlord's perspective, the net gain for the land component is highest in castor followed by cotton and pearl millet. For the water component, the gains are highest in castor and are lowest in mustard. This is mainly because castor is an eight-month crop spread between the monsoon and winter seasons, and needs around eight irrigations. The winter crops of cotton and wheat need a similar number of irrigations. Hence, the water charges are almost the same for castor and other crops, while the cash value of castor is much higher.

How can these figures be used to understand the rationale behind sharecropping? The contract includes a two-third share for the landlord, as the land and water rights rest with him. Therefore, the profit for the landlord includes the net gain from the land and water components put together (see Table 7.1). The tenant gets the return from the labour days invested into agriculture through sharecropping. The market rate of the irrigation price includes the cost of electricity, maintenance of the tube well, and a profit margin for the shareholders. In the present analysis, the irrigation cost is calculated taking the prevailing market rate including the profit of the shareholders, which is calculated separately. If that cost were to be calculated, the net profit would further move towards the landlord.

The tenant's rationale

There were two motives for tenants to take up sharecropping despite being aware of the terms, conditions and the situation that favours the landlord. First, the present tenants are smallholders (and former water buyers) who have been pushed out of water markets due to the significant reduction in hours of electricity supply and decline in tube well yields. With a large-scale reduction in the functioning of water markets, the total irrigated area in the village decreased and the people who depended on it needed alternative work to survive. These former water buyers were a ready work force for sharecropping.

Second, taking up sharecropping also assures employment over the contract period, a situation that is in favour of tenants. Within the prevalent situation, it was the most convenient arrangement for the tenants to tie-up with the landlord. Hence, sharecropping was preferred.

TABLE 7.1 Per Hectare Net Gain for the Components of Labour, Land, and Water in Sharecropping, 2002–3

Crops	Gross Production per ha Equivalent (in INR)	1/3rd Share of Gross Production in INR	Labour Component Investment		Labour Component Net Gain (in INR)	Landlord's Investment (in INR)	Land Component Net Gain (in INR)	Irrigation Costs (in INR)	Water Component Net Gain (in INR)
			Cash (in INR)	Labour Days					
	a	b = a*0.33	c	d	e = b – c	F	g = b – f	H	i = b – h
Castor	50,666	16,720	3,026	82	13,694	1,926	14,794	4,709	12,011
Cotton	29,989	9,896	3,081	98	6,815	3,494	6,402	4,514	5,382
Mustard	17,956	5,925	1,850	38	4,075	1,255	4,670	4,496	1,429
Wheat	25,272	8,340	2,311	63	6,029	3,547	4,793	4,613	3,727
Pearl millet	23,590	7,785	2,345	72	5,440	1,529	5,820	5,880	1,905

Source: Field data collected in 2001–2.[15]

TABLE 7.2   Wages for Tenants under Sharecropping, Per Hectare

Crop	Average Labour Days Incurred by Tenant	Average Financial Benefits of Tenant (in INR)	Wages for Tenant (in INR)
	a	b	b/a
Castor	82	13,694	167
Cotton	98	6,815	70
Mustard	38	4,075	107
Wheat	63	6,029	96
Pearl millet	72	5,440	76

Source: Field data agricultural seasons, 2001–2.

As Table 7.2 shows, the wages equivalent of income gained by tenants under sharecropping range from INR 70 to INR 167 per day, averaging approximately INR 100 per day. This is double the wage for a casual agricultural worker in Sangpura in 2002, which provides a rationale for engaging in sharecropping.

The landlord's rationale

The rationale for the landlord to offer land for sharecropping is twofold. First, one-third sharecropping is the best option for the landlord, as it combines the land and water share components. Hence, he appropriates two-thirds of production. Further, the average cost share for labour, land, and irrigation comes to 45, 17, and 38 per cent respectively. Among all costs, the percentage of labour costs is the highest, followed by the cost of water. The tenant, who shares 45 per cent of total costs, gets around 22 per cent of the profit. The landlord, who shares 55 per cent of the total costs, receives 78 per cent of the net profit. This profit sharing makes the one-third share-cropping system the preferred mode of renting out land. Further, the landlord also saves the costs otherwise incurred in supervision and management. The crops are largely looked after by the tenants, including the day-to-day management.[16]

Second, from 1990 onwards, Sangpura and surrounding areas in Mehsana District experienced large-scale long-distance migration of the upper class farmers, mostly Patels. In Sangpura also, out of around 200 families of Patels, 100 families have close relatives like brothers, fathers, or first cousins in the United States (US). Most of them have migrated illegally, paying a huge amount of money to agents. The process started in the early 1990s and is still operational despite the stringent policies of the American government, especially after the 9/11 attack on the twin towers in 2001.[17] The elaborate social network of Patels in America easily introduces new persons into their community and thus the chain goes on. Of late, people from the Prajapati caste have joined their Patel counterparts. In 2003, two out of the 70 Prajapati families had migrated to the US. This process made many families independent of agricultural operations in the village, as migration secures a considerable income from their kin in the US through remittances.

Feminization of the Labour Force

Another issue is the feminization of the agricultural work force in the village. Sangpura is located in the central part of the industrialized area along the Mumbai–Mehsana highway that is also called the 'golden corridor' of Gujarat. It has a number of small and medium industries spread out along a 500-kilometre stretch. The industries in the vicinity create a demand for unskilled labour on a daily wage basis. The average wage outside the village was around INR 100 per day in the 2001–3 period, while the agricultural wage labourer in Sangpura got only INR 50 per day. Thanks to lucrative options for wage workers available outside agriculture, it was convenient for male labourers of Sangpura to search for non-agricultural employment. Women do not work outside the village due to their day-to-day household responsibilities such as childcare, cooking, and looking after the needs of other family members. This situation has led to increased participation of women in agricultural operations inside the village. In the past, for example, transporting the grain to the thresher was the responsibility of men. This is now undertaken mostly by women. Other male jobs include field preparation and making borders. This is now largely done by means of tractors by subcontracting the task to the tractor owners. Of all manual work in agriculture, only irrigation remains clearly a male responsibility; other tasks are slowly shifting towards women. The irrigation component includes the per day wage (done by men), which is largely based on the number of irrigations rather than on daily wages. For convenience, it is estimated here on

TABLE 7.3   Gender Division of Labour, Per Season

Crop	% of Man Days Involved	% of Woman Days Involved
Castor	42	58
Cotton	31	69
Mustard	35	65
Wheat	55	45
Pearl millet	42	58

Source: Field data agricultural seasons, 2001–2.

an hourly basis.[18] Hence it may overestimate the number of actual working days for men. In Sangpura, agricultural wage rates are the same for men and women.

Table 7.3 provides the gender-segregated labour requirement for five major crops. Cotton is a highly female labour-intensive crop related to the need for labour to pluck cotton buds. It is followed by mustard, castor, pearl millet, and wheat. Wheat and castor require threshing, which is largely the responsibility of men. However, threshing is fully mechanized for these two crops and the only task for men now is to collect the grains and put them into the bag. Harvesting of pearl millet and mustard is done manually and is largely done by women. Therefore, even though the labour days for cultivating wheat are more for men, in physical terms the work is heavier for women. An overall segregation of labour days shows that the responsibility of men and women are in the proportion of 40 to 60, respectively.

UNEQUAL CROP SHARING AND THE NEW FACE
OF WATER MARKETS

The analysis above showed that there is a transfer of surplus through sharecropping contracts from the tenants to the landlords. The issue becomes important given the prevalence of these contracts and their contribution to overall village economics. In Sangpura, the incidence of sharecropping closely followed the changes in the mode of agricultural production. With the introduction of irrigated agriculture and well mechanization, large areas came under agriculture. The technological innovation for water extraction in the form of tube well irrigation resulted in an expansion of the area under irrigation from a single extraction point, and gave rise to water markets. During the 'golden years of agriculture' in Sangpura, sharecropping touched an all-time low and water sales were reported to be the highest. The holders of water rights found it easier and more lucrative to sell their water than to put their land under sharecropping and share the booming produce with tenants.

However, this situation did not last more than a decade because of the deepening of the water table, over-cultivation of the land, and a sharp reduction in the supply of electricity. Agriculture was becoming a costly occupation while profits were decreasing. Water costs comprised a large share of the input costs. In the end, even the upper classes no longer saw agriculture as a lucrative business. The reduction in electricity supply also curtailed the tube well yield, as a consequence of which water markets started to shrink. Small and marginal farmers who had earlier been dependent on other people's water sales for their irrigation, had to abandon agriculture or shift to rain-fed monsoon crops. Once again, Sangpura experienced a rise in sharecropping contracts that shared the produce for the land, labour, and water components under the most popular one-third sharing system. In the 2001–3 period, around 80 per cent of the irrigated area was under sharecropping.

Thus, sharecropping had become the new face of the water market. Water was sold to tenants in return for a one-third crop share. Technically speaking, in the one-third system, combining the share for land and water should leave at least 33 per cent of the benefits for the tenant. The analysis showed that the average net profit share between the landlord and the tenant is in the proportion of 78 : 22. This is against the share of average investment, which is in the proportion of 55 : 45 for the landlord and the tenant, respectively. This ratio does not include the cost of supervision and monitoring on the part of the landlord. This is reduced to a minimal level when land is given in sharecropping. On the contrary, the supervision and monitoring cost lies with the tenant; it is not calculated here as an investment cost. Thus the labour days invested by the tenant are much higher than those of the landlord. Under sharecropping, the risk of agriculture is shared with the tenant in an unequal way. The profit share is unequal against an almost equal input sharing. This is an economic relation of inequality where surplus is generated by the landlord through controlling the labour process under sharecropping by virtue of the property rights in land and water. This surplus flows towards the landlord through the unequal distribution of profits in a situation of extreme scarcity of water, in a period when agriculture in the village is characterized by lower productivity and negative growth.

THE POLITICAL ECONOMY OF GROUNDWATER IRRIGATION IN NORTH GUJARAT

As described, Sangpura's case clearly demonstrates the changing relations between production technology and socio-political institutions. The rise in groundwater irrigation was an immediate response to technological changes in agriculture, which created a demand for irrigation. Numerous wells and tube wells were installed to meet this demand. Water markets further mediated this process of providing initial access to the resource.

However, as discussed, groundwater markets not only built on existing social relations of production but also reproduced inequality in the access to and distribution of resources. The control of groundwater by the privileged few was not limited to the agricultural production system but spilled over into other arenas of social life, thus defining the course of agrarian change. Feeding back into political linkages and working against regulatory mechanisms for checking groundwater overexploitation, it helped in consolidating the power position of the dominating classes. These factors define the political economy of groundwater governance of north Gujarat, to the detriment of its limited groundwater resources.

In the following section, I summarize these trends while looking at the interlocking triadic control of social power (which defines access to resources for different social groups), socio-ecological variables in crop production systems (which determine the level of surplus generation for households), and domains of state functioning (which reproduce the unequal and unsustainable pattern of groundwater utilization). These interlocking control systems are detrimental to the livelihoods and survival strategies of the poor and marginalized sections of the rural population of north Gujarat.

SOCIAL POWER AND THE DISTRIBUTION OF GROUNDWATER

Politics is the process through which the social relations of power are constituted, negotiated, reproduced, transformed, or otherwise shaped. In the context of resource use, four levels of power can be identified. The first level consists of the everyday struggle over access and use of resources. The second level includes the political nature of policy formulations contested by different interest groups. The third level deals with the state and party politics related to water resources, or hydropolitics. The fourth level refers to institutions, agreements, and conventions at global level that shape and influence water use (Mollinga and Bolding 2004). I found a great deal of interlinkage between the first, second, and third levels of groundwater politics, where contestations over groundwater resources are linked to everyday village politics and political power, which extend beyond the structures of domination in the village into the larger state politics. The democratic process operates at different territorial levels and works through the social structure linking various structures of power occupied by different groups in society. It also leads to organized articulations of social, economic, and political power in favour of the dominant classes. The politics of domination enhance the control over resources by powerful social groups. The distribution of groundwater in the village is intertwined with the distribution of power. This leads to the mining of groundwater, caused by the ability of dominant classes to generate a surplus from agriculture and other activities and reinvest in drilling deeper down the aquifer. The fact that Sangpura is situated in an alluvial area with a large deep aquifer enables the process of exploitation to continue unhindered, but with disastrous long-term consequences.

One of the important questions leading this analysis concerns the mechanisms that reproduce class and caste domination, and provide power and legitimacy for dominant groups to define the level of surplus extraction. Groundwater irrigation in Sangpura built on the existing inequality in land distribution, which provided economic power to some social groups over others to access a productive resource. There is a very strong and well-established link between landownership and tube well ownership. This linkage defined the level and distribution of household surplus extraction from the booming agriculture and water markets in the early stages of their growth. The large farmers, mainly belonging to the Patel caste, were able to mobilize caste ties to access institutional finance and inputs needed for commercial agriculture. The unbridled access to groundwater, along with the mobilization of material, financial, and symbolic resources, reinforced the economic and political domination by the upper classes.

However, the arena of access and control over groundwater resources did not go uncontested. Other social groups challenged the dominant alliances in more than one way but did not succeed. This is mainly due to two reasons. First, the upper classes have constantly strategized ways to reproduce their domination and tried to have control over the process by various means. These strategies also included sharing of limited profit from institutional arrangements like sharecropping, so that the conflict does not reach the level of organized contestation. The relations of exploitation in Sangpura are embedded in the class and caste structure, and in the ability of dominant classes to control the labour process. The strategies of upper castes are quite visible in various alliances that can shape collective forms of consciousness, behaviour, and organization for achieving and defending their interests. The upper castes form dominant alliances to transform a 'class in itself' into 'class for itself'. Their alliances result in conscious strategies, which have both planned and unintended consequences in everyday politics and day-to-day decision-making for keeping control by dominant classes. The choices made are also part of a much broader political agenda, closely coinciding with the present social characteristics of the state. The control over groundwater resources is a result of this larger consciousness-making.

Second, the cultural dimension of caste diffuses economic inequalities and subordination, as they are seen more as 'given' or 'natural' facts. This implies that the castes lower in the social hierarchy do not form a 'class for itself'. However, this does not downplay their strategies in devising plans for resisting this process of domination. In fact, their resistance includes various exit options and a search for alternative livelihoods rather than confronting and questioning the unequal distribution of resources.

SOCIO-ECOLOGICAL VARIABLES AND CROP PRODUCTION SYSTEMS

The ecology of Gujarat has a defining role for processes of agrarian change. The very existence of a deep aquifer that can be mined in

Sangpura provided avenues for not confronting the ecological prob-lem of a declining water table. In 2002, Sangpura was situated in a 'dark' zone of groundwater utilization, with an ecological crisis being manifest in declining productivity of agriculture in recent years. The crop production base in the village was reaching a threshold, due to intensive agriculture and other ecological factors, which undermined dry land agriculture.

However, this declining profitability has not led to agriculture becoming uneconomic, at least not for the dominant and economically powerful classes. These people still derive a surplus, even in case of declining agricultural growth. This is caused by their ability to control the labour process and increases in input use under new production systems. The large area that came under sharecropping was also used as an instrument to control labour and to ensure that the water markets take a new form. In the early stages of water market development, the water charges were based on crop sharing. This gave way when profit from the sale of water superseded the profit from crop sharing. Crop sharing also meant the sharing of risk in agriculture between tenant and landlord. When the productivity of the land was highest, this risk was minimal; hence, water charges were collected on an hourly basis and in cash. With the decline in agricultural productiv-ity and shrinking of water markets, the risk in agriculture went up many-fold. The rise in the institution of sharecropping in this period was a response to share this risk by creating a stake for tenants. Water charges were again calculated against crop share, while labour costs were minimized by using lower-caste women tenants and keeping the wage rate constant for over a decade. This power of the dominant classes to shift the risk and pricing of groundwater whenever it suits their class interest throws light on the social relations in Sangpura.

A pertinent question is what strategies by the farmers can restore fertility, and why new technological changes are not adopted in Sang-pura. Of all measures to maintain fertility and land productivity, using compost is most popular among farmers. This is largely because of its availability in the village thanks to dairy farming activities, and its time-tested result in maintaining productivity without harmful effect to crops. Regarding new technological choices, there has been little innovation suited to the requirements of present-day agriculture in Sangpura. Even if the technology exists, information is not widely available. In the era of productivity decline only a few rich farmers have been able to experiment with different varieties of crop. Some of them, who experimented in the past, have benefited. Experimentation is limited to a few people and to their 'capability' to take the risk to experiment.

The increasing irrigation cost does not lead to the use of water efficiency devices such as drip irrigation systems. This is chiefly because under the one-third sharecropping system, the gain from the land and water component goes to the landlord. A drip irrigation system reduces the labour cost for irrigation, but the cost of installing a drip system is much greater than the cost of total labour charge in irrigating the field. Under the flat rate electricity system, accessing water is still economic in combination with cheap labour under sharecropping and profits from milk production systems. Hence, landlords take a rational decision where they prefer to use manual labour over drip irrigation. These aspects show how technological adaptation is the product of changing social relationships. Policy prescriptions to popularize a particular technology can only have very limited effects if they are not backed by a sound understanding of prevalent social relations and the ability to change them.

The declining productivity does not have the same effect on all social groups, as the case shows. The large farmers could cope with this situation through partial diversification from agriculture. In this way they have still been able to derive surplus. It is the small and marginal farmers that have been at the receiving end of these changes. This aspect makes a case for devising strategies to protect the interests of historically marginalized groups in the wake of extreme scarcity situations. The answers to the productivity question lie in devising strategies and promoting technology that consider the present resource constraints, and innovation suiting the particular ecology. This may include water-use efficiency measures and recharging shallow water aquifers. The concept of water use efficiency needs to be broadened to incorporate issues of productivity, equity and sustainability in resource management (Hussain *et al.* 2001). North Gujarat has numerous ponds and tanks that have silted-up over the years due to neglect and non-dependability. Restoration of these tanks could be used as a long-term strategy for recharging shallow aquifers. Another aspect to the improvement of productivity is to innovate and sustain seed varieties. While high-yielding variety (HYV) seeds in Sangpura no longer seem to show their higher yield potential due to poor seed management and uncertain water use, traditional wheat provides an almost similar yield with less external inputs. There is a dire need for research and development (R&D) in ecologically suitable agricultural technology for input efficiency, especially for water-starved areas like north Gujarat. Stabilizing agricultural production and the environmental system is fundamental: if they are negatively affected in an irreversible manner, this will have considerable social consequences.

* * *

In this chapter, I have dealt with some of the problems faced by marginal and near-landless households who take up sharecropping for their livelihood in Sangpura. An important driver behind these processes and changes in the domain of agrarian relations is the unsustainable and inequitable path of development of groundwater exploitation. The decline of water markets in an ecological context of falling groundwater tables (which requires increasing investments in pumping technology) draws an increasingly sharp boundary between those able to create or maintain access (largely the upper-caste/upper-class landowners) and those who get excluded.

Sharecropping interlinked land, water, and labour markets in an attempt by the landowning elite to share or devolve the risks of agriculture and the increase in irrigation costs. Large farmers adjusted to the situation of declining profit from agriculture through long-distance migration and the appropriation of surplus through sharecropping contracts. The marginal and near-landless farmers were forced to take up land on a sharecropping basis to cope with their increasingly difficult situation. The new situation has created a divide between those who can afford to partially move out of agriculture, and those who are forced to bear the consequences of declining productivity in agriculture, due to their inability to seek employment outside the village.

The chapter has illustrated how sharecropping is part of the changing social fabric of village society, due to changes in systems of production. Under the boom period of high productivity and low risk for richer farmers, the multidimensionality of patron–client relations was reduced in favour of market-driven prices and wages, leading to a change in the basis of class relations. However, when ecological and co-variant electricity supply conditions moved agriculture towards lower productivity and higher risk, a shift in labour contracts emerged. This shift is driven by water pricing and the difficulties of accessing groundwater as the water table is chased downwards. The shrinking water markets led to a shifting of risk through the sharecropping contracts. Former water buyers, who had been excluded from the purchase of groundwater for irrigation, were a ready labour force who could absorb this risk for the upper classes. Thus, sharecropping in Sangpura is a new face of the exploitative water market, and not a mere technical arrangement for allocating productive resources. In fact, it is a product of varying social relationships that provides an instrument for the dominant classes to transfer the burden of resource depletion to other people lower down in the socio-economic hierarchy.

In Sangpura, groundwater distribution and control are based on the ability of individuals to afford the costs of investing and deepening the tube wells. This gives economically powerful sections of society an edge to monopolize control over groundwater, which has wider socio-political consequences. At the local level, this manifests itself in control by the dominant classes over various arms of the state institutions like the panchayat, the dairy cooperative, and the credit cooperative (Prakash 2005). This is not a unidirectional agenda for controlling groundwater, but part of the broader agenda of dominant social groups to manipulate material, financial, and symbolic resources, which helps in reproducing their socio-political and economic domination.

These cases show the link between the local power structure and the social characteristic of the state, suggesting a wider relevance of this analysis for north Gujarat as a whole. The lack of effective state enforcement through strict laws and regulations to check groundwater exploitation rests largely on its political unwillingness. The political leadership of the state does not want to go against the large farmers' lobby, which is an important vote bank. This sustains a political interest among politicians and policymakers in the continuation of practices of mining deep aquifers. This process has been part of populist agricultural policies all over India, driven by the need to mobilize vote banks. 'While farmers have long been dependent on the state for essential inputs into agriculture, in recent years politicians in the state governments have now also become dependent on farmers' (Dubash 2002: 251). This shows that resource exploitation is not apolitical. A way to deal with this is to work on approaches based on balance of power concepts, with explicit recognition of the political nature of management (Moench 2000). As shown, dominant classes in Sangpura have been effective in controlling productive resources without much affirmative action by the state to provide a safety net for people lower down in the class and caste hierarchy.

To sum up, the DTWs in Sangpura were introduced as the outcome of changes in production relations, while their introduction is also a driver of further changes in these relations. The initial expansion and later demise of water markets and the level of surplus generation changed social relationships and an individual's access to groundwater resources. Dominant social groups devised strategies to control material and financial resources through various means. This control did not go uncontested. However, the contest was heavily weighted in favour of dominant classes, chiefly because of their ability to cope with changes.

Groundwater has been an important part of the interlocking control, as agrarian institutions, socio-ecological variables, and domains of the state functionally interacted with the means to extract it. This is chiefly because of the characteristics of groundwater, which transformed agriculture on the basis of an individual's level of access to it. Further, this accessibility got translated into new ways of surplus extraction for the farming households. The social characteristics of the state provided an opportunity for the local power holders to control access to groundwater. Together, these led to unchecked exploitation of groundwater, which only benefited the small minority of upper class farmers. Through the shift in labour contracts in accordance with groundwater pricing, the risks of agriculture shifted away from the landowners to the sharecroppers through the sharecropping contracts. As groundwater accessibility decreased and the market shrunk, the upper classes maintained their relative social and economic positions and brought the working classes into a new dependency relationship, defining the course of future agrarian change in north Gujarat.

Notes

[1] The introduction of HYVs after 1965 and the increased use of fertilizers, chemicals, agricultural equipment, and irrigation (the Green Revolution) provided the production increase needed to make India self-sufficient in foodgrains. This has led to a sharp increase of agricultural production. The Green Revolution has often been criticized as unsustainable, as it requires immense amounts of capital for equipment, irrigation, and fertilizers. In addition, the crops require so much water that water tables in some regions have dropped dramatically. See http://en.wikipedia.org/wiki/Green_Revolution_in_India (accessed 12 February 2012).

[2] Groundwater markets are actually irrigation service markets that emerged in Gujarat around 1910. They have become sophisticated economic institutions that stimulated the emergence of a class of pump irrigation entrepreneurs who invested in DTWs to sell pumped irrigation water to a client basis for profit (Shah 2009: 98–9).

[3] DTWs are drilled wells that operate at depths typically beyond 40 metres and require submersible pumps operated by electricity or diesel. In 2003, the depth of DTWs in the study area was 100 metres.

[4] Groundwater irrigation is available throughout the year, enabling intensified agricultural operations, reducing risk of crop failure and creating year-round wage employment. These factors do influence the economy. However, there is an increasing trend to assume simple correlations between groundwater irrigation and poverty decline (Moench 2002; Narayanamoorthy 2001). These studies ignore the other poverty-related variables at work. The availability of groundwater is locality specific and its use usually subject to the non-availability of surface water resources. Access to groundwater depends on access to land and capacities of individuals to invest in tube wells. Second, due to lack of strong regulation, there is a danger of overdevelopment if the policy prescription of 'groundwater development for poverty alleviation' is followed.

[5] Hardiman puts forward the class/caste nexus in the distribution of tube well water in Gujarat. According to him,

water controlling cartels do not require a formal union or constitution; they operate according to a precise local knowledge in which each family is all too aware of its standing within the hierarchy, its corresponding entitlements, and the bounds beyond which they cannot step. In the past, dominant classes controlled subordinate groups within a village with an iron hand, often involving violence. Today their political control is challenged, with members of lower castes voting for politicians of their own community and attempting to assert themselves in a whole number of different ways. Increasingly, therefore, dominant groups have to rely on their economic power both to control subaltern classes and prevent them from gaining any economic powers themselves. This is the real rationality of the system. (1998: 1541)

[6] For criticism of Shah's work, see Palmer-Jones (1994) and Dubash (2002).

[7] A pseudonym of a village in Mehsana district, Gujarat. A taluka (sub-district) is the lowest level of government administration above the village.

[8] According to the Central Groundwater Board of the Government of India, in overexploited subdistricts the level of groundwater abstraction is more than 100 per cent of average annual recharge. In dark blocks this level is between 85 and 100 per cent. Grey blocks have between 65 and 85 per cent, and white blocks less than 65 per cent.

[9] For data on poverty, I have used the government data available at the village record. All the below poverty line (BPL) families are listed in the village record and supported though various government schemes.

[10] I follow the definition by Cheung:

[S]hare tenancy is a land lease under which rent is a contracted percentage of the output yield from the tenant per period of time. As a rule, the landowner provides land, and the tenant provides labour; other inputs may be provided by either party. Share tenancy (or sharecropping) is thus share contracting, defined here as two or more individual parties combining privately owned resources for the production of certain mutually agreed outputs, the actual outputs to be shared according to certain mutually accepted percentages as returns to the contracting parties for the productive resources forsaken. (1969: 1107)

[11] Kos is an irrigation device operated manually using draught animals.

[12] 'Bhagiyo' is a Gujarati word that means sharecropper ('*bhag*' means share or division).

[13] 'Dhani' originates from dhan (wealth). Dhani is someone who possesses wealth. Landlords in Sangpura and elsewhere in north Gujarat are addressed by this term.

[14] I interviewed 85 farmers engaged in a sharecropping contract, selected through stratified random sampling. The input–output data of five major crops were taken in 2002–3. Interviews were held between July 2002 and May 2003, to cover monsoon, winter, and summer.

[15] For land, the net gain was calculated by deducting total cash expenditure from the share of produce. For labour, the total cash expenditure was deducted from the share of produce. This figure does not include the price equivalent of the number of labour days invested as it is seen as overall return from the produce of labour, which is different from the concept of working as wage labourer. For water, the irrigation schedule and the amount paid for the water was looked at for each farmer. This cost was deducted from the water component share so as to calculate the net gain. At the time of fieldwork, mustard, wheat, and pearl millet had very different yields, and for some farmers, especially smallholders, these crops failed. Thus, this analysis

does not reflect the normal situation for these crops, as yields are normally higher. In 2001–3 the value of 1 USD ranged between INR 47.5 and INR 49.

[16] Compared to the tenant, the landlord's time investment in the sharecropping arrangement is minimal. The only time investment is during harvesting of the crop and selling it on the market. However, it was observed that the landlord keeps a vigilant eye on the whole process through an elaborate social network.

[17] During my fieldwork in 2001–3, two people from the Patel caste and one from the Prajapati caste migrated to the USA.

[18] In reality, irrigation is carried out by the hour in Sangpura. To calculate the percentage of men days, the record of all hours spent in irrigation have been taken, divided by eight to get the number of days put in.

8 TRANSFORMATION OF TANK IRRIGATION POLICY AND TECHNOLOGY ON THE INTERFACE OF A RECURSIVE STATE–SOCIETY RELATIONSHIP

Esha Shah

The nature of the state's role in shaping irrigation in South Asian societies is intensely contested since the seminal work of Wittfogel (1957; see also Leach 1959 and Ludden 1979). In the two decades since the publication of Wittfogel's book, several scholars refuted one of the main arguments of Wittfogel: the large-scale creation of irrigation infrastructure needed a centralized bureaucracy and a despotic state to control coerced labour (Hunt and Hunt 1976). The debate that followed inaugurated the study field of the community's role in constructing, managing, and using irrigation resources. This debate gradually shifted its terms of reference from state to communities, from construction to maintenance and management, from water control to local agriculture, and from past to present.

As a corollary to these debates, some influential scholarly works of our time give a powerful discursive space to the idea of community that is tied up with the imagining of traditional or pre-modern knowledge systems. James Scott, for example, in his influential book *Seeing Like a State* (1998), proposes the Greek concept of *metis* to contrast the local, implicit, embedded, practical experience of knowledge with the abstract, general knowledge deployed by state agencies. Another influential scholar, Stephen Lansing, proposes that a holistic cultural approach attentive to indigenous systems of order can solve many contemporary political predicaments. Calling it a 'perfect order', Lansing (2006) argues that the centuries-old traditional system of the water temple in Bali achieved a balanced ecology without recourse to centralized state power. Lansing argues that 'culture', with its deep roots in pre-modern times, is an alternative rationality to the pervasive economization of modern times sponsored by a state.

In the case of development policies for tank irrigation in south India, the shift of narrative frame from state to community, culture, and traditional knowledge is not only academic but also relates to policy. A number of policy-related studies operate with the assumption that the traditional tank irrigation system fell from the heights of its pre-colonial grace as a result of the intervention of the colonial state and subsequently because of the apathy of the post-colonial state (Agarwal and Narain 1997; Janakarajan 1989; Mukundan 1988; Sengupta 1985; Shankari 1991; Tippaiah 1997). Deterioration and degeneration of tank resources as a result of the loss of community, custom, and traditional knowledge is the fundamental assumption upon which much of the current policy dialogue rests. The retrieval of the lost knowledge traditions of community institutions are central pillars of tank development policies and practices funded by institutions like the World Bank (see World Bank 1981) and the European Union (EU). The ideas of traditional knowledge and culture are central in these academic and policy shifts of narratives from state to community.

This chapter challenges both state-centric and community-worshipping approaches to understanding the changing irrigation knowledge and resources, and demonstrates the recursive—mutually transforming—roles played by state-society interactions (see Mosse 1997b). The chapter follows three strands of arguments, which are elaborated in three subsequent sections. First, the boundaries between state and society are considered blurred. Although those who control state power are in a position to take decisions with far-reaching socio-economic consequences, and although these decisions may at times reflect the interests and pressures of other powerful actors both at home and abroad, the choices made by the state are ultimately political choices that respond to demands from society. When state organizations come into contact with various social groups, they clash with and accommodate various moral orders (Migdal *et al.* 1994: 12). This process of shaping the moral order involving agrarian relations, as a result of state-society interactions, is explored in the next section. It discusses the ways in which changing agrarian relations since the introduction of the Green Revolution, and the resulting new farmers' movement, have transformed agricultural policy at national level. Section three takes this argument further, with specific reference to tank irrigation policy in the south Indian state of Karnataka.

The following two sections, thus, discuss how different social actors participate in development processes in different ways. The acts of collective protest adopted by rural elites in the form of the new farmers' movement are powerful forms of social and political response to state-sponsored development processes. The state, on the other hand, by accommodating multiple strands of demands not only derives higher legitimacy, but also makes political alliances with different sections of rural society. The chapter thus challenges the mutually exclusive spheres of state and community and argues for more inclusive approaches to understand forms of political interaction between state and society. Second, the chapter demonstrates that knowledge and technology are the domains where the contestation of power between state and community and different sections of community happens. Scott (1998) contrasts the state's way of seeing with that of communities. Challenging such unified categories of state and community, this chapter shows that by designing, using and managing technology the social field of power is created, contested, transformed, and reproduced by both state and society actors in close interaction with each other. Pursuing this argument, the fourth section explores the transformation of tank technology designs in the Shimoga district of Karnataka, in response to changing agrarian and state–society relations since the introduction of the Green Revolution in agriculture.

STATE–SOCIETY INTERACTION: FARMERS' MOVEMENT AND NEW AGRARIANISM

Since the introduction of the Green Revolution in the 1960s, the emerging agrarian politics at national and regional level culminated in a landed farmers' movement in many parts of India over the next two decades. This divided Indian society along sectoral—urban versus rural—lines rather than class lines, and generated a significant point of departure for both agrarian and irrigation policies. This phase, often described as a wave of 'new agrarianism', was led by a hegemonic class of owner–operator landholders that swept national and regional politics from its start in the 1980s and continued through the 1990s. This phase was represented by several agrarian-minded regional parties. Examples are the Shetkari Sanghthana of Maharashtra, the Karnataka Rajya Raita Sangha (KRRS), the Tamilnadu Agriculturists Association, the Bhartiya Kisan Union led by Mahendra Singh Tikait in north India, and the Kisan Union led by the *kisan* (farmer) ideologue Charan Singh in Uttar Pradesh (UP) (Rudolph and Rudolph 1987: 2, 49–54). The main and most consistent demands of the movement have been to increase the prices of agricultural outputs and reduce the costs of inputs. The fixing of a procurement price for agricultural outputs has been a crucial concern of the movement. Other issues included the lowering of prices of agricultural inputs such as electricity and irrigation water, betterment levies, and taxation of agriculture. Non-payment and waiving of government loans has also been a strong issue (Brass 1995; Gupta 1998; Varshney 1998).

The nature of the movement has been populist and oppositional, in the sense that the demands are made on behalf of all rural and farming communities irrespective of class, caste, religion, ethnicity, and gender, and the chief target of the demands has been the state government. The movement's main manifesto is based on the premise that the state's policies with an urban bias have failed to deliver the development promise to rural farming communities (Gupta 1998: 80). The farmers' movement and its populism stress that the interests of urban, industrial 'India' have systematically undermined the well being of agrarian, rural 'Bharat'.

There has been considerable consensus among scholars that all agrarian-minded organizations have heavily mobilized owner-cultivators for their core support (Gupta 1998: 80). Even when small farmers are an important constituency of the movement, it is still heavily biased towards the landed sections of the peasantry and even more towards those farmers who have a relatively large marketable surplus. The movement has rarely raised the concerns of landless labourers and, more importantly, never questioned the differential distribution of land and water resources between the landed and landless peasantry. In this way, the movement is not only biased towards the landed class, but has a fundamental connection with Green Revolution technology that tied the cultivators to the market in an unprecedented way.

New Agrarianism in Karnataka

The agrarian context of Karnataka presents a slightly different picture from that of India as a whole. The non-Brahmin agricultural castes of Vokkaligas and Lingayats are together the largest landholding groups and politically the most powerful groups in Karnataka. The socio-economic and political dominance of the Vokkaligas and Lingayats has consisted, among other things, of the control of a substantial proportion of fertile land around villages, possession of the post of village headman, and leadership of patron–client networks, often through money lending (Manor 1989: 332). Since the colonial period, no region-wide or imperial government could make its authority significantly penetrate the dominance of these castes at village level (Manor 1989: 327). As a result, the dominant forms of land control and local power structures continued without much disruption during and after the British period (Manor 1989: 328).

However, it has been argued that Karnataka has a more equitable distribution of resources than other parts of India (Manor 1989). In terms of the landholding pattern, Karnataka has a higher percentage of owner-cultivators than the national average. Before the second wave of land reform began in 1974, the percentage of owner-cultivators had increased from 70.2 per cent in 1961 to 88.8 per cent in 1971 (Manor 1989: 343); 92.8 per cent of the cultivated area was wholly owned and self-operated (Rajan 1981: 57). According to the 1971 agricultural census in Karnataka, the total area under tenancy was 7.4 per cent (against a national average of 8.5 per cent) (Rajan 1981: 57). This pattern of landholding has not significantly changed, despite two phases of land reforms initiated at the state level (Kohli 1987; Pani 1997).

The owner-cultivators, a majority of them Vokkaligas and Lingayats, have thus emerged as the most influential class of landowners in the social, economic, and political history of Karnataka. This dominance, which in the past manifested itself mostly at village level, has acquired a new face. The interests of the landed in Karnataka found new political alliances and articulations at the state level, in line with their counterparts elsewhere in India. Since the 1980s, beginning with the agitation for remunerative prices for agricultural outputs, the landed farmers in Karnataka have also organized into a regional political group—the KRRS—arguably the state's most 'powerful farm lobby' (Kripa 1992: 1182). The new farmers' movement not only emerged from a pioneering base among sugarcane growers of Shimoga district, but also has a vast following among paddy growers, especially in the surplus rice growing areas of the state, including Shimoga district (Assadi 1997; Nadkarni 1987).

The phase of 'new agrarianism' has achieved a significant gain in terms of prices and assured markets for output and reduced prices for inputs for farmers. The movement succeeded in putting pressure on the state government to eliminate the procurement levy on rice[1] and relaxation of the restriction on the movement of foodgrains.[2] These demands of the paddy-growers were in fact the main demands of the movement in the 1980s and 1990s (Assadi 1997: 79; Nadkarni 1987). The third and most important gain, which began in the 1980s and has continued throughout the 1990s, is the penetration of the rice and wheat growing farmers' agendas in the foodgrain price policy at both national and regional levels.[3]

Several scholars commenting on the current agrarian crisis, sharpened by several farmers committing suicide, have acknowledged a considerable political clout of farmers' organizations (Assadi 1997; Rao and Gopalappa 2004; Rao and Suri 2006). The defeat of ruling governments in both Karnataka and Andhra Pradesh (AP) in the 2004 state elections, for instance, is attributed to their respective apathy to farmers' issues. The Congress finally formed a coalition government in Karnataka (and did not survive long), but the chief ministers of previous governments, Chandrababu Naidu of Telugu Desam in AP and S.M. Krishna of the Congress party in Karnataka, were famously known for their urban and information technology bias. In the Karnataka elections of May 2008, the KRRS gave a call and a slogan to boycott anti-farmer political parties (see Shah 2008a).

There have been other means by which the input costs have been reduced for rice- and wheat-growing farmers, in addition to the encouraging price policy. In Karnataka, several tax concessions were granted in the phase post the new farmers' movement. In addition to the concessions on betterment levy, water rates, and electricity charges, sales taxes on fertilizers and insecticides were reduced in 1983 and 1984–5 from three to two per cent and from four to three per cent respectively (Nadkarni 1987: 132). This upward trend of favourable price policy of both output and input, coupled with the assured market thanks to government procurement, and an assured and stable trend of prices beginning in the early 1980s as a result of farmer lobbying, has made paddy growing an attractive enterprise in Karnataka.

The farmers' movement thus has gained significantly in terms of a shift in price policy, but the relevant question for this study is: what do these politics imply for the day-to-day management of other crucial resources like tank irrigation water at the local level? Another, related, question is: how has this new agrarian politics influenced tank irrigation policy at the state level in Karnataka? Changing agrarian power equations at the regional, national, and local levels since the Green Revolution, as discussed earlier, generated a point of departure for management of tank irrigation resources. The rise of centrist politics together with the rise of a politically ascendant class of owner-cultivators, who have acquired an influential position at both the national and regional levels, is one of many factors that has undermined old forms of authority at village level. The remainder of this chapter discusses how this shifting locus of agrarian relations and the corresponding shift in state–society relations produced perceptible changes in tank irrigation policy at the state level and tank irrigation designs at the local level.

NEW AGRARIANISM AND TANK IRRIGATION POLICY IN KARNATAKA

In the early 1970s, administrative responsibility for tank maintenance was transferred from the revenue department (RD) to the public works department (PWD). The management of tanks, the power to collect water and irrigation charges,[4] and judiciary powers to penalize cultivators were retained by the RD. However, the village-level revenue officers—Patel and Shanbhoga—remained in charge of tank management even after tanks were formally transferred to the PWD in the early 1970s. During the subsequent decade, the PWD repaired several tanks, as part of larger agrarian reforms initiated at the national and regional levels. For the first time since independence, replacement and repair were taken up on this scale and the distribution network, sluices, and damaged bunds attended to (for further insights on how these replacements and repairs accompanied a marked shift in tank designs in various parts of Karnataka, see Shah 2003). In some parts of the state, these design changes preceded the introduction of Green Revolution varieties of paddy in the 1970s. In other parts, these changes accompanied the shift in cropping pattern only in the 1980s. Either way, the repairs provided new economic opportunities to the landed class in tank-irrigated areas. The choices made by this landed class influenced the way tank resources have been managed and maintained in the last three decades. Not only did a major shift in cropping pattern occur during this time, but also a whole new techno-managerial system emerged, especially in the command areas of paddy-irrigating tanks.

In terms of distribution of financial resources, the magnitude of investments made for minor irrigation in Karnataka during the Third Plan period (1961–6) was 157.9 million, up from INR 50.8 million in the Second Plan. The Fifth Plan had double the provision of the Third Plan for minor irrigation, but the greatest increase in its provision can be noticed during the Sixth (1980–5) and Seventh Plans (1985–90). The increase in investment during the 1980s suggests that this was a crucial time for tank irrigation policy. The rise in investment in minor irrigation works, however, was accompanied by a state of confusion about which government department should be in charge of tanks. Since the early 1980s, the responsibility for tanks has been transferred several times between various government departments. A quick review of the history of the minor irrigation department (MID) supports this point.

After the reorganization of Karnataka in 1956, the PWD was constituted with two wings: communication and buildings, and irrigation and public health engineering. Separate chief engineers were appointed for each branch, although the lower level of engineering staff remained common to both branches. This means that the lower level of engineering staff looked after all the public works, although each branch had separate higher staff. In the history of PWD, this phase is known as the undivided PWD. In this period, there were constant complaints from staff and farmers that the public works of irrigation were neglected and given low priority. It was only in the 1980s that a separate irrigation department was formed. It was further split into major and minor divisions. In 1980–1, the department of public health engineering was also constituted separately. Almost all tanks, barring a few large ones defined as major or medium irrigation schemes, were handed over to the MID. In 1987 there was a further shift, when Karnataka constituted Zilla Parishads (ZPs)[5] with the aim of decentralizing and devolving power to lower administrative levels at district and below. After the formation of ZPs, tanks with an irrigated area of less than 200 hectares were handed over to the engineering section in the ZP. As a result, the MID was left with only a few tanks to look after. However, within a couple of years, tanks with an irrigated area of more than 40 hectares were handed back to the MID. During this period, the paper manifestations of tanks, their files, shuttled between the PWD, MID, and ZP, often degenerating into a state of oblivion.

This confusion about which department is responsible for what and which tank (based on size) rests with which department, suggests that the 1980s were a period of crisis, generated by expectations that the state would invest in the management and maintenance of tank irrigation resources more than it had done until then. The state bureaucracy had inherited from its British predecessors a large volume of technical data like detailed contour maps, supply statements (discharge data) of all important rivers and their tributaries, a census of tanks, and detailed maps of the tank-irrigated areas. However, the state bureaucracy inherited no institutional arrangements to manage the tanks on an ongoing basis. As already discussed, the British did not interfere much with local forms of tank management dominated by local elites.

After independence the same social arrangements for the management of tanks continued, in which the village revenue officers (Patel and Shanbhoga) played an important role, although the RD and PWD were officially responsible for tanks. The main mandate of the PWD and MID remained the engineering aspects of repair and maintenance. There were no institutional structures created in or funds allocated to these departments to manage tanks on a daily basis. This point can be explained by the fact that the PWD and its successor, the MID, have mainly engineering staff trained primarily to handle civil engineering work. Each of the engineering staff is in charge of a few hundred tanks spread over a certain geographical area. Even visiting each of the tanks, once in a while, and attending to the major repair and maintenance tasks demanded time and energy from the engineering staff, to the extent that expecting them to partake in day-to-day management of each tank was impractical. Only for a few well-functioning and bigger tanks there was a person appointed by the MID to open and close sluices and watch physical structures. These persons were appointed mostly from the service castes from nearby villages, and were not considered regular MID staff. Evidently, the farmers of irrigated areas themselves attended to most crucial decisions about operation of physical structures and water distribution. These included bearing the costs for operation, management, and maintenance of tanks, except the occasional repair and maintenance done by the MID and previously the PWD. The managerial and financial involvement of the state bureaucracy in tank management on a daily, routine basis has therefore been considerably limited.

This crisis from the beginning of the 1980s signified another point of departure. With the rise of populist politics of farmers in the early 1980s, the lack of state participation in the management of rural infrastructure began to be seriously questioned. The political process that began in the 1980s and culminated in the formation of ZPs expected the state machinery to invest a sizeable amount of resources to build, manage, and maintain rural infrastructure, including irrigation resources. Thus, for the first time since the colonial period, the state was expected to invest financial and human resources for routine repair, maintenance and management of tank resources. Shifting small tanks from department to department is perhaps a symptom of the confusion these expectations have generated.

The pressure on the state to invest in tank resources can be understood in two ways. First, it can be interpreted as a sign that the rural elites want to shed the inherited responsibility of tank management, which had started to become a burden once there were other means to expand political and economic gains. With an intensifying and diversifying cropping pattern since the Green Revolution, controlling tank management was no longer the only means by which rural elites could gain economic power. While intensification of paddy cultivation, supported by favourable price policies, generated unprecedented gains for rural elites, diversification of cropping patterns also presented them with other opportunities to expand economic power. At the same time, investment in tank irrigation, which earlier reproduced their political and economic power over resources, had become increasingly non-remunerative with changing social and economic relations.

Shifting conventional social arrangements for management and maintenance of tanks is the second explanation for the increased pressure on the state. Whatever the process of decision-making at the tank level, the actual tasks of operation and maintenance cannot be carried out without the labour input of non-landowning and service castes members. Conventional arrangements for both decision-making and execution of management and maintenance tasks have been rooted in social relations of power at village and at tank levels. These arrangements have come under increasing pressure in the last two to three decades with changing caste and economic relations, as also corroborated by Janakarajan (1989) and Mosse (1997a). Not only have the decision-making processes regarding tank resources been significantly transformed, but, more importantly, the mobilization of labour from non-landowning caste members has become increasingly difficult. Lower-caste tenants, for instance, are refusing to partake in canal cleaning. Some of them have acquired their own land and are questioning non-participation of higher caste farmers in maintenance of physical structures. Neergantis (workers who distribute water in a tank command, operate the system, and clean canals) are employed somewhere else, and hence water is not distributed field to field.

Therefore, the crisis faced by the state bureaucracy reflects a larger process of shifting power and economic dynamics at the village level. The political economy of agrarian change has created new opportunities for rural elites who now expect the state to invest, financially and managerially, in the upkeep of tank resources. This demand is also buttressed by the fact that the conventional social arrangements that ensured labour input for management and maintenance of tank resources cannot be reproduced in their entirety. This push and pull has culminated in an overall crisis, in which the state is expected to invest but has no institutional arrangements to do so, while rural elites are exploring greener pastures and demanding the state to take over the financial responsibility of tanks, and in which the non-landowning peasantry is disrupting earlier arrangements of labour contribution. This crisis ultimately culminated into the act on decentralization and formation of ZPs, and the allocation of considerable financial resources for upkeep of irrigation infrastructure. As a result a large number of tanks with less than 40 hectares of command area were transferred from the MID to the ZP.

The formation of ZPs reflects the impact of the new farmers' movement on ruling politics. It was an election promise of the Janata Dal government, the first non-Congress government ever acquiring power in Karnataka. It came to power in 1982 after a landslide victory against Congress. Their election slogan was 'we want to rule not from Delhi but from the village'. On winning the election, Janata Dal recognized that holding on to power would require major concessions to rural constituencies, especially to the landed class. Decentralization and devolution of power to the district and lower levels were soon taken up. In subsequent years, as per the stipulation of Act 27, departments were surrendered by the state government to the ZPs.[6] This was not the first time that an attempt was made to decentralize power to lower levels (Chandrashekar 1984: 683). It was Janata Dal's successful appropriation of the populist rhetoric of the farmers' movement that made it possible to convert the bill on decentralization in the Act.

The dual nature of the country's development, which has created rural 'Bharat' and urban 'India', has been the central feature of the discourse of influential leaders of the farmers' movement such as Sharad Joshi of Shetkari Sangathana in Maharashtra and Charan Singh of the Kisan Union of UP and Nanjundaswamy of KRRS. This division between Bharat and India was intended to highlight the urban bias of development policies which, according to this view, resulted in a widening gap between urban dwellers who work for the state and industries, and rural people who are sustained by agriculture. Gupta argues that this duality of Bharat and India, also expressed as 'they' versus 'us', 'collapses a variety of dissatisfactions experienced by different segments of the rural population into a unitary framework' (1998: 80). The appropriation of this unitary framework by the ruling Janata Dal in 1983 was indicative of the fact that the landed farmers' agendas had penetrated deeply into populist politics at the state level. The roots of decentralization of power to ZPs during the Janata Dal regime, which resulted in the transfer of a significant number of tanks to ZPs and a significant rise in the financial resources provided for the planned development, grew in the soil of this populist politics of the landed classes.

How easy it has become for the landed class to access the resources available at the lower levels of administration is an issue for future research. At a normative level, the intervention of the state bureaucracy to provide services has freed the dominant actors at the local level to a considerable degree from the need to reinvest their extracted surplus back into the local system.

Even after the formation of ZPs, confusion over institutional identity of tanks was not yet over. After shifting tanks between government departments a few times in the 1980s, the state government has recently sought to hand them over to the community, back to where they were believed to belong. A World-Bank-funded project initiated in the state in the early 2000s sought to develop an approach to community-based management and improvement of tank resources. The project aimed at transferring one-third of the existing tanks to farmers. Management and operation of the tanks with an irrigated area of less than 40 hectares that were looked after by ZPs were scheduled to be transferred to communities in 2002, when the first write-up of this research began (Shah 2003). Tanks with more than 40 hectares of irrigated area were supposed to be handed over to farmers' organizations. The project was conceptualized as a part of the globally initiated irrigation reform policy supported by multilateral funding agencies; in the case of Karnataka, the World Bank. This reform policy forms part of other global initiatives that envisage a reduced role of the state in the affairs of civil society.

The previous discussion shows that tank policy in Karnataka has faced an apparent contradiction with respect to initiatives to transfer tanks to communities. On the one hand, farmers are demanding a higher level of state intervention and participation in creating, managing, and maintaining rural infrastructure, which has been articulated vociferously through farmers' movements. On the other hand, the state is developing an approach which will make communities responsible for tank management and maintenance. There is another side to this contradiction. In spite of state departments officially being in charge of tanks, in reality only communities have managed and maintained tanks, a point that even the state bureaucracy acknowledges. The state bureaucracy's failure to appropriately manage local resources is put forward as a justification to return tanks to communities. Tanks have been managed by communities all through and hence the state policy that advocates development of an approach to make communities manage and maintain tanks sounds a bit like a hoax.

In the next section, I discuss the changing designs of tank technology since the Green Revolution at the interface of state and society in the Shimoga district. Though some empirical details (based on fieldwork conducted between 1999 and 2000) may be relatively dated now, the larger point about technological designs as sites for interaction and contestation among actors from society and state still remains relevant.

SOCIAL DESIGNS: TRANSFORMATION OF TANK IRRIGATION TECHNOLOGY

It may not have been sheer coincidence that water distribution infrastructure like sluices and distribution canals in many tanks of Shimoga district began either to disappear entirely or were severely modified since the early 1970s, with more modifications undertaken in the 1980s and 1990s. Shimoga produces the largest quantity of rice and has the second largest area under rice in the state. High-yielding varieties (HYVs) were first introduced in Shimoga in 1966–7. According to a survey of two rice-growing areas of the district, all farmers had adopted new varieties by 1972 (Krishna Murthy 1975: 121), much earlier than in other parts of Karnataka. A majority of landowners of the tanks discussed here adopted new varieties from the early 1970s. However, it took another decade for a complete shift to intensive rice cultivation in tank commands, followed by the policy shift on procurement price at the national and regional levels. The shift in designs of irrigation infrastructure, therefore, followed not only the introduction of Green Revolution varieties, but also the intensification of rice cultivation since the early 1980s.

The most apparent shift has been the disappearance of water distribution canals. A paddy-growing command area (*atchakat*) is usually so designed that it facilitates the movement of water from field to field. Upstream and downstream fields are provided a relative level difference so that they can be irrigated in succession from head to tail. This design of fields, the position of canals and their slope thus facilitate movement of water in the atchakat, but also favour some lands in the atchakat over others. The direction of movement of water also produces grades of different types of land in the atchakat. It assures supply more to certain patches than to others. Most often these patches are located in the head reach. The favoured patches acquire certain features, owing to their submergence under water for longer duration, which make them progressively more suitable for paddy cultivation. These grades of lands relate to the landholding pattern in the atchakat. Almost always the best land in the atchakat belongs to the historically privileged groups of landowners.

Another major shift in tank designs since the intensification of rice cultivation has been the disappearance of the sluice operating mechanism. Conventionally, tanks irrigating paddy in Karnataka have the 'plug and pole' type of sluice. Tanks in a wet region like Shimoga, with average annual rainfall between 1,500 and 2,000 millimetres, are relatively small and have lower bunds than tanks in the drier or mixed regions with average annual rainfall between 600 to 900 millimetres or less.[7] The sluice apertures in the tanks of the wet region are located in embankments. In the bigger tanks in the mixed region they are located in the water storage area away from the embankments. Sluices in tanks in the wet region have conventionally been operated from the platform located over the opening in the embankment. In two tanks in Shimoga district discussed below, the sluice operating mechanism—plug and pole—was stolen or suddenly disappeared in the 1980s. They were never replaced, barring in two tanks, where they again went missing.

Take, for example, the case of the old Saulanga tank. Three canals—two distribution canals located on the extreme edges of the atchakat connected to two sluices, and one drainage or seepage canal located in the middle—existed in the atchakat until a decade ago (Figure 8.1). After the introduction of high-yielding paddy in the atchakat, the section of the canal became narrower and finally disappeared completely in the course of the 1980s. The seepage canal was used to collect drainage water from land on both sides. It also served as a carrier canal that supplied water to the tail-end. Now irrigation in the atchakat is entirely done from field to field, from head to tail. While earlier, water could reach the tail-end in a few hours, now it takes at least seven to eight days. As a result, the amount of tail-end land receiving irrigation has drastically decreased, and the choice of crops has been severely curtailed. Tail-end farmers plant either semi-dry crops or early maturing paddy varieties that are broadcasted.

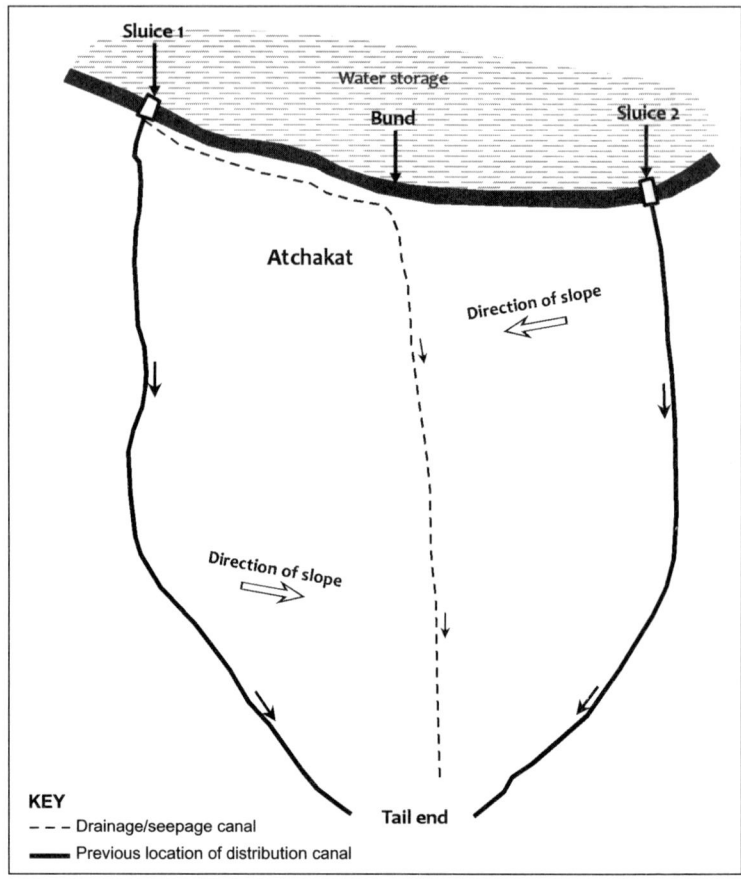

FIGURE 8.1   Saulanga Tank Atchakat

Source: Adapted from Shah (2003).

In another tank called Chinnikatte Taverekere (Figure 8.2), the left bank canal (LBC) on the extreme edge still survives, but the seepage canal in the middle, that used to take water to the tail-end, has been heavily encroached upon and silted up. Tail-end farmers say that its carrying capacity has been reduced to one fourth of the original. One more distribution canal, marked right bank canal (RBC) in Figure 8.2, has completely disappeared as a result of encroachment. Sluice 1 of this tank, which used to provide water to the RBC and is located at a higher level than the rest of the atchakat, has gone out

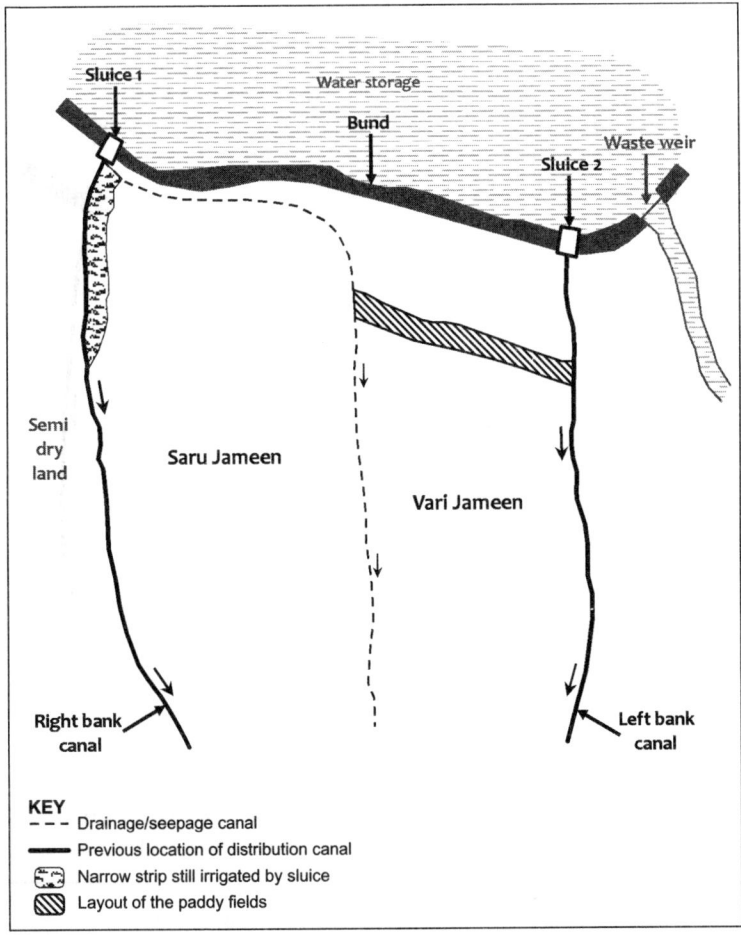

FIGURE 8.2   Chinnikatte Taverekere Atchakat

Source: Adapted from Shah (2003).

of use and now feeds only a narrow strip of land on the extreme left edge. The entire atchakat now receives water from the LBC. Water is first distributed from field to field to a part of the atchakat called *vari jameen* (upper land). The drainage from vari jameen is collected in the seepage canal and then distributed to lower parts of the atchakat called *sara jameen* (lower or seepage land). The tail-end now receives water once in eight days, while it needs it every day. Although vari jameen practically forms the head reach in this tank due to its proximity to the canal, it is actually the higher end of the atchakat, given its higher level compared to the sara jameen.

Similarly, the plug and pole arrangement of the sluice disappeared 20 to 25 years ago, soon after it had been repaired and replaced by the PWD. Before this period the sluice had a heavy plug, which used to be opened three to four times in the irrigation season after permission was granted by the *patel*. Each time the sluice was opened, it was kept open for a few days until the entire atchakat had been irrigated, and again closed until the next round of irrigation. The plug and pole of the sluice existed until only three decades ago. The sluice is now stuffed with gunny bags and paddy stems before the rainy season, opened in July, kept open for the entire paddy season, again stuffed in October, opened in January, and closed before the rainy season.

Yet another tank, Kumsidoddakere, has neither distribution canals nor sluice operating infrastructure. The wooden plug and pole existed until the early 1970s, while the tank was managed by the village-level revenue officers (patel and *shanbhoga*). In 1971, the tank was handed over to the PWD, when officially the patel and shanbhoga stopped being responsible for the tank. This happened at almost the same time as the new paddy varieties were introduced in the atchakat. The distribution canals existed at places shown in Figure 8.3. Tail-end farmers of this tank alleged that, soon after takeover by the PWD, the powerful farmers of the head reach, including the patel and shanbhoga, first encroached upon the canals and later even destroyed the remaining part. This happened after the introduction of transplanted paddy, when the atchakat started to face water shortage. Destroying canals ensured that water is first supplied to the head reach and reaches the tail-end only if it is allowed to.

Another example is a tank called Sorturuhosakere (Figure 8.4). This tank is roughly 300 years old. It had 80 to 120 hectares of atchakat that has now come down to 12 hectares, all owned by one extended family of the Lingayat caste. There is an intense struggle going on between tail-end lower-caste Kuruba farmers and head-reach, higher-caste, Lingayat farmers. Tail-end farmers are not only prevented from taking water from the tank but also from acquiring land in the head-reach. Even violent means were adopted to prevent one Kuruba farmer from purchasing a plot in the head reach. A Keladi

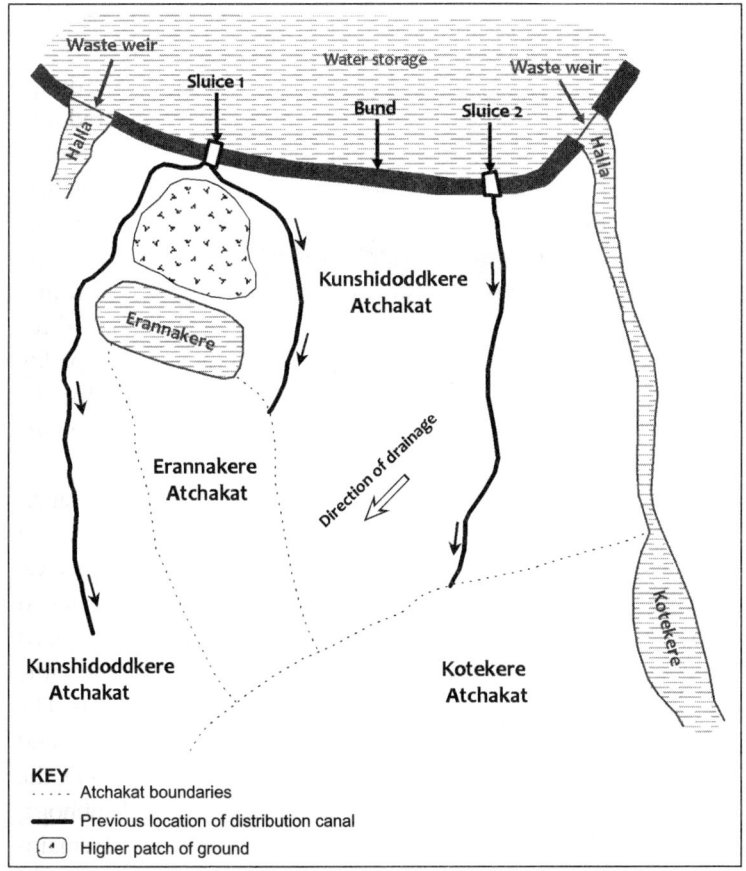

FIGURE 8.3   Kumsidoddakere Atchakat

Source: Adapted from Shah (2003).

Naika, two centuries ago, donated the whole of the atchakat to one
Lingayat family whose descendents now own land in the head-reach.
One of the descendents, the most powerful farmer in the atchakat,
is a civil contractor, who has also taken construction contracts from
the MID. Thanks to his influence the sluice was replaced and repaired
and canals were extended and lined three decades ago. But even after
the repairs by the MID, only farmers related by caste and kinship,
who own the 12 hectares of head-reach land, have been allowed to
take water from the tank. In case the tank receives more than three

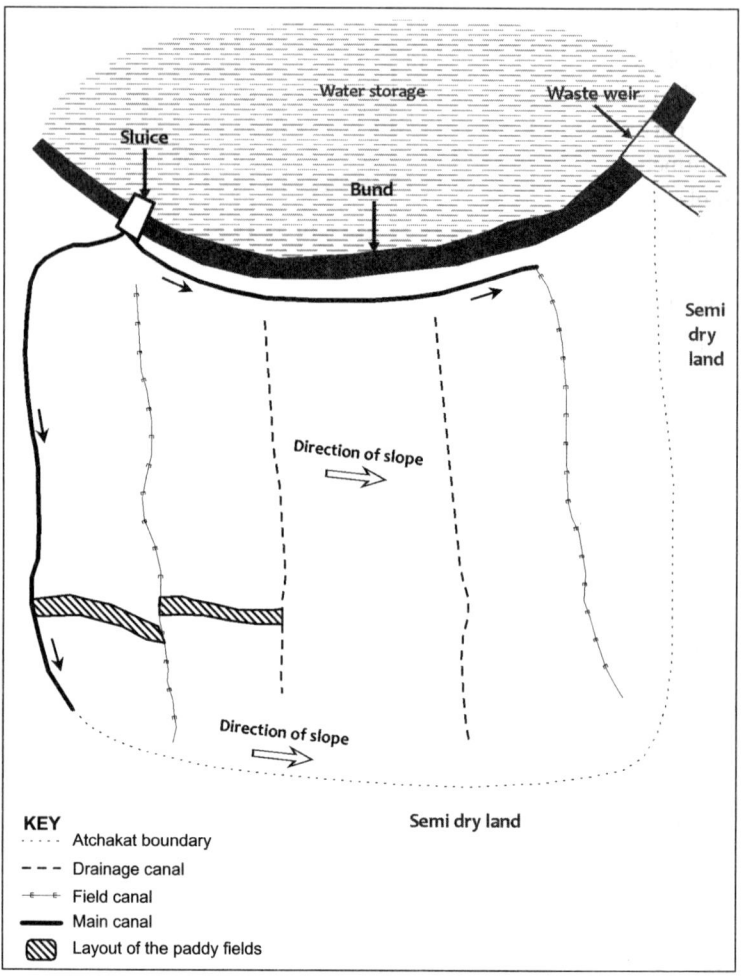

FIGURE 8.4   Sorturuhosakere Atchakat

Source: Adapted from Shah (2003).

metres of water, the tail-end farmers are allowed to irrigate semi-dry crops a couple of times but not to take water for paddy cultivation.

This shift in the crucial water distribution infrastructure in the tanks discussed above does not seem to be a case of the deterioration of physical structures as a result of bad management or maintenance. Instead, in all tanks, the disappearance of canals and sluice operating mechanisms was sudden, and related to the introduction of Green Revolution rice varieties and the shift in state irrigation policy. It is hard to comprehend why it would be difficult for farmers to replace this part of the tank infrastructure if needed. The plugs and poles are routinely replaced by farmers in other parts of the state, and once replaced, can easily last for twenty years. When I asked farmers how difficult it would be to replace the plug and pole, the answer was that 'it would be futile to do so as they will be stolen again'. Cleaning and mending of especially earthen canals do not require either sophisticated technical assistance or capital investment. Repair or reconstruction of any type of earthen or masonry work such as the embankment, the stone revetment, the sluice platform, or the waste weir superstructure are difficult operations, hard for farmers to handle on their own. Desiltation of the tank submergence area is another difficult operation, which may need an organized form of technical assistance. Farmers, however, routinely handle repairing and cleaning canals, and replacing plugs and poles of sluices.

The disappearance of essential infrastructure signifies an erosion of social relations of power that previously maintained them. Nadkarni, discussing the class character of the farmers' movement of the 1980s, noted that 'the landlords had a sense of security in the feudal order, which today's rich peasants do not necessarily have' (1987: 3). The disappearance of crucial physical structures marks the breaking down of the old managerial order, in which the elite class seems to no longer have a stake. Their earlier involvement with the management and maintenance of the physical structures earned them a right to be part of the decision-making authority, which in turn ensured that their interests were protected. Now, as long as they have land located in the privileged part of the atchakat that receives assured irrigation, they have not only been apathetic to management and maintenance but also, as the examples of tanks in Shimoga district show, embarked on destroying some of the structures to retain their privileged positions.

The erosion of the local power structure also has another side. On the one hand, the farmers' movement of the 1980s intensely affected the corridors of power at the state level and has since also influenced the ruling governments. On the other hand, it has resulted in the loss of external government authority even to nominally interfere and create a normative structure of order at the village level. This was perhaps most evident in Shimoga, the heartland of the new farmers' movement. One of the important strategies of the movement in the 1980s was to expose corruption in public life and oppose the seizure of farmers' property by government officers in lieu of loan recovery. This culminated in government officials being prevented from entering the village (Assadi 1997: 58–62).[8]

The agitations were a response to the high-handedness of government officers and their corrupt practices. The success of the actions further increased the power of the landed and elite farmers, especially sugarcane and paddy growers. This alliance, in my opinion, modified and suspended the normative structures of state authority at the local level. Irrigation infrastructure has been modified in other districts too, but not to the extent it has happened in the Shimoga district. The extreme case of Sorturuhosakere, where the atchakat has been reduced to landowners related by caste and kinship, eventually came to the notice of one MID officer. Thanks to his initiative, revenue officials visited the village and entered seven hectares of tail-end land in their *pani* books (land registers) as wetland. After this event, some of the tail-end land was allocated water for semi-dry cultivation. However, the tank is still monopolized by farmers from one extended family.

These changes in the distribution infrastructure are as much a result of shifting power dynamics at the local level as an outcome of the political economy of a changing cropping pattern; as discussed ahead, these are closely intertwined. There are two aspects of changing cropping patterns that play out significantly towards transforming the local power dynamics and eventually resulting into modifications of irrigation infrastructure. These will be discussed now.

Advent of the Green Revolution: Shift from Broadcasted to Transplanted Paddy

Farmers I interviewed narrated the change in terms of 'before, when only broadcasted paddy was cultivated, and after, when all farmers began to grow transplanted paddy of new varieties'. I maintain the same distinction to narrate the process of change over time.

As already discussed, tanks in the wet region of Shimoga have a small capacity compared to the tanks in other regions because tanks here fill up several times during one monsoon season, which is also the irrigation season. Further, tanks in this region have sufficient capacity for irrigating their atchakats once. Earlier, paddy was usually sown in the last week of May by throwing dry seeds on the land that sprouted after the arrival of the first rain. The sluices were opened by a *talwar* (a village servant) one and half months after sowing, with permission from the shanbhoga, and that too was postponed if there was enough rain. Water was supplied three times during the whole season. It was thus distributed intermittently and not continuously, and rotated between the tail-end and head reach. The presence of canals was absolutely crucial for rotation. Almost all tank-irrigated areas also had a seepage or drainage canal that carried excess water from the head reach to tail-end, rotated water between the head and tail-end, and carried irrigation return flow to the downstream tank. One round of irrigation took 15 to 20 days, depending upon the size of the atchakat. Sluices were operated during the irrigation round— opened in the morning and closed in the evening. For intermittent rounds of irrigation, the sluice operating mechanism was kept in order, as each round of irrigation would practically empty a tank. Sluices also had to be closed for storage of inflow till the next round of irrigation.

Irrigation now is provided as per the requirements of transplanted paddy. Sluice outlets are stuffed with gunny bags, paddy stems, stones, and mud before the beginning of the rainy season. They are opened at the time of land preparation and transplantation, when broadcasted paddy in the tail-end would be one and half to two months old and require water. After completion of the first round of irrigation and depletion of water in the tank, sluices are again closed with gunny bags, stones, and paddy straw until the second round. This method of sluice control is fairly labour-intensive, at times it even takes two days to entirely remove all the material stuffed inside the sluice openings. Only farmers growing transplanted paddy close and open the sluice outlets with the help of hired labour, whom they pay from their own pocket. Once opened, sluices are kept open for the rest of the season because closing them under water is very difficult. Transplanted paddy needs continuous supply and drainage during the maturing season except in the first month, or so farmers believe. Hence sluices are closed only at the end of the season. The disappearance of the canals and the sluice operating mechanism ensures that water is supplied to the tail-end field to field, only via the head-reach and according to the requirements of transplanted paddy, completely negating the possibility of rotation between head reach and tail-end.

Thus, the disappearance of crucial water distribution structures ensures that water is supplied according to the requirements of transplanted paddy grown in the privileged patches in the atchakat. This shift in tank designs emerged along with the change in social relations in the context of intensification of paddy cultivation.

POLITICAL ECONOMY OF AGRARIAN CHANGE AND TANK DESIGNS

Paddy-growing farmers in Shimoga derive their economic power from land other than tank-irrigated land. The groundwater level in the region is comparatively high (available at 50 metres) and instances of borewell failure very rare. The tanks I studied are relatively barren of borewells: one or two borewells in roughly 120 to 160 hectares of land irrigated by a tank. However, the number of borewells in the region outside the tank atchakat has been steadily going up.

The reason is that almost all landowners in the tank-irrigated area also have a piece of dry land, locally known as *hankalu*. Conventionally, subsistence crops such as *ragi* (millet), *jowar* (sorghum) and horse gram were grown on hankalu land, but in the last decade or so the cropping pattern on these lands has radically changed. Those who could afford it have invested in borewells on hankalu land to grow a variety of cash crops such as vegetables, cotton, maize, groundnut, betel nut, coconut, and banana, in addition to conventional subsistence crops such as coarse ragi, jowar, and horse gram. The process began almost 20 years ago when DCH 32 cotton first replaced subsistence crops. However, cotton at that time was rain-fed, given the fact that this region has a well-distributed and high amount of rainfall.

A more diversified cropping pattern based on extensive use of groundwater has developed only in the last decade or so. Most new crops are grown for the market and, as one landowner said that 'the real farming activities now have shifted to hankalu land'. A piece of land in the tank-irrigated area still fetches a much higher price than hankalu land because it assures at least one crop of paddy. Nevertheless, as my respondent said, the focus of agricultural activities has shifted to hankalu land as it provides more opportunities, especially after the arrival of Indo-American seed varieties of maize and vegetables. A favourable price structure—higher prices for output and lower prices for input—as a result of farmers' pressure at regional and national level has also made it possible, even for small farmers, to invest in cash crops (Nadkarni 1996), at least in the wet region.

Those who could afford it would install a borewell on hankalu land instead of on atchakat land, where only one or at the most two paddy crops could be grown. Consequently, two distinct cropping regimes have emerged in this region, one dependent on tank water and the other on rain and groundwater. These two cropping regimes remain separate in terms of their agricultural activities and water utilization patterns, unlike in other parts of Karnataka and South India, where borewell irrigation and tank irrigation have overlapped and clashed.

There is nonetheless a significant economic interrelationship between these cropping regimes. Income earned from cash crops grown on hankalu land is reinvested in tank-irrigated areas. More detailed research is needed to understand how the shift in cropping pattern on hankalu land has influenced tank irrigation practices, but prima facie it looks as if tank-irrigated paddy land provides insurance of one crop for subsistence and for the market, while hankalu land provides new economic opportunities.

* * *

In the debates on tank irrigation in south India, the colonial and postcolonial state is often described as guilty of both intervening and neglecting tanks, destroying them either way. The postcolonial state has been criticized for neglecting indigenous irrigation resources in favour of modern (medium, major, and well irrigation) schemes. Whether or not it intervened too much in or neglected tanks, the modern state emerges in these arguments as pervasively powerful and capable to drive society. Such state-driven approaches, culminating in the advocacy of community-based resource management, not only pre-suppose state and society as two distinctly separate and undifferentiated spheres, but as a result also tend to visualize social change as unidirectional. Such approaches strip the various components of society of their agency and oversimplify struggles for domination spread through society's multiple arenas by portraying them as struggle between (an oversimplified version of) community and an all-powerful state. Furthermore, the idea of community is increasingly tied down to the idea of traditions, customs, and pre-modern forms of knowledge.

A number of scholars have criticized such notions of community adopted by irrigation reform policies and backed by international agencies as only remotely based on the realities of social arrangements around resource utilization. These critics (for example, Harriss 2002) show that, in stark contrast to the World Bank's 'community' as an abstract group with social capital (solidarity, interpersonal trust), community arrangements are embedded in hierarchical, discriminatory, and conflicting social relations. Discussing tank irrigation in south India, Mosse (1997b, 1999) regards such idealistic approaches to customs and traditions as 'imagined' and shows how the colonial government invented the idea of community management for its policy on tank administration.

Notwithstanding the fact that these critiques draw attention to a fundamentally important point, they often replace solidarity and trust among communities with hierarchies and differential access to resources. With some exceptions (for example, Mosse 1997a), these studies rarely discuss the component of knowledge as part of community formation. The critique of the social composition of community, while important to acknowledge, does not sufficiently help to comprehend the processes by which patterns of natural resource utilization are created, transformed, and reproduced, and the role that knowledge plays in these processes. Both the simplistic notion adopted by the irrigation reform policy and some of its critiques assume community as an administratively and culturally bound space, which acquires its identity only when contrasted with the space occupied by the state. The idea of knowledge similarly tends to be static, either generated through intrinsic traditions and customs or imposed through external state-sponsored processes. This chapter has gone beyond such bounded notions of community and state and static ideas of knowledge by arguing that both irrigation policy and tank technology acquire their characteristics and meaning in the dynamics of interaction spread through society's multiple arenas, including various knowledge traditions.

I have shown in detail that landed farmers from rural India have interacted with political processes of development in the most organized fashion in the form of the new farmers' movement. This has had far-reaching consequences for agrarian policy at the national and regional levels and for the way tank technology is modified and used at the local level. I have argued that these newer and potent forms of political interaction spread over multiple arenas of state–society interaction can only be recognized by challenging the unified and contrasting notions of state and community.

Furthermore, this chapter demonstrates that not all farmers are passive recipients of policies developed by the state agencies. It corroborates the argument that a politically expedient group of farmers has been actively driving agrarian change in a particular direction by means of making specific productive and knowledge choices (Shah 2008b). In fact, it is in the making of these everyday forms of technological and other choices, derived from multiple sources and traditions of knowledge, that the relationship between various sections of community and between community and state is sharply materialized. The chapter, thus, demonstrates the ways in which technological designs of tank irrigation are the sites where mutual interaction and contestation between different social groups and the state, as well as between different knowledge traditions, take place.

Notes

[1] The procurement policy of 1966 was meant to procure foodgrains below the open market price, to support and subsidize the availability of cheap food mainly to the urban poor through the public distribution system (PDS). In this policy, paddy cultivators had to surrender part of their produce to the government. For further discussion, see Mooij (1998).

[2] Apart from a procurement levy on rice growers, the government also imposed a restriction on the movement of grains from surplus-producing areas to deficient areas, either within one federal state or between states. Enforcement is easy in the surplus rice-producing areas: if supply is larger, the prices are lower and procurement easier (Mooij 1998: 93).

[3] The Agricultural Price Committee (APC) was constituted in 1964–5 to advise the government on price policy and to recommend a minimum support price (MSP) after taking into consideration the cost of production. Its formation preceded the introduction of Green Revolution wheat and rice varieties. The APC was later replaced by the Committee on Agricultural Costs

and Prices (CACP). Under the political pressure exerted by agitating farmers in Karnataka and elsewhere procurement prices have increased over the last decades. Even in the 1990s they continued to increase and are consistently set higher than those recommended by the CACP (Dev and Mooij 2002: 64).

[4] A fixed rate per acre is collected as an irrigation charge from all cultivators receiving irrigation, whereas the water charge is collected every season depending on the crop cultivated.

[5] The ZP is the elected district-level political and administrative unit of a three-tier structure of local government.

[6] The act mentioned in the text was called 'Karnataka Zilla Parishad, Taluq Panchayet Samithis, Mandal Panchayets and Nyaya Panchayets Act 1983'. It came into effect in 1985. After the 73rd constitutional amendment, it became the Karnataka Panchayati Raj Act in 1993.

[7] The categories of wet and mixed regions are constructed to include the interaction between ecological and historical characteristics (see Shah 2003).

[8] Local bureaucrats, especially from the irrigation and (RDs), have been targets of farmers' fury during these agitations. Their offices and files were burnt at times as part of the agitation (see Nadkarni 1987: 92). The government officers who came to sequester farmers' property to recover loans were heckled routinely, and at times locked up following the strategies adopted by Punjab farmers. Many villages were declared as no-entry zones for government officers (Assadi 1997: 58–63). Pitched battles were fought between the KRRS squad and government officers. At times some officers refused to enter certain villages (Assadi 1997: 62).

9 PROPERTY RIGHTS, WATER RESOURCES, AND TECHNOLOGY: THE MISSING ECOLOGICAL LINK

Jyothi Krishnan

This chapter discusses how property rights over land and water combine with the use of modern water extraction technologies to result in unsustainable and inequitable distribution of water in southeastern Palakkad district in the state of Kerala, India (Figure 9.1). This region has witnessed an intensification of paddy cultivation, especially with the introduction of modern reservoir-based canal networks since the 1960s. There has also been diversification into commercial production of coconut, bananas, and areca nut. Large canal systems have become increasingly dependent on water supply from distant reservoirs facilitated through inter-basin transfers, as water supplies in local water sources like rivers, streams, and ponds are dwindling. Increasing individual control over water sources and water-lifting technologies has also resulted in unsustainable, inequitable, and ecologically harmful water use patterns. Water scarcity is manifest in drinking water shortages and frequent crop failures.

Taking the Varayiri watershed as a case, this chapter looks into the interplay between property and technology in the management and use of water resources, and the lack of an ecological orientation in development, which has resulted in increasing water scarcity. The findings presented here are based on a qualitative survey of changing land and water management practices undertaken in the Kollengode and Elavenchery panchayats in Chittur taluk in 2000 and 2002. Data were derived from detailed farmer interviews on land and water management practices, supplemented by cross-season field-level observations of water management, and irrigation distribution practices around different water sources. Given the skewed nature of land holdings in the area, farmers from all landholding categories were interviewed.[1]

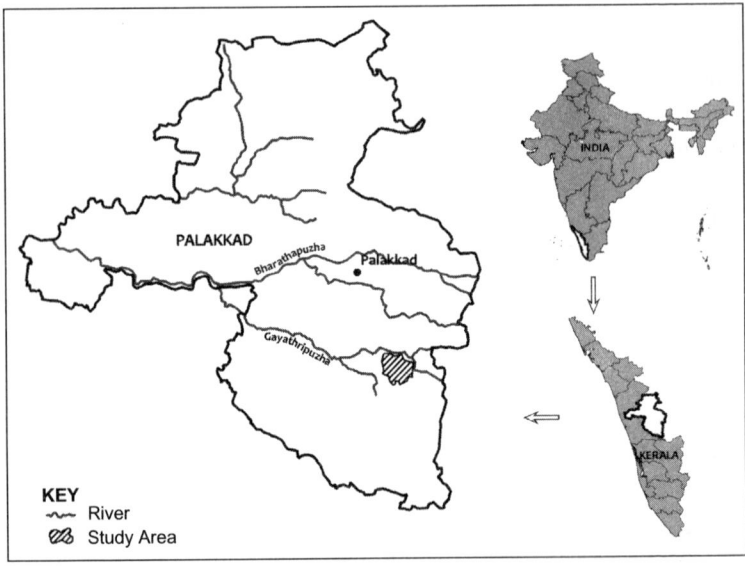

FIGURE 9.1   Kerala, Palakkad District, and the Study Area

Source: Author.

The chapter first introduces the Varayiri watershed, describing briefly the changes in irrigation technologies, cropping patterns, and broader land use. This is followed by a discussion of the interplay of property rights and ecology as they are seen evolving in water use patterns under distinctive and diverse sources of water found in the watershed, ranging from traditional ponds and shallow wells to more recent large-scale irrigation systems and tube wells. This discussion examines the operationalization of private and public property in the case of water, and the role of individual control over energized pumping devices in strengthening private control over water resources, resulting in both inequity and unsustainability. Finally, the chapter looks at possibilities for reform in the context of ongoing decentralization initiatives in the state, ending with a short conclusion.

THE CHANGING WATER SCENARIO IN KOLLENGODE AND ELAVENCHERY

The Varayiri watershed, covering some 1,450 hectares, drains into the Gayatri river, a tributary of the Bharathapuzha river. It is located in the northern parts of the panchayats of Kollengode and Elavenchery in Chittur taluk in southeastern Palakkad district, which is considered one of the rice bowls of the state. Water scarcity has been reported from both panchayats, the government having had to compensate paddy farmers for losses incurred from failed crops on a number of occasions.[2] Water scarcity in this part of the district has long been taken as a given, being correlated with the unique agro-climatic features of the region.

The eastern part of Palakkad district is located in the Palakkad Gap, a unique geological formation where a 45-kilometre-wide abrupt break occurs in the otherwise unbroken run of the Western Ghat mountain range (900–1,500 metres) that marks the eastern state boundary. The Western Ghats create a barrier to the eastward-blowing moisture-laden winds from the Arabian Sea, stimulating rainfall except in this gap region. Not only does this area receive less rainfall and higher temperatures than elsewhere in the state. It is also subject to continuous strong dry winds that funnel through the Gap from the neighbouring Coimbatore plains in Tamil Nadu. The annual rainfall recorded at Kollengode during the study period was only 1,459 millimetres, compared with the state average of 2,500–3,000 millimetres annually (mainly June–September during the southwest monsoon). Maximum summer temperatures recorded at Palakkad have often crossed 40 degree Celsius (Government of Kerala 2003).

Seasonally concentrated rainfall (mostly during the southwest monsoon), in addition to the relatively lower amount of rainfall, makes water economy crucial. Early trials with irrigation in this region become manifest in a high number of ponds or tanks, referred to as *kulam*. Kollengode is the panchayat with the largest number of such traditional water harvesting structures (363) in Palakkad (Kerala State Land Use Board 2001). These kulams typically consist of an embankment built across the line of drainage so as to hold back surface run-off. On average, they cover an area ranging between 0.4 and 1.2 hectares, with a central depth ranging between 2 and 3 metres (6–10 feet). The embankment or the main bund known as *varambu* is supported by two other bunds on either side leaving the remaining side open along the slope. Sluices are placed on the main bund to regulate the outflow of water. The kulams allow storage of intermittent surface run-off (for later use for irrigation, watering cattle, and domestic purposes) and enable percolation of water to supply locally excavated depressions and pits known as *kuzhi*. A total of 150 such kulams had been carved out in the Varayiri watershed alone, on average one pond every 100 square metres.

The ponds were an integral part of the agricultural landscape and earlier irrigation history of the district. Water use in the command area of each pond was determined by the existing agricultural land classification system, which was in turn defined by the drainage pattern in each micro catchment. Paddy lands were classified into two broad categories, *kalayi* and *potta*. Kalayi lands refer to double or triple cropped paddy lands located in the valley bottoms, while potta refers to single cropped paddy lands located on gentle slopes.[3] Kalayi lands had access to water stored in the pond for additional crops of paddy whereas potta lands did not.

The sluices were designed in a way that suited this land classification system. Water stored in the pond was supplied through sluices (*ovu*) placed at higher and lower levels, called *mele ovu* (higher sluices) and *kizhe ovu* (lower sluices). The high-level sluice is placed on one of the two supporting side bunds of the pond. When a pond is full of water, water flows out through the mele ovu to the fields located on either side, which are mostly potta lands, located on par with the water level in the pond. Once the water level recedes, water will not flow out through the mele ovu into the potta lands. Hence, if potta lands enjoyed water rights at all, it was only to the water flowing through the higher sluice of the tank. As a result, potta lands were cultivated with dry crops during the second cropping period. In some cases, lifting water was permitted, but since it was manual lifting, this option was resorted to only in cases of great urgency.

The lower sluice, on the other hand, located on the main bund, opened out into the kalayi lands located directly below the pond, along the main drainage line. Moisture-retaining kalayi lands had access to a greater amount of water and double cropping of paddy was confined to kalayi lands alone. This helped to minimize the incidence of crop loss and maximize the chances of a good crop on kalayi lands. With ponds storing only a limited amount of water, adhering to this land classification system was important in ensuring a second crop of paddy. An early commencement of the second crop in order to ensure an early harvest (before the onset of the summer months) was also a critical feature. Kalayi–potta was not just a system of land use classification, but also

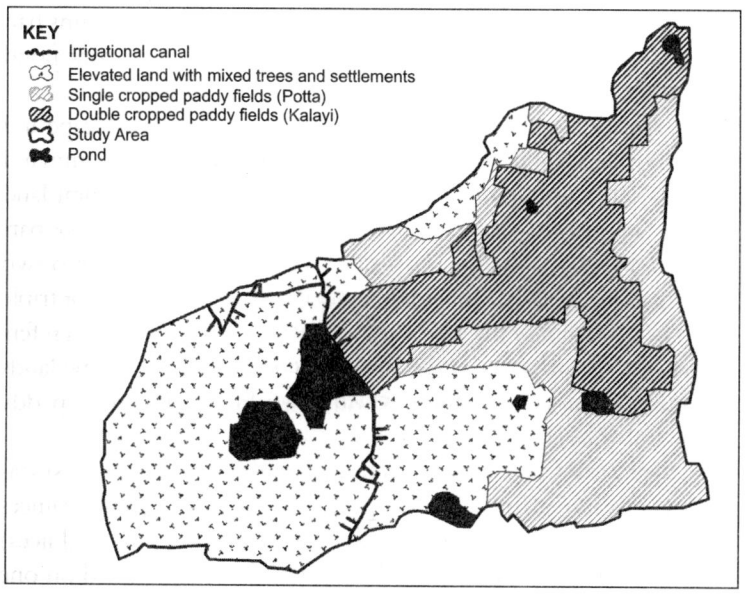

KEY
〜〜 Irrigational canal
⌇⌇ Elevated land with mixed trees and settlements
▨ Single cropped paddy fields (Potta)
▨ Double cropped paddy fields (Kalayi)
◌ Study Area
▰ Pond

FIGURE 9.2   Catchment and Command of the Velanganpadam
Pond, Kollengode Panchayat
Source: Author.

a system of allocating water rights. As an example, Figure 9.2 shows
the land use configuration around the Velanganpadam pond which,
despite changes in cropping patterns and water supply, still shows this
kalayi–potta distinction in land use and water rights.

This pattern changed significantly with the introduction of the
Chulliar dam and its canal network [part of the Gayatripuzha irrigation
system (GIS)] in the area in the late 1960s, and with the intensified use
of energized water lifting mechanisms from the 1990s. The introduction of large-scale irrigation targeted the expansion and intensification
of monsoon-based paddy cultivation into double cropped lands. The
left bank canal (LBC) of the GIS traverses the southern ridge of the
watershed, which comes under the command area of four distributary
canals.[4] The introduction of energized lifting devices also enabled
lifting of water from ponds on to potta lands in other seasons. The
kalayi–potta distinction no longer mattered in access to irrigation
water, except for pond water. In addition to surface sources, tube wells

now constitute an important source of irrigation. Between 1960 and 2004, the area sown more than once under paddy grew from 31 to 63 per cent of Palakkad district (Government of Kerala 1971, 2005), with the greatest increase occurring between 1960 to 1975, when canal systems were introduced in the district.

The other important change with respect to irrigation has been the marked increase in area cultivated under perennial crops such as coconut, banana, and areca nut[5] since the 1970s, particularly under lift irrigation. In 2001, these perennial crops covered approximately 34 and 20 per cent of the total agricultural land in the Kollengode and Elavenchery panchayats, respectively. The cultivation of these crops requires intermittent irrigation, which intensifies during the relatively dry period of December to April, when surface water flows and storage are at their lowest. As canal and lift irrigation has extended water access, production of these tree crops has spread into the upland *parambu* lands (lands upstream, which could not be irrigated) and can replace paddy crops. Banana cultivation is mostly undertaken on paddy land, resulting in permanent change of land use and multi-season water demand.

While the availability of pumps has facilitated the cultivation of these crops, their cultivation has in turn intensified the need for pumps, fuelling a vicious cycle of changing cropping patterns and increased water consumption, particularly during the second crop season for both paddy and perennial tree crops. Although the advent of the reservoir-based canal network initially supported double-cropping of rice over larger areas, since the 1990s, water supply has become increasingly unreliable. As a consequence, pumping of water from ponds, streams, and wells is increasingly used to meet the water requirements of the second crop. Where canals flow near ponds, water may be diverted into the pond, or lifted onto nearby land that was once potta land.

With these changes, Palakkad is now the chief paddy-producing district, and has seen the least decline in paddy production across Kerala as other perennial crops have boomed. However, despite intensification of paddy over a larger area since the 1950s, this cropping pattern has also declined to make way for other crops, as shown in Table 9.1 (see Krishnan 2009).

In the two panchayats studied, there has been limited conversion to banana stands, but in Elavenchery, some 6 per cent of paddy

TABLE 9.1   Area (in hectares) under Major Crops in Palakkad
District, 1975–2005

Crop	1975–6	1995–6	2004–6
Paddy	185,182	135,630	118,701
Coconut	16,994	48,336	55,533
Banana	587	4,413	10,705
Other plantains	3,483	3,409	6,871
Pepper	851	3,460	7, 305
Ginger	383	1,239	936

Source: Government of Kerala (2001, 2007a).

land has been converted to coconut and other crops (a mix of areca nut, banana, coconut, peppers, vegetables), perhaps because of its tail-end position in the irrigation system. However, as mentioned earlier, cultivation of tree crops has expanded with possibilities to lift water from canals, streams, and ponds, to make up one-third of the cultivated area of Elavenchery and one-fifth of Kollengode. Figure 9.2 gives an example of contemporary land use in the catchment and command area of a pond.

While changing irrigated agriculture provides insight into the manifestation of increased water use and water scarcity, the larger ecological context, which is also undergoing rapid change, is often overlooked. Deforestation has been the most notable factor. Early in the nineteenth century the now almost dry areas of the Palakkad Gap were covered with rich forests (Nair 1991). The numerous scattered hillocks in the Palakkad plains were still covered with dense vegetation as recently as 1970, prior to the government takeover of privately owned forests. The Varayiri watershed is no exception. The nationalization of private forests in the 1970s, the spread of housing clusters on sloping terrain that was hitherto thickly vegetated, and the increasing spread of banana and coconut cultivation have cumulatively reduced the tree cover in the area. While there are no quantitative estimates of deforestation for the concerned panchayats, broad estimates for the region show a reduction of forest cover from 60 per cent of the total area in 1905 to about 6 per cent in 1973 (Nair and Chattopadhyay 1985, cited in Sooryamoorthy 2003).

Coupled with deforestation, other notable land use changes with an impact on the water regime of the area have been the use of paddy land for non-agricultural purposes (notably house-building), the bulldozing of scattered hillocks and levelling of sloping lands, and sand mining from paddy fields and pond beds. These changes are widespread in the study area, Palakkad district, and much of Kerala.

The cumulative impact of all these changes on the water resources of the region has not been investigated so far, and such a quantitative study was beyond the scope of this study. Yet understanding this changing ecological milieu is important, as local irrigation sources have undergone degradation, visible now in silted up ponds with resultant reductions in storage capacities, reduced flows in streams and rivers, and falling water levels in wells. Canal supply during the second crop period, when water requirements are high, originally partially buffered farmers from degradation of local water sources. However, the increasing unreliability of canal water supplies during the past decade has brought the deterioration of local water conditions to light. Pond water levels are now at an all-time low during the summer months, and well-to-do farmers have invested in tube wells as a solution. The panchayats have also initiated schemes and projects to redress water scarcity, most of which focus on lift irrigation projects exploiting available ground or surface water, and check dams to help recharge. All of these are desperate short-term efforts to cope with the water crisis, while no attempts are being made to redress the wider ecological degradation that has set in. Just as irrigation expanded without real reference to traditional systems, there are still no integrated measures to restore the ecological integrity that could support these systems. By showing the changes in property rights allowing this land use change and ecological degradation, this chapter hopes to encourage quantitative studies of changes in the water balance that also relate to equity of access to water.

WATER USE PATTERNS: PROPERTY AND ECOLOGY

This section discusses water use patterns from the viewpoint of property rights that mediate the interaction between people and water, focusing specifically on the impact of property rights on the sustainability and equity of resource management. Amongst the institutional arrangements that mediate the human use of natural resources, property has been recognized as one of the most significant (Benda-Beckmann 2001; Berkes 1989; Hanna *et al.* 1996; McCay and Acheson 1987), determining not only who may use which resource and in what ways, but also shaping the incentives people have for investing in and sustaining the resource base over time (Meinzen-Dick and Pradhan 2001: 10). Given the long historical interrelationship between land and water rights, it has been argued that efforts at addressing agriculture-related water needs must take into account the complex interlinkages between land tenure and water rights (Cotula 2006). In most customary systems of land and water use, the right to use water has been dependent on the property rights to land (Hodgson 2004). The riparian law of water rights operational in many countries is, for instance, dependent on the property rights to land adjoining rivers and streams. Similarly, rights to the use of groundwater are linked to the rights to land under which it is located. Land, therefore, has been considered as the most important property variable that determines the property character of irrigation water (Abeyratne 1990: 20). Abeyratne also suggests that, while research-ing property in water, especially in irrigation, a 'back door' approach through research on land rights be adopted.

Farmers in the study area meet their irrigation requirements from multiple irrigation sources: canals, ponds, streams, shallow wells, and tube wells. Public and private property classifications of water in different irrigation systems determine the extent of control that individual farmers exercise over these water sources. These public–private categories over water are largely derived from the existing system of land rights.

Ponds and wells are located on privately owned land, and therefore classified as private property. The stream is considered as common or public property. The canals are clearly classified as public property, mostly because it is the state that invested in the irrigation infrastruc-ture of reservoirs and canals, in the process acquiring the land on which these are situated. This situation is further complicated by the fact that water shifts from one property regime to another as it moves through interconnected surface and groundwater irrigation systems.

Property categorizations have been critiqued for creating an inadequate understanding of the ecological properties of the resources concerned (Klug 2002). Being rooted in the superiority of human entitlements, they pay little attention to the need for long-term protection of the resource (Freyfogle 1996; Klug 2002). With regard to land rights, Freyfogle (1996) points out that rights to a parcel of land are not framed with respect to its specific features like soils, terrain, and vegetation. As a result, the land use options pursued by the landowner need not take into consideration the natural features of the land, and can therefore result in ecologically unsustainable land use patterns. This holds true for water as well. The existing public and private classifications of water and the differential access they sanction pay no attention to the unique hydrological properties of the resource.

Water supplied from the Chulliar dam through the network of main and distributary canals and brought to fields through field channels is considered public water, or *sarkar vellam* (literally, 'government water'). In terms of access to water, this implies that all farmers in the command area are entitled to receive a share, the infrastructure (reservoirs, canals, and regulatory devices) being the property of the state irrigation department. Misappropriation of this water can, therefore, be viewed as a violation, and can be contested by the deprived users. Water in the pond or tank is regarded as the private property of those who own land in its command area. Public or government-owned ponds are few in number. Ponds are public property only when their ownership has been surrendered by the concerned owners to the government. This did take place in the case of a few ponds during the land reforms in the 1970s, when large landowners surrendered land on which ponds were located to adhere to the stipulated ceiling limit.

In the case of privately owned ponds, rainwater, when channelled into the pond, becomes the private property of those who own land in its command area. These ponds are better classified as shared private property or common property of the concerned landowners. They become individual private property when the land in the *ayacut* (command) area of the pond is owned by one single individual. The right holders of water in the pond have the right to exclude non-members from taking water for irrigation. Like ponds, shallow pits (kuzhis) and wells (both shallow and deep) are also treated as private property. However, unlike ponds, usually these irrigation sources belong to a single owner or to a family.

The stream, which constitutes yet another important source of irrigation, is both common and state property. The stream can be considered a common property resource of riparian landowners, for they stake the first claim to the water in the stream. Non-riparian landowners can take water from the stream only through the lands of riparian landowners, for which they require the latter's consent. The stream can also be considered public or state property, as the use of water from the stream for irrigation purposes is regulated by the state. While riparian landowners are entitled to use the water in the stream for irrigation (by diversion), existing state irrigation laws forbid the pumping of water from streams and rivers. Farmers evade this restriction by digging wells on riparian land and pumping water from them. Theoretically, water in a well is private property as it is located on private land, but in reality it is water from the stream that seeps into the well. Farmers, therefore, acquire the necessary departmental sanction for the installation of electric pump sets by showing the well as the water source, and not the stream. In addition, mobile diesel pump sets are indiscriminately used by riparian landowners to pump out water directly from the stream, as their installation requires no departmental sanction.

Farmers thus evade existing rules by appropriating common waters within wells located on their privately owned holdings. The same is observed when canal water is stored in ponds. While water supplied through the canal network is considered public property, once this water is diverted into ponds it is transformed into the private property of the concerned right holders. Many farmers engage in this practice of discretely diverting public canal water into private ponds. This practice gives more water to those with access to ponds, but further deprives those without. Therefore, the most coveted irrigation sources are privately owned ponds and wells, primarily because they ensure timely access to water. While canal water meets the water requirements of paddy (if supplied as per schedule), timely access to canal water is dependent upon many factors, including bureaucratic functioning and the scale of misappropriation by wealthy farmers able to divert the water (see Krishnan 2009). Ponds and wells may not meet the entire water requirement of a paddy crop, but they provide critical interim relief, particularly when canal supply is unreliable. More importantly, owners of these sources are free to extract water as they please.

In a nutshell, these property rights, and the differential access that they provide, do not take into account the hydraulic interconnectedness between different water sources, nor the cumulative impact of water extraction from them. I return to this later in the chapter.

THE SILENT (PUMP) REVOLUTION

The advent of energized water lifting devices in the area since the 1970s has extended possibilities of resource extraction and strengthened private control over resources. These modern technologies lead to unrestricted exploitation of water from privately owned water sources like tube wells and ponds, and are capable of transforming public water sources like streams into open access sources. Pump sets have facilitated individual access to water, and provide much needed flexibility and individual control over application of water compared to the bureaucratically managed canal water distribution practices of the Chulliar system. Individually operated pump sets are more numerous here than group lift irrigation schemes, which irrigate fields belonging to a group of farmers.

Kerosene or diesel pumps, which appeared around the 1960s, were soon followed by electric pump sets in the 1970s. Both are used to lift water from surface streams and ponds, and groundwater from wells (both shallow and deep) and kuzhis. In the 1960s, generous state subsidies were made available for installing pumps, at the rate of 25 per cent for electric engines and 50 per cent for diesel engines (Frankel 1972). In Palakkad district, the number of pump sets rose from 79 in 1964–5 to 1592 in 1967–8 (a 20-fold increase in three years). In 2001, there were 245 energized pump sets for irrigation purposes in Kollengode and Elavenchery panchayats, ranging from below 1.5 horsepower to above 5 horsepower (Kerala State Land Use Board 2001). To promote lift irrigation from smaller water sources (ponds, wells, streams, and rivers), a scheme for the free supply of pump sets was introduced in the 1970s (State Planning Board 1976: 1). This was justified on grounds of increasing food production by assuring a third crop, as well as enhancing the utilization of under-utilized water resources, implying that water sources like ponds and groundwater were indeed under-utilized.

The State Planning Board (1976: 4) also recommended the urgent electrification of pump sets, calling for special budgetary provisions by the Board, if necessary.

Currently, the electricity costs are borne by the government. The farmer has to bear the initial installation costs (of the electric motor and construction of a pump house). Departmental sanction is required for this, with the agricultural officer testifying that the water source to be pumped is privately owned. This certificate is required before agreement is given by the Kerala State Electricity Board for connection of an electric pump set.

Farmers report that they have had to pay bribes to get this certificate and also for speedy processing of formalities. Some farmers incur the expense of installing additional electricity poles if their holdings are far removed from the existing distribution network. Since the initial installation costs are high, it is mostly the large and well-to-do farmers who invest in these connections. Medium and small farmers invest in diesel engines or hire these as needed. The installation of diesel power sets is now far less common, given costs of pump sets, diesel fuel, and approval to operate. Installation of tube wells requires a sanction from the revenue division officer (RDO). While it is easy to obtain a sanction for pump sets under 3 horsepower, larger pumps face more difficulties.

The digging of tube wells is unregulated and left at the discretion of the individual landowner. While one or two tube wells appeared in the 1970s, they have come to prominence for irrigation only since 1990. Before1990, only two hectares of land in the Kollengode panchayat and none in Elavenchery panchayat were irrigated by tube wells (Census of India 1991). By 2005, the number had increased to 119 and 149 tube wells, respectively, in these panchayats (Government of Kerala 2006).

The revolution brought by energized pumping of surface and groundwater resources can be seen from the areas irrigated. District-level data on net irrigated area for Palakkad district shows that during the 1995–2005 period, while the area irrigated by government canals and ponds had declined, the area irrigated by private canals, private wells and tube wells had increased (Government of Kerala 2001, 2007a). Though corresponding figures at the panchayat level are not available, the trend could well be the same.

ECOLOGICAL IMPLICATIONS OF
WIDESPREAD WATER LIFTING

This section looks at these changing impacts on water sources, as experienced by farmers interviewed. During the second crop season, pumps work non-stop in the lower reaches of the Varayiri watershed, where canal water supply is inadequate (being located in the tail end), to lift water from the Varayiri stream and the Gayatri river. Prabhakaran, a farmer from the Valluvakundu area remarked how during the second crop season one can see a row of diesel engines along the stream, pumping the stream till it is dry. The increasing intensity of water withdrawals from streams and rivers is often justified by the misplaced logic, endorsed by many a riparian landowner, that water that flows downstream is 'wasted', and hence should be used as much as possible. As Mallika Rajan, a farmer from Peringotukavu, put it: '[I]n all streams and rivers, water is running wastefully into the sea, and it is of no use to anybody.'

This reveals the lack of appreciation of the ways in which streams, rivers, and groundwater are hydraulically connected. Neither the panchayat (in whose control the streams have been placed) nor the minor irrigation department (MID) have initiated any assessment of the cumulative impact of individual withdrawals from the stream. In addition, the implementation of lift irrigation schemes irrigating between 50 and 100 hectares now exerts added pressure on declining flows in the river network.

The pumping of water from ponds irrespective of the topography-based land use classification system is reported to have reduced water levels in many ponds. Potta lands now enjoy the same extent of water access as kalayi lands. However, the soils of potta lands require more frequent irrigation during the second crop of paddy, increasing water consumption. This intensified pumping not only reduces storage in the ponds, but also affects water levels in nearby wells and kuzhis. Where canal water can periodically supply ponds, the reduced storage is partially masked during the second crop season when the need for irrigation is at its peak. Elsewhere, storage in ponds declines more quickly, as in the poorly-served tail-end or outside of the canal network.

With regard to groundwater exploitation, farmers from certain areas report how drilling of tube wells in proximity has lowered water levels in shallow wells in that area. Madhu from the Tachakora area in Kollengode panchayat, for instance, remarks that the water in the open well in his house compound never used to dry up until these tube wells were dug in the vicinity. Similarly, Mukundan in the low-lying Velampotta area observed that the digging of two tube wells by a neighbouring landowner to irrigate his coconut plantation, led to falling water levels in his pond and well, so much so that he had to dig another well for drinking water purposes.

While there are no clear estimates of groundwater withdrawals from the panchayats studied, the adjoining Chittur block panchayat has been notified as overexploited by a recent estimate of groundwater withdrawals in the state (Bijoy 2006). Groundwater exploitation could also impact stream flows, and vice versa too, when streams are hydraulically connected to groundwater aquifers (Burke *et al.* 1999; Glennon 2002). In such cases, water will move between stream and aquifer depending on the water table in the surrounding aquifer, with streams recharging groundwater when water levels in the aquifer are lower than the stream, and receiving water when water levels in the aquifer are higher. Pumping from either of these sources therefore affects water levels in the other. In the Varayiri watershed, more work is needed on how many streams are hydraulically connected with local aquifers, and on the impact of intensifying groundwater extraction on declining base flows in streams and rivers, particularly during the summer months.

There are neither estimates of the total abstractions from ponds, streams, and wells, nor a quantitative understanding of how pumping from one source affects water flows in others. The neglect of small water sources like streams and ponds in flow measurements makes it difficult to assess the kind of pressures that might be building up on these sources, which has important implications for local water scarcity. In the study area, extraction of water from ponds and wells (both shallow and deep) is left to the discretion of the concerned owner(s). Even in the case of a common property resource like the stream, there is little regulation.

Assessing locally available water supplies and the total demand for water in each region is becoming increasingly relevant, not just in the specific study area but in the entire Palakkad district, to ensure that water consumption does not threaten the long-term sustainability of the water sources concerned. In all parts of the district that are irrigated by water supplied from distant reservoirs, the deteriorating functioning of reservoirs casts doubts on the long-term availability of 'external' water. Locally available water resources therefore assume critical importance for meeting the water requirements of each region.

INEQUITIES IN LAND OWNERSHIP: INEQUITIES IN ACCESS TO WATER

A survey of water access patterns in the area revealed that access to water was unequally divided between landownership categories. As mentioned earlier, farmers were classified into three broad landholding categories: large, medium, and small (see Note 1). A sample survey of water access patterns in these categories revealed that large farmers had greater access to privately owned sources of irrigation than farmers in the other categories. Taking the case of access to ponds alone, 89 per cent of the large farmers had access to ponds, as against 48 per cent of the medium farmers and 23 per cent of the small farmers. In the case of access to groundwater, 35 per cent of the sampled large farmers owned tube wells, while only 6 per cent of the medium farmers and none of the small farmers did. Of the 35 per cent of large farmers who owned tube wells, 18 per cent had dug more than one tube well.

The linkage of water rights with land rights and the resultant development of private reserves of water has led to a situation wherein the majority of large landowners have access to multiple water sources. A larger proportion of large farmers had access to more than one privately owned source of irrigation (a pond and a tube well, a pond and a shallow water pit, or a pond and a shallow well). While 58 per cent of the large farmers benefited from such access, only 8 per cent of the medium farmers and none of the small farmers did. This indicates the existence of a high bias between farmer groups with regard to access to water.

It is pertinent to note that present-day inequities in access to ponds is a result of the implementation of land reforms in the area in 1970. The issue of redistribution of water was ignored in the land redistribution, as a result of which the former landlords and large tenants retained land that had access to ponds. The land that small farmers received as a result of redistribution rarely had access to ponds (see also Krishnan 2009). This inequity was aggravated as larger landowners were also able to invest in tube wells and pumps. The phenomenon of storing public canal water in privately owned ponds reproduced and increased this inequity.

INDIVIDUAL CONTROL OVER WATER LIFTING TECHNOLOGIES

As pumping is expensive, inequities in water use arise from the differences in purchasing power between well-to-do farmers and those less well off. Since not all farmers can be 'self providers' with their own pump sets (Wood 1999: 782), inequities result (Wilson 2002). The 'self-providers' in the study area are not a homogenous category: owners of electric pump sets are at an advantage compared to owners of oil engines, as the running costs of the former are lower. Most medium farmers own oil engines, which are fuelled by diesel or kerosene. Most small farmers who own oil engines can do so only because of government subsidies. Many medium and all small farmers hire oil engines rather than own one.

The costs incurred by farmers who hire oil engines ranged between INR 50 (roughly 1 USD in 2002) and INR 70 per hour of pumping, which included the rental charges for the pump sets and fuel cost. Most medium farmers and all small farmers hiring oil engines were found to calculate whether this amount was worth spending. As a result, all small farmers invest money in pumping only when it is certain that without it the paddy crop will fail. Most often pumping was resorted to at the flowering stage, when water availability was critical. Not a single small farmer pumped water in order to apply fertilizers on time. While they knew that not doing so would reduce their yields, they felt that the costs of pumping did not justify the benefits. Paddy yields in the area average 3,000 kilograms per hectare, with wide variations between large and small farmers. Many small farmers suffer poor yields due to the non-availability of water, with yields falling to a mere 1,000 kilograms per hectare (550 kilograms per acre). The field survey reveals the resultant inequities: while 94 per cent of the large farmers resorted to pumping to irrigate their crops, only 59 per cent of the medium farmers and 28 per cent of the small farmers did so.

This differentiation in water security is shown by the fact that, while only four per cent of the large farmers and the medium farmers suffered crop loss in the second crop season (that is, they lost their entire crop), 28 per cent of the small farmers suffered crop loss. The large and medium farmers who suffered crop loss were concentrated in the tail-ends of the irrigation canals, while the small farmers were more spread out over the study areas. Similarly, while no large farmer suffered from excessively poor yields, four per cent of the medium and 43 per cent of the small farmers did. A poor yield is defined here as a yield of less than 550 kilograms per acre (roughly 1,300 kilograms per hectare). These 550 kilograms comprise one truckload of harvested grain, and farmers were found to refer to one truckload of paddy per acre as the lowest or minimum relevant yield.[6] Energized lifting of water, therefore, has not only revolutionized the quantity of water available to farmers, it has also widened the gap between 'resource-rich' (Raju *et al.* 2004: 270) and resource-poor farmers in terms of access to water.

PRIVATE ENCLOSURES OF WATER VERSUS DEPRIVATION

Hence, while the existing public and private categories of property rights help explain skewed access to water resources, the picture is complete only when one understands the socio-economic conditions in which these rights are embedded. Benda-Beckmann's (2001) distinction of categorical and concretized rights is illustrative in this regard. Categorical rights refer to the broad conceptual legal categories assigned to resources, and concretized rights refer to the way in which the former are embedded in the immediate socio-economic context. The latter include rights over the means of appropriation (money, technology, and so on) and the existing distribution of wealth.

In this context, Ribot and Peluso's (2003) argument in favour of distinguishing access from rights is also useful. While property, according to these authors, refers to rights to things or resources sanctioned by law, access is a broader term that refers to both legal and extra-legal processes and relationships, and thus to a wider range of social and power relations that constrain or enable people to benefit from resources. In the present case, a combination of property rights and relations of access helps to explain the existing inequalities. The following examples make this clear.

Landed farmers able to invest in pumps are found to invest in creating an interlinked system of ponds, tube wells, and shallow pits on their private land holdings. Ponnunni, the owner of 10 hectares of land, for instance, has dug a tube well to a depth of almost 100 metres (300 feet). During the second crop season, his 10 horsepower motor pumped water continuously from a pond to which he has exclusive water rights. This pond was brimming with water while ponds around display their dry beds. Bhaskaran, a large farmer who owns land in the tail-reaches of the Peringotukavu distributary canal, disillusioned by the repeated failure of canal water, was in the process of digging two tube wells in 2002, dug to a depth of 73 and 82 metres. He invested a total of INR 200,000 in sinking two tube wells, pumps, and pipelines that transported the tube well water into his ponds. He had ownership rights in three ponds, in two of which he had exclusive rights. The increased storage in one of these was expected to raise the water levels in the adjoining shallow pit as well. He hoped that by pumping water from his three ponds and two tube wells he would be able to grow a second crop of paddy. Such investments create enclosures of water that are protected by the existing rights regime, wherein enclosures of water on private land are regarded as private property.

In contrast, the case of Velan illustrates the travails of the resource-deprived small farmer. The 60 cents (0.24 hectare) of land that Velan's wife inherited is located close to the Kuttikadu pond. The owner of the pond had not raised a second crop. Hence there was some water in the pond. Velan, along with another small farmer named Krishnan and a medium farmer, Chentamarakshan, who owned about 0.8 hectares of paddy land, requested the caretaker of the Kuttikadu pond to allow them to pump water from the pond. The three had hired an oil engine and purchased some kerosene. Velan's land was located furthest from the pond, so the fields of Krishnan and Chentamara were irrigated before the water reached his field. Since the water level in the pond had receded, they had to dig a pit in the pond bed before pumping the water. Chentamarakshan ridiculed Velan for being able to mobilize just one bottle of kerosene. Velan later remarked that the bottle of kerosene was all that was left of the cooking fuel at home. Since he was unable to share the cost of the kerosene, Chentamarakshan made him do the manual work.

After all this effort, Velan got three sacks of grain (about 170 kilograms of paddy) from his plot. A small farmer like Velan resorts to pumping of water only when he is convinced that it is absolutely necessary to save his paddy crop. To do this he used his last reserves of kerosene, otherwise to be used as cooking fuel. Meanwhile, large farmers like Bhaskaran or Ponnunni run their pump sets continuously, filling their ponds to the brim. While this kind of resource use is a critical determinant of water scarcity in the study area, the cases also illustrate that scarcity is a relative concept crucially linked to broader issues of inequality.

The differential ability to invest in energized lifting devices also enables the landed and wealthy farmers to disproportionately exploit common property resources. Ramachandran is a large farmer owning five hectares of land in the lower parts of the Varayiri watershed. His land adjoins both the Gayatri river and the Varayiri stream, being located close to their confluence. He has installed an elaborate system to fill up his pond: an electric pump set that pumps water from the Gayatri river via the well that he has dug on land adjoining the riverbed. This water is piped to the Varayiri stream, from where it is pumped up again into yet another well. This water is then taken to his own exclusive pond located on higher ground, irrigating his paddy fields. Located at the tail end of the canal network, his green fields stand out amidst the surrounding parched yellow fields during the second crop season of paddy.

THE FUTURE OF PRIVATE OWNERSHIP AND CONTROL

The unrestrained extraction of water from the different irrigation sources in the area highlights two important issues. First, the classification of water sources as private property promotes overexploitation. This is particularly so in the case of privately owned wells (both shallow and deep) and ponds. Second, even when the resource is classified as common property, as in the case of the stream, manipulation of these legal boundaries coupled with the use of pumps creates de facto private access and control. The current operationalization of rights to water not only creates inequalities but also promotes unsustainable use of a scarce resource. During the past two decades it is increasingly recognized that an unregulated private regime encourages indiscriminate consumptive utilization of water and leads to rapid depletion (Torori *et al.* 1996).

Alternative property regimes for water are increasingly recommended, particularly in the case of groundwater use, which has been firmly rooted in the private domain. Of late, surface water sources are also being subjected to such reconsiderations. The law of riparian rights has been critiqued for confining the right to use flowing water to those owning land alongside a stream or river (Orindi and Huggins 2005; Singh 1992). This is illustrated in the study area as well, where farmers who do not own riparian land have no right to convey water through the land of riparian owners in order to irrigate their land. They must have permission, and may have to irrigate the riparian farmers' land before their own.

In this context, Orindi and Huggins (2005) argue that public access points or rights should be provided to those who do not own riparian land, to prevent such exclusion from being passed on to future generations. Similar suggestions have emerged with regard to ownership of ponds and wells. In the study area, farmers without access to ponds have argued that all ponds should be declared government property and that all farmers should have a right to the water in ponds. Interestingly, Bhatia (1992) has reported that small and marginal farmers in Gujarat have come up with a similar suggestion as far as ownership of tube wells is concerned.

Private property rights have also been argued to be inappropriate for the management of natural resources from a perspective of sustainability. Private property rules have been critiqued for treating natural resources as temporally and spatially bounded commodities (Ojwang and Juma 1996). The basic feature, however, of water resources is that they cannot be fixed in time and space (Sick 2007). McKean (2000) notes that granting individual private rights to such natural resource systems leads to their chopping up into bits and pieces, which is inappropriate from an environmental perspective. This is true of the existing distribution of rights to water sources in the study area. While some own tube wells, others own shallow wells or ponds (individually or as a small group), and yet others enjoy access to the stream. All of them extract water from these varied sources as though they are unrelated.

It is argued here that the unique properties of water should be central to the formulation of water rights. The extent of access granted by property classifications and modern lifting technologies should take into account the total availability of water and the impact of current use patterns on its long-term availability—not just for irrigation but also for other essential purposes such as drinking water. In this regard, the earlier kalayi–potta classification is worth reviewing. This classification of water rights, whereby rights were restricted to kalayi land during the second paddy crop, had the conservation of existing water supplies as a high priority. New lift systems give equal rights to kalayi and potta lands, in a less efficient use of existing water supplies. The present water crisis demands a fresh look at the existing constellation of water rights and land rights.

OPPORTUNITIES FOR REFORM

Scarcity of water for drinking and irrigation has assumed serious proportions in many parts of Palakkad district over the past decade. As a result, scarcity redressal has been taken up by both state and local governments (the village panchayats). The emphasis on decentralization in the state during the past ten years has placed local water resources management under the initiative of the panchayat as an important strategy in solving scarcity problems. Accordingly, the management of public water sources like streams, irrigation and drainage channels, lakes, tanks, and wells was vested with the panchayat. Panchayats were also required to come up with local strategies and plans for each sector like water resources, agriculture, and soil conservation.

Reviewing panchayat performance in this respect between 1997 and 2002 (the Ninth Plan period, when decentralization was introduced in the state) reveals that, while panchayats drew up local water resource management plans, the focus remained on structural interventions. This comprised constructing check dams (low dams measuring three to five metres in height) across streams and rivers, stone lining of ponds and stream channels to prevent side erosion, and installing lift irrigation schemes. While water conservation was a priority for most panchayats in the district during the first phase of decentralization, water conservation was equated with restoration of ponds alone, again focusing on stone lining of sidewalls. In not a single case was the restoration of the pond catchment given due emphasis. The pond was viewed as a mere storage reservoir, not a related part of a watershed-based complex of water-harvesting measures. Non-infrastructural measures that promote recharge by enhancing the vegetative cover and in situ moisture conservation were not considered. The panchayats, however, cannot be held solely responsible for such a narrow vision. They have only followed the approach adopted by irrigation departments so far, one focused on civil engineering structures and not on the ecological milieu that determines water availability.

Similarly, while attempts are made to abate the current water crisis, the existing resource distribution pattern is ignored. The suffering of tail-end farmers in the Chulliar ayacut is accepted by farmers and irrigation officials alike. Solutions have focused on implementing further inter-basin transfers that augment the storage of the Chulliar reservoir, leaving untouched once again the issue of distribution. This overlooks how those with access to interlinked sources of irrigation and pumps are well buffered from water scarcity.

Reducing present demand levels is considered an important strategy in abating the water crisis (Rogers and Hall 2003). For this, a reconsideration of existing property rights formulations is an important first step. While there is little clarity in formulating alternative property regimes, greater collective rights with legislative backing has been advocated as a way forward (Bhatia 1992). Other measures include imposing an upper ceiling on water extractions. While land ceilings have been an accepted method to prevent concentration of landholdings, a similar concept in water use could help restrict its uncontrolled exploitation. Orindi and Huggins (2005) argue that fixing an upper limit on individual extractions is an important means of regulating water withdrawals. Van Koppen et al. (2004) suggest that a water tax be imposed on large-scale water users, not only to help finance water management services and generate net income, but also to provide incentives to use water prudently. Water charges and water permits, and restricted power supply and increases in electricity tariff rates are some of the suggested measures to control water use in general and groundwater consumption in particular (Saleth 2005). However, all of the above measures assume some form of external enforcement and monitoring, which is challenging in this environment. The effectiveness of restricted power supply has been undermined by the ability of rich farmers to invest in large pumps and multiple wells, and by availability of diesel pump sets (Saleth 2005; see also Prakash, this volume). Therefore, voluntary compliance with such regulations is critical in ensuring their effectiveness (Sick 2007). Formulation and enforcement of such measures at the community level have been advocated to instil a sense of responsibility for the conservation of community resources (Bhatia 1992).

The latest decentralization initiatives in the state offer space to test out such institutional initiatives. In line with contemporary national and global policy emphases on a river basin approach to manage water and natural resources, state government guidelines for the ongoing Eleventh Plan (2007–12) emphasize a watershed-based approach to local planning (Government of Kerala 2007b). The guidelines propose that all development activities undertaken by a panchayat, particularly in the realm of agriculture and irrigation, fall in line with a larger watershed plan. These watershed plans, to be prepared by the concerned panchayats, are to be part of larger river basin plans as well.

While this initiative raises questions of spatial 'fit' that arise from managing a biophysical system along political–administrative boundaries (Moss 2006), the need for an integrated approach to manage water and natural resources is beyond doubt. A comprehensive assessment of water resources with a focus on management and distribution issues can be a starting point for panchayat-level debates on sustainable and equitable use of an increasingly scarce resource. Understanding of the functioning of hydrological systems, of resource depletion and sustainability of current water use patterns is an important precondition for any of the institutional reforms discussed. In practice, to date, little has been achieved, with panchayats confining their activities within administrative boundaries, not venturing to think in terms of river basins and watersheds.

* * *

Measures to redress water scarcity need to be grounded in the locally and regionally specific ecological and socio-economic contexts. A sound appreciation of these dimensions is critically amiss in all measures used to date to develop and manage water resources for food security and rural development. This gap has serious consequences for the ecology and economy of Palakkad district and other regions of Kerala. The ecological factor has to be taken into account in the management and distribution of water, both in general, and in understanding of its diverse yet interrelated forces. In Palakkad, the degradation of local water sources has been masked by the periodic availability of water through the canal system, although the limitations of this system are now showing. In addition, remedial measures focus on either inter-basin transfers of water or infrastructural measures like check dams that do not ensure the long-term sustainability of the resource. Collective action in this regard has been confined to lobbying for increased supply of water, never addressing the issue of control of access or demand.

Looking to the diverse relations of the ecological dimension, a critical mediation of water distribution can be found in the existing property rights regimes, which sanction unrestrained extraction of water from privately owned water sources. Controls over technology development and management are insufficient to prevent water defined as public from being taken into the private domain. Lack of attention to defining and managing the public–private divide means that little attention is paid either to the hydraulic connections between the so-called private and public water sources, or to the ecological impacts of unrestrained extraction from one water source on another. Skewed access to these privately owned water sources aggravates both the injustice of this situation and the negative ecological impact of fragmented yet over-exploiting use of diverse water sources. This transforming privatization of water has been driven by agricultural policies enabling intensification and commercialization of production, but with little attention to the sustainability, durability, and equity of technologies selected and water supplied. First, intensification of rice production and, later, growth and shift into perennial cash crops, have changed the spatial, temporal, and quantitative demand for water. Both technologies and institutions have enabled a shift in water use that has also been a shift in the nature of property rights.

The present political climate in Kerala, with a strong focus on decentralization and local-level natural resource management planning, provides an opportunity to initiate locally relevant measures to redress scarcity. Yet there remains a need for the state to debate agrarian and agricultural policies alongside these new policies for resource management, if new local management is really to address this ecological dimension of water problems. As this chapter has argued, this ecological dimension requires attention to the relations and institutions that mediate between people and water, not only to water. This requires understanding and recognition of the property rights and production relations mediating between people and water through technology, as well as attention to the ways in which the physical cycle of water availability is being changed by technology. While the results so far are not very promising, the availability of a critical space for local-level interventions needs to be taken seriously.

Notes

[1] The state-level landholding classification identifies large, medium, and small farmers as those with holdings above 2 hectares, between 1 and 2 hectares, and below 1 hectare, respectively. For this study, small farmers were classified as those with holdings less than 0.4 hectare and not 1 hectare, as a majority of the small farmers in the study had holdings below 0.4 hectare. A total of 144 farmers were surveyed, with 48 farmers in each of the above landholding categories.

[2] Between 1995 and 2002, farmers from Kolumbu, Karinkulam, Vattekad, and Panangattiri in Elavenchery panchayat lost their second crop of paddy on four occasions and were compensated by the Krishi Bhavan (the Agricultural Office). The area affected by crop loss covered approximately 400 hectares. (This data was provided by the Agricultural Officer at the Elavenchery Krishi Bhavan. This is not published data, but derived from the compensation that was paid to paddy farmers who lost their second crop of paddy.)

[3] Paddy is normally cultivated in two cropping seasons: the first (the autumn crop between April/May and September) largely coincides with the southwest monsoon. The second (the winter crop) is cultivated between October and January, when rainfall is less adequate and external water supplies are required.

[4] The Kollengode, Payilur, Peringgotukavu, and Karinkulam distributaries are supplied from the second reach of the LBC of the Chulliar dam of the GIS. The GIS is supplied by the Chulliar and Meenkara dams.

[5] Rubber cultivation has not been taken up in the study area, unlike other areas of Kerala, probably due to climatic conditions.

[6] Yields were found to vary between less than 1,000 kilograms and 4,000 kilograms per hectare, reflecting differential access to water at critical stages of crop growth.

10 THE POLITICAL ECONOMY OF TANK MANAGEMENT IN TAMIL NADU

R. Manimohan

Monsoon rain is an important source of water for agriculture in the semi-arid regions of India, of which temporal and spatial variation in rainfall are major characteristics. In response to this climatic variability, people and societies have created numerous surface storage structures to conserve rainwater for further use. These form part of a culture and history of adaptations transforming nature into productive resources. Tanks are examples of this nature–society interface in India (Ludden 1985; Sengupta 1993). Many tanks are concentrated in south India. Particularly, the states of Andhra Pradesh (AP), Karnataka, and Tamil Nadu account for about 60 per cent of the area irrigated by tanks in India, holding almost 120,000 tanks out of the national total of 208, 000 tanks (Sakthivadivel *et al.* 2004: 3521). While in 1989–90 tanks still irrigated 1.8 million hectares, a little over 20 per cent of the net irrigated area in India (Vaidyanathan 2001: 7), this area fell from 2.43 million hectares to 1.53 million hectares between 1961 and 2000 (Narayanamoorthy 2007: 194). Thus, in south India tanks are inherently part of the physical landscape and social memory. However, as these figures show, tank systems are in decline, particularly since the 1960s, in terms of both area and governance systems ensuring the functioning of tanks (Janakarajan 1993; Narayanamoorthy 2007; Shankari 1991). This narrative of tank decay has become so influential that it has almost become a defining element of tanks.

Being associated with decay and neglect, tanks become ammunition for those who make polarized arguments in debates about the role of dams and surface irrigation in development. One school of thought idealizes tanks as alternatives to large-scale canal irrigation systems, glorifying community traditions as an alternative to bureaucratic meddling. In contrast, another school of thought is pessimistic about the relevance of surface irrigation systems (both large-scale and small-scale, including tanks) in India today, eulogizing the role of groundwater in sustaining agricultural and economic growth. This view argues for the conversion of tank systems into groundwater recharge structures, instead of keeping them as surface irrigation infrastructure. By their characteristic arguments and their perceptions of problems and solutions, the former could be labelled as the 'new traditionalist school' (NTS)[1] and the latter as the 'new institutionalist school' (NIS).[2]

However, the proponents of both schools share a similar static view on the decay of tank commons, perceiving the problem 'through a particular lens in which they see local institutional breakdown, communal disintegration and social apathy, but not social action and conflict' (Goldman 1998: 44). Both fail to acknowledge the different trajectories of tank use, or what Nagaraj and Jayaranjan (2004: 48–94) describe, in the context of Tamil Nadu, as the process of 'reordering' in irrigated landscapes.

Then, how should one envision the current state of irrigation tanks? How could one capture the dynamism exhibited by tank systems? This chapter attempts to answer these questions by employing a political economy approach integrating landscapes, material artefacts, and institutions that overcome the limitations posed by NTS and NIS analysis. Political economy investigates the social relations of production, locating production in the labour process (Marx 1974: 173–92). In India, the modes of production debate,[3] which engaged in identifying the conditions of capitalism in agriculture during the Green Revolution, rigorously employed a Marxian political economy framework. However, three basic criticisms have been raised of the debate. First, it does not take into account the role of irrigation technology in shaping agrarian change (Bolding et al. 1995). Second, the debate has focused on the role of private property but omitted the importance of the commons and their ecological dimensions in explaining agrarian change (Agrawal and Sivaramakrishnan 2001). Third, it has failed 'to take cognisance of the role of caste' in the analysis of class relations in Indian agriculture (Chakravarti 2001: 108). This chapter incorporates these three elements in the study of the political economy of tank management, using 'socio-technical landscapes' as a frame of reference, explained further in the next section.

This chapter focuses on changing use of land and water resources in a tank-irrigated agrarian landscape in Pudukkottai region in Tamil Nadu, south India. It examines whether and how specific patterns of social relations and material artefacts can be related to, and explain, persistent environmental crises in tank cascades. Sections of the chapter illustrate the nature of agrarian and environmental change in this region. Two distinct landscapes have evolved as wet and dry zones, respectively, characterized by commercial farming in the former and subsistence farming in the latter. These agrarian social characteristics, along with the natural properties of the landscapes, shape the process of redesigning irrigation and drainage artefacts in both zones. This transformation has been driven by several forces that have challenged the functioning of interlocking tank cascades, as described in the final sections.

This chapter reports the major findings of a larger research project on the changing use of tank cascade landscapes in Pudukkottai district between 2000 and 2002, with subsequent short visits in 2005. Tanks are the major source of irrigation in this largely semi-arid region. There are a few rivers or jungle streams like the South Vellar and Agniar, which bring water during the monsoons. With the increasing number of wells and borewells, cultivation is carried out using both tank and well water.

To explore the changing patterns of tank use, the study uses data of several kinds. It combined archival research with qualitative and quantitative methods. Information on land use, agricultural and irrigation practices, and resource conflicts during the colonial period were gleaned from archival sources in the Pudukkottai Darbar Records Office, now part of the revenue divisional office in Pudukkottai. Detailed information about post-independence land use, agricultural practices, and irrigation management has been collected primarily through document reviews, participant observation, in-depth interviews, and a survey of tank users in two tank cascades (wet and dry). The wet cascade is located in Alangudi taluk, and the dry cascade stretches across two taluks: Kulathur and Pudukkottai (Figure 10.1).

SOCIO-TECHNICAL LANDSCAPES AND TANK IRRIGATION IN TAMIL NADU

The concept of a socio-technical landscape is a frame of reference for this study, integrating social and material contents of irrigation

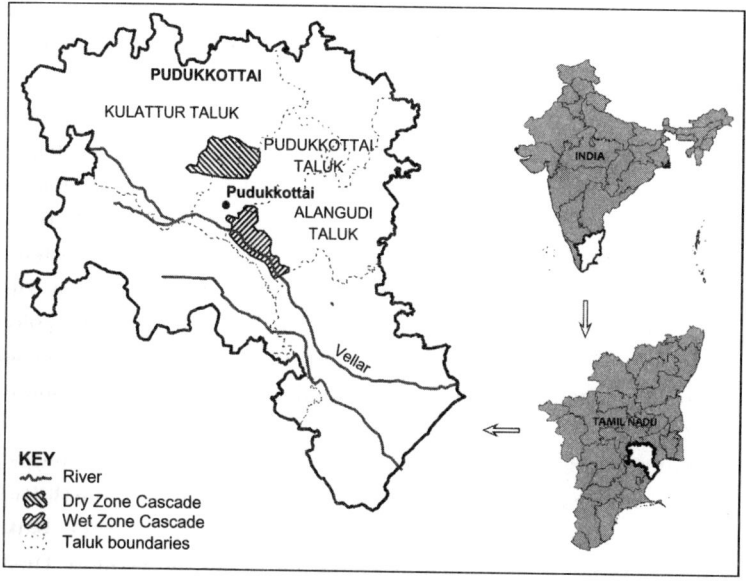

FIGURE 10.1   Pudukkottai District and the Study Area

Source: Author.

artefacts and physical landscapes of the irrigation system within the broader social organization of agricultural production it is part of. A point emphasized here is that space matters in water control, and social relations always have a spatial form (Massey 1984). This chapter, based on insights from political economy, integrates artefacts, landscapes, and social relations in the study of tank management, in which the spatial dimension is emphasized and integrated with the framework of 'water control'.

Water control consists of technical control, organizational control, and socio-economic control (Mollinga 2003; also see the Introduction to this volume). These dimensions are connected with the concept of power. Mollinga argues that 'management institutions and technical artefacts can be understood as the embodiments of particular social relations of power, and, the other way round, socio-economic and political power in irrigation takes concrete shape in particular forms of organisation and technologies' (2003: 40).

Pudukkottai is not a typical climatic region of Tamil Nadu. Pudukkottai is basically a dry tract with high temperatures (40 degree Celsius in May) and high potential evapotranspiration. It receives an average annual rainfall of 920 millimetres, which is scanty (Gandhi 1983). Most rainfall is concentrated in two monsoon seasons: the southwest monsoon (July–September) and the northeast monsoon (October–December). The northeast monsoon, which accounts for about two-thirds of the annual precipitation, is unpredictable and erratic, and frequently associated with cyclones and short spells of heavy rainfall (Ratnavel and Gomathinayagam 2006). In most parts of the year, except for the northeast monsoon period, more water is lost to evapotranspiration than is supplied through rainfall runoff (Athreya *et al.* 1990). Society has adapted to this climatic variability by manipulating the undulating topography and developing irrigation tanks. These are water bodies or reservoirs along the natural line of drainage to capture rainfall and surface runoff. Thus, tank irrigation technology is 'customised to a particular combination of local ecological conditions, social objectives and institutional capabilities' (Ambler 1994: 266).

The topography of Tamil Nadu slopes gradually downwards from the western mountains to the eastern coastal plains. In Pudukkottai, a series of rivers traverse the gently sloping terrain. These rivers and streams (*kattaru* or jungle streams) are dry for most of the year but flow full during a good monsoon. A tank is formed by throwing an earthen embankment across the drainage line of a valley to capture the rainfall and surface runoff. Tanks are generally found in series or cascades in these valley landscapes, draining their surplus flows into rivers or other tanks downstream. The river flows are often diverted to feed tanks. Besides, there are hundreds of rain-fed tanks too. Both are common features of agrarian landscapes across Tamil Nadu as well as in Pudukkottai. Discussions of cascade tanks elsewhere in India and in Sri Lanka can be found in Sakthivadivel *et al.* (1996, 2004).

The shape and size of tank embankments vary, as determined by topographical features and social and political choices. However, most tanks are crescent-shaped, in correlation with the gradual fall of slope contours. The height of the embankment is greatest at the valley bottom and is generally less on the sides of the valley due to the gradient of the terrain. The semi-circular shape of embankment helps to diffuse the force of standing water impounded behind it.

Tank water is released for irrigation to fields located below the tank. Sluices (*kumili*) are structures constructed in the tank embankment for regulating water released from the tank for irrigation. They draw tank water through hollowed rectangular stone or concrete barrels which sit beneath and within the embankment. Sluices are constructed at varying levels, depending upon the contours of the tank bed and the elevation of the *ayacut* (tank command area) lands. Specialist operators (*nirani*) open and close the sluices during fixed time intervals to release water. Water moves under the force of gravity. A network of channels, taking off from the sluices, carries the water to the ayacut and fields in the command areas of the tanks. Spillways are placed at the end of tank embankments to discharge heavy inflows during the monsoon rains. The surplus flow is carried away by the surplus discharge course (*marukal*).

As a runoff storage structure, every tank has its own catchment area. As mentioned, tanks are mostly not found in isolation but in groups or cascades (see Figures 10.2 and 10.3). These interconnected tanks are formed within the boundaries of the cascades or meso-catchments. In physical terms, a tank cascade or meso-catchment includes a main

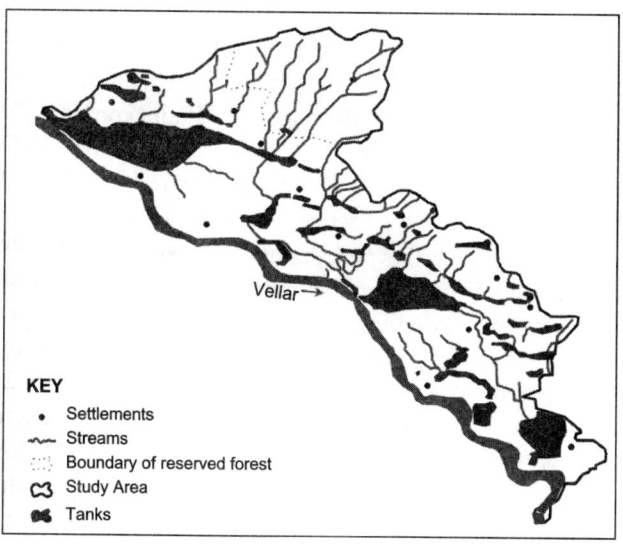

FIGURE 10.2   Vallanadu (Wet Zone) Tank Cascade

Source: Author.

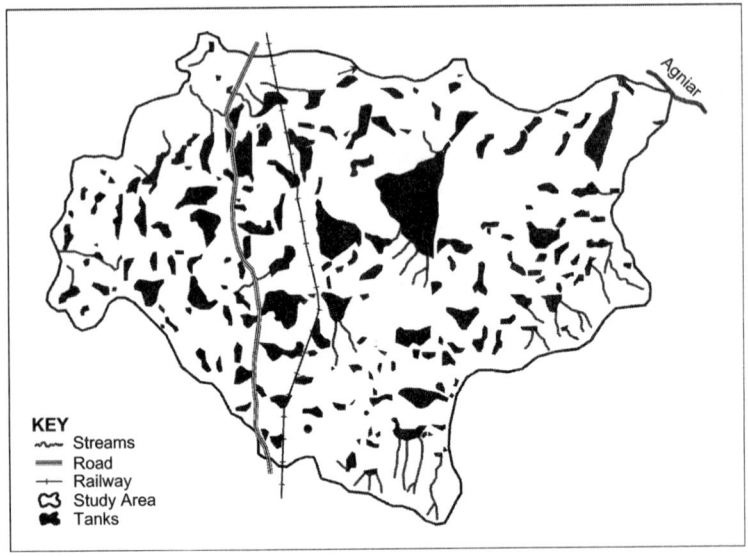

FIGURE 10.3   Sathyamangalam (Dry Zone) Tank Cascade

Source: Author.

valley with a main drainage line flowing through it and a few or several smaller valleys on the sides, with minor ephemeral streams (*odai*) which converge into the main drainage flow. Pudukkottai has records for some 4,000 tanks and 146 interlocking tank systems, with tanks supporting some 80 per cent of the irrigated area.

In cascades, tanks form a series from the higher elevation of the valley catchment until the last tank drains into a stream or seasonal river. The surplus flow of one tank in the cascade is released to the next tank downstream through a drainage canal. Moreover, in a valley landscape, bigger tanks are found along the main drainage line. Smaller tanks are dotted on side valleys at relatively higher elevations. They are also formed in series and the surplus flows through them are drained into a bigger tank, the location of which is on the main drainage line. In addition to the surplus flows from heavy rainfall events, the irrigation return flow of a tank is also conveyed to the next downstream tank. Consequently, a considerable amount of water is reused. Hence, a series of tanks located at varying levels share the entire flow of a cascade.

The tanks in cascades are intercepted by different land uses. Forest (*kadu*) and dry lands (*punchei*) are common in the higher levels of a shallow valley. Irrigated 'wet' fields are carved out at the lower level of the valley in front of each tank so as to be able to withdraw water. In this way the closing end of the command area of a tank usually forms the foreshore of the immediate downstream tank. Well-irrigated garden lands are found somewhere between the dry lands and wetlands but their location is in the close vicinity of the tanks. Human settlements are often found at higher locations, sometimes within the irrigated areas.

The integrated system of tank cascades in a sloping terrain is made possible by spillways. As already mentioned, the stream course that carries away surpluses (marukal)—normally a natural drainage stream—simultaneously feeds the tank located at lower elevation. Thus, the spillways—in this case called calingulas—regulate water flows in tanks located in series from higher to lower elevation. Calingulas act simultaneously as storage and drainage mechanisms in tanks in cascades.[4]

A brief description of the technical artefact now follows to show its role in this specific socio-hydrological context. The overflow weir has square shaped vertical cut stones fitted at equidistance on a masonry body wall that forms the crest of the weir (Figure 10.4). Generally, these cut stones are 6 inches × 6 inches (0.1524 metres) square and their height varies from 0.3 to 0.6 metres (1–2 feet). Such cut stones are called dam stones (*anaikkal*). The distance between the dam stones also varies from 0.6 to 0.9 metres (2–3 feet). The bottom level of the dam stones or the level of the body wall is mostly designed at half tank level of the tank. The top of the dam stones equates with at full tank level (FTL).

Calingulas not only act as safety valves protecting tanks in risky situations of heavy flood but also act as 'points of contact' (Mollinga 2003) for many actors and are a historically negotiated public space in the social ecology of tank cascades. The technical features of calingulas clearly exhibit certain elements of social organization and property rights attached to them. When floods recede during the monsoon rains, *ayacutdhar*s (irrigators in the command area) close the calingula by making clay and turf bunds up to the top of dam stone level. By this act they are able to store water to the FTL without affecting the downstream water rights.

FIGURE 10.4 Spillway (Calingula) Showing the Cut Dam Stones

Source: Author.

In the past, downstream farmers had the right to fill their own tank up to the sill level of the weir. Only after that could the immediate upstream tank ayacutdhars raise the clay and turf bund up to the top of the dam stones or at FTL. On normal rainy days, as soon as each tank had filled up with water to crest level of the weirs (half tank level), the downstream farmers put up the turf bund between dam stones on the crest first. Then successively the others, moving upstream, also did so in order to store water in the tanks up to FTL. The practice is different when there is an abundant or additional inflow followed by heavy rain. Then farmers in the upstream tank take away the earthen bund first, and subsequently successively the other bunds going downstream, to drain the water. In a situation of extreme flood, these practices provide in-built flexibility for quick disposal of flood flows by automatically washing away the turf bunds put between dam stones when water overtops them. The surplus weir comes into operation without the FTL being exceeded to a significant degree. In this way, calingulas are material expressions of water rights between upstream and downstream irrigators in cascades.

Cascade water management included regulation of water use practices at the level of individual tanks too. It was a principle and practice that, until the tank located at the lowest end of a series was filled with water to its FTL, the upstream tanks should not be opened for irrigation. This was ensured by giving privileged rights to the tank at the end of a cascade. These rights consisted of a tax-free grant to royal kinsmen or to temples. These privileges to downstream temples and elites gave them political control over upstream tank users in a cascade. This can be understood as a mechanism to ensure adequate water flow throughout a cascade and mobilize labour and resources for feeder channel maintenance (Gomathinayagam 1995: 12).

Now these interlocking tanks are struggling to survive amid processes of transformation in agricultural technology, land use, and social institutions. A combination of factors is found responsible for these transformations in tank cascades, including state-initiated land use changes in forest cover and cropping, and shifts in labour relations during the colonial and post-independence periods, detailed further in later sections.

WET AND DRY TANK CASCADES: AN INTRODUCTION

As a result, two distinct landscapes have evolved as wet and dry zones, in this otherwise dry region. This section summarises their variable responses to the environmental crisis, as characterized by different patterns of water management and agrarian change (Table 10.1). These

TABLE 10.1 Reordering of the Socio-technical Landscapes of Tank Cascades

Characteristic Features	Wet Cascade	Dry Cascade
Number of interlocking tanks	19	84
Source of water to tanks	Anicut (diversion dam), rainfall and runoff	Rainfall and runoff
Percentage of tanks having calingulas as spillways (number of tanks in parentheses)	52% (10)	28% (24)
Topography (in relative terms)	Flat	Steep
Land use in upland catchments	Mono-cropping plantations; red soil and laterite mining	Mixed cropping plantations; granite mining and stone quarrying
Tank capacity loss due to siltation	26%	31%
Degree of inter-linkage of tanks	High	Low
Intensity of cultivation and water use	High	Low
Rate of groundwater extraction	High	Low
Degree of commoditization	High	Low
Cropping pattern	Paddy, sugarcane	Paddy, pulses, oilseed, millets, vegetables
Functional role of spillways	Drainage	Storage
Agrarian social character	Commercial farming	Subsistence farming

Source: Fieldwork 2000–2.

not only imply changes in agronomic practices, hydrological conditions, and the level of soil moisture, but also transformations of the social character of the agrarian economy. Both the wet and dry zones exhibit change, but of different forms, in water management practices, cropping patterns, and degree of commoditization. Consequently, the rate of accumulation and the level of class differentiation are higher in the wet zone than in the dry zone, characterized by commercial farming in the former and subsistence farming in the latter. These agrarian social characteristics, in turn, shape rights to water in a manner that suits the requirements of the new agrarian regime and accommodates the interests of its potential claimants. Further evolutions in the socio-technical landscapes of tank cascades are explained in the final section, after more history is given on the wider changes in landscape and society.

Within the constraints and possibilities posed by ecology and different patterns of socio-economic development, wet and dry cascade villages register variable levels of adoption and growth of modern water extraction technologies. In both the wet and dry zones, resourceful farmers invested in borewells and wells to escape from the crisis posed by soil erosion and sedimentation (which reduced the effective capacity of tanks). However, well owners in the dry zone face insufficient water supply from their wells as the terrain overlies hard rock aquifers, which limits their farming activities. The groundwater potential is greater in the wet zone and is becoming the major source of growth.

Small and poor peasants in both zones have reacted differently to the ecological crisis. In the dry zone, most peasants continue to practise farming as a subsistence option, combining it with non-farm employment of lower remuneration. They cultivate a single subsistence crop, mainly paddy, during the northeast monsoon season depending on tank water. Men migrate mainly due to the adverse environmental consequences. Consequently, a substantial amount of the population has left agriculture in the dry zone villages.

In the wet zone, small and poor peasants use their farmland intensively along with resourceful farmers. Landholders who do not own a well 'buy' water from well owners. A market for water has emerged, which is embedded in the prevailing labour relations. Groundwater-led agrarian development has paved the way for growing socio-economic differentiation and commoditization.

What are the implications of these agrarian and environmental changes in tank water management? The consequences differ in the wet and dry zones. In the wet zone, the utilization of water and its social management are critical factors in the new cropping regime.

Tank systems often come to be considered as sources of groundwater recharge rather than surface irrigation systems, though a significant number of non-well-owning farmers use tank water for irrigation. The emergence and consolidation of commercial farming in the wet zone has radically transformed the way tanks were managed in the past. New social institutions are crystallizing which embody elements of the interests of the class of commercial farmers. Through these processes the subordinate classes are brought into the fold of the dominant interests through economic coercion and ideological hegemony. Thus, tank management institutions in the wet zone, in general, become a powerful vehicle to carry forward the professed interests of the commercial farmer class.

In the dry zone, tank management institutions are reshaped along with the emergence of new elements of subsistence farming, also increasingly articulated within the scope of commercialized agriculture. This process indicates the significance of continuity and change in aspects of tank management practices that evolved and changed through the centuries.

How do these transformations affect the functioning and use of tank cascade landscapes in both zones? The direct effect is the redesigning of the technical dimensions of calingulas. As already noted, the tanks in series are technically connected by calingulas which act as storage and drainage infrastructure simultaneously. Soil erosion and sedimentation reduces the depth of tanks, and, in direct correlation, the functioning of calingulas. In interlocking tank systems, this situation generates tensions between upstream and downstream tank users. Farmers who cultivate land close to tank foreshores generally have a concern to protect their crops from submergence. Those who plough fields in the tank ayacut are interested to retain more water in tanks to maximize crop yield. Both these interests stand in conflict. In order to safeguard their crops, tank foreshore farmers clandestinely break calingulas at night, which is discomforting for the farmers of the downstream tank. This drives the actors into negotiations over alteration in spillway designs.

One of the most important results of this process in recent times is that calingulas are often being replaced by other types of spillways. While in the wet zone cascade, 52 per cent of the tanks retain calingulas as spillways, in the dry zone only 28 per cent have calingulas.

The agrarian social interests and the spatial characteristics of the landscapes in both the wet and dry zones act together in shaping this transformation. This shows how tanks and their users respond to the broader societal, economic, and environmental change, discussed in more detail in the next section.

LAND SETTLEMENT AND TANK CASCADES

The building of tank irrigation infrastructure in Tamil Nadu went hand in hand with the establishment of agrarian social institutions in the various historical periods. Agrarian institutions that flourished along the drainage basins in Tamil country were known as *nadus*.[5] Inscriptional sources suggest that the density, size, boundaries, and subdivisions of nadu settlements were defined in relation to water resources either in tank-irrigated zones or in river-irrigated core wet zones (Stein 1980; Subbarayalu 1973). Studies suggest that reservoir building with stone sluices and weirs dramatically changed the South Indian agrarian landscape between AD 700 and AD 1000 (Gunawardana 1984; Gurukkal 1986; Heitzman 2001). Provision of calingulas on the side of tank embankments eased hydrological linkages along a drainage basin, which facilitated water transfer in micro-environments through institutionalized social practice. The development of these integrated designs or interlocking systems of decentralized tanks has thrived over time along with the development of institutions of property, governance, and a fiscal system.[6] By the ninth century, this system of property, governance and fiscal relations became the foundation of the medieval agrarian order which sustained the interlocking tank cascades, and in turn, the latter reinforced the former.

However, these systems fell into crisis during the late nineteenth century, when the colonial state introduced certain structural changes in the fiscal system and governing mechanisms by implementing land settlement schemes in Pudukkottai.[7] These altered the redistribution system of agricultural production in such a way that it could no longer function as a supportive mechanism for interlinked tank networks.

Three settlement schemes were prominent. The first scheme implemented was the Amani settlement begun in 1878–9 and completed in 1881–2. It replaced grain-shares with cash rent. The second was the Inam Settlement that started in 1888 and came to a close in

1896. In addition, the Revenue Settlement of 1910 had an important bearing on the region's agrarian regime. The implementation of these settlements thus stretched from 1879 to 1912.

The implications of these settlement schemes for tank management were disastrous, weakening its local authority structures. In addition, the establishment of the police and court system, and revenue, irrigation, and forest departments structurally replaced the local governance system that previously oversaw local disputes and conflicts, including intervillage tank water disputes. Moreover, the state initiated certain revenue-augmenting measures in the management system of its forests and uncultivated lands that formed the catchment of the interlinked tanks, which, until then, had been commons, also shaping runoff generation into the interlocking tanks. These measures included sale of uncultivated dry lands for cultivation and auctioning of forests for commercial exploitation. These changes disturbed the basis of property relations and the functioning of the interlocking tank systems. Inter-tank disputes in the pre-colonial period centred around upstream and downstream water rights that were underwritten in the design, but the changes brought by colonialism started shifting the focus of disputes from water rights to design elements themselves.

The immediate result of these structural changes was destabilization of the existing institutions for tank management, particularly the organization of desiltation of tanks and channels. As a result, silt started accumulating in the tanks. This, in turn, gave rise to renegotiations of irrigation and drainage rights. As the tanks silted up, irrigators increasingly realized the reduction of the effective capacity of tanks. They demanded that the government increase the level of the calingulas to a level sufficient to compensate for the lost capacity due to siltation.[8]

As a result the state initiated a scheme of 'enlargement and repairs' which replaced calingulas by broad-crested weirs.[9] The scheme was implemented mainly in large tanks located on the main drainage canal within cascades, where the interests of both the colonial state and the agrarian elite converged. It also matched with the objectives of the new land revenue settlement programme of the state that prescribed a land revenue assessment procedure according to irrigation source.

Thus, the older social, economic, and ecological values inscribed in the design elements of calingulas did not match with the new values attached to the colonial land and forest settlement schemes and new property and fiscal arrangements. After independence the tendency to change calingulas into other types of spillways continued.

ORGANIZATION OF AGRICULTURE UNTIL THE 1970S

The policies pursued in the late nineteenth and early twentieth centuries in Pudukkottai brought several qualitative changes in the organization of agricultural production. However, the elements of farm production were not completely destroyed and some persist even today.

CROP CYCLE

Given the variable rainfall and uncertain climatic conditions, agriculture in Pudukkottai evolved to correspond with the cycle of monsoons. Crops were grown in two seasons locally known as *kalam* and *kodai*. Kalam—the first season—normally begins in July or August and cultivation commences in accordance with rainfall. The two monsoon seasons characterize the kalam (the southwest monsoon and the northeast monsoon). Hence, kalam extends over six to seven months into January or February. Kalam is the main cultivation season, in which paddy was the principal wet crop in tank ayacuts (*nanchei*). Kodai (summer) cultivation commences in February–March and ends in July–August. Various farming strategies were adopted to work with the tank environment. In kodai, short-term paddy, millets, and vegetables were cultivated in the tank ayacut in accordance with water availability and field soil moisture levels soon after the harvest of kalam crops. Kodai crops were mostly protected by irrigation from sand-bed shallow wells, operating swing baskets.[10]

Millets, oilseeds, and pulses were major crops in the rain-fed lands (*punchei*). Vegetables and garden crops were cultivated in garden lands (*thottam*) supported by wells.[11] Both in garden and dry lands mixed cropping and crop rotations were generally practiced. This maximized the diversity of production and spread risk. Soil fertility was improved through spreading of withered leaves from plants grown on the fields, which also protected soils from erosion.

AGRARIAN STRUCTURE AND SOCIAL MANAGEMENT OF WATER

Agrarian structure consists of different classes in society. The role of different social classes in agricultural production and their claim over the distribution of the production is based on their relation to property. The management of water in Pudukkottai is shaped by the prevailing property relations along with landscape and water conditions.

In agrarian contexts, land and water are important objects of property: in Pudukkottai the hierarchy in land control paralleled the caste hierarchy. Land ownership and control were dominated by Kallars (landowners and cultivators), and by Brahmins and Nattukkottai Chettiyars as absentee landlords who leased out their land to Kallars and subordinate castes like Valaiyars, Pallars, and Paraiyars who were agricultural labourers. Many Kallar landowners also leased out land to their relatives and to their Valaiyar, Pallar, and Paraiyar tenants. Dominant Kallar landowners cultivated their land with *pannaiyal*s (attached labourers) and day labourers.[12] Thus, the landlords extracted surplus from land through their ownership. Muslim landowners were not uncommon in the region. Peasants cultivated their land mostly with family labour and exchange labour combined with occasional wage labour.

Landownership and inheritance determined water rights. Landowners in tank ayacuts alone were entitled to tank water. Tenants enjoyed the rights of their landowners. Contribution of labour to system maintenance by the landholders was an obligation to continue their rights. The work of tenants and farm labourers was generally attributed to landowners. Water rights were exercised through a system of rotational turns (*murai*) and shares (*pangu*). Priority was given to certain ranked lineages (*karais*) located in the villages. Kallar *ambalars* directed and supervised the implementation and smooth functioning of water rights.[13] They organized maintenance of tanks in series and their channels, in order to facilitate agricultural production. This work was generally carried out by agricultural labourers and tenants.

The important categories of agricultural labourers in Pudukkottai were pannaiyals and *attha coolies* (day labourers). The pannaiyals were permanently attached to dominant landed households. They were paid by their employers from the annual grain harvest and fed by the households of the landlords. In some cases, they were also permitted a piece of land for their own benefit. They stayed in homesteads given by the landlords. Such 'gifts' and 'payments' were telling signs of the social position and location of the 'gift giver' and the 'gift receiver'. Thus, food and housing were mechanisms to keep labour in perpetual servitude, powerful weapons to control labour and its mobility. The system was most iniquitous and exploitative. The tank-irrigated agrarian landscape was, thus, built through these material and symbolic attributes of labour.

ENVIRONMENTAL CHANGE AND AGRARIAN RELATIONS AFTER 1970

Land use changes initiated during the colonial period in the tank catchments still have effects, but further significant shifts in land use occurred from the late 1970s, particularly in the upper catchments, transforming agriculture and irrigation practices in Pudukkottai. The role of the state was crucial in bringing about these changes. Post-independence governments 'inherited' the colonial model of natural resources management and governance (Ludden 1992), albeit tempered with changing approaches and funding for local develop-ment planning. It is a top–down technocratic model of planning and implementation, containing features of commercial interests.

Land Use Transformations in Catchments

Land use changes in tank catchments in Pudukkottai stem particularly from five interlinked measures around forestry plantations and social forestry on tank foreshores, intensification of farming, soil conserva-tion measures, mining, and quarrying, and social and legal changes affecting tank maintenance, are discussed further.

Plantations

In 1972, the government of Tamil Nadu converted 87 per cent of forest land in Pudukkottai into plantations: some 19,700 hectares (76 square miles) of forest land. It was given to the Tamil Nadu Forest Plantation Corporation (TAFCORN), a public sector undertaking, to establish eucalyptus plantations to supply wood mainly for paper industries. It was a policy shift towards commercial forestry. These older forests were part of catchment areas of tanks and provided farmers with resources for agricultural activities and livelihoods. This created conflicting claims on the use of land and water between TAFCORN and tank irrigators, and also exacerbated soil erosion in the upper catchment forests and sedimentation in downstream water bodies.

Many supply channels flowing through the forests bring runoff to the tanks during monsoons, particularly in the northeast monsoon period October–December. Farmers in the downstream areas had the right to receive the runoff from the forests to their tanks, and the responsibility to maintain the channels. In the early 1970s the state government paid little attention to the interconnectedness between forests, tanks, and farmers' rights, when it decided to clear the original vegetation and release the land to TAFCORN. Then TAFCORN ploughed the land by tractor and planted trees both on land and in the courses of supply channels. They also erected contour bunds for soil and water conservation measures in the plantation area. Farmers protested that the contour bunds arrested the runoff over which they claimed rights. The farmers uprooted planted saplings and broke the contour bunds to protect their water rights. There were no compensation arrangements or reconciliation efforts. These cycles of bund creation and destruction continue across the monsoon seasons, and the flash floods during monsoon bring heavy sediment loads from the plantations as the earth bunds are broken and mixed with the flow. Occurring every monsoon season, the erosion and sedimentation has been a severe problem in the catchment and tanks, respectively.

Social forestry

Social forestry in tank foreshore areas has also reduced the water holding capacity of tanks.[14] Social forestry evolved into a movement in Tamil Nadu with the aid of the Swedish International Development Authority (SIDA) from 1981 onwards. The original objective of the project was to plant 260,000 hectares on tank foreshores—nearly half of the total tank area. Despite increasing difficulties in securing land, in the mid-1980s tree planting covered 158,000 hectares of common lands, 132,000 hectares of which was on tank foreshore lands, 24 per cent of the total tank area in Tamil Nadu. In the total tree cover of the social forestry programmes, 80 per cent of the area planted was on tank foreshores (Arnold *et al.* 1988: 13–14). Hence, social forestry became an integral part of the tank landscape.

Despite being considered an economic resource, social forestry plantations are also considered a potential source of ecological imbalance in the tank landscape. Importantly, tank foreshore plantations prevent bullock carts from entering into the tank bed to remove the silt—a recurrent issue raised by farmers during the early 1980s when farmers used to apply tank silt to fertilize their fields. Over time, the leaves and branches of plants fall into the tank water and settle on the tank bed with the suspended silt load. This substantially increases the rate of siltation in tank water and reduces the effective capacity of tanks.

Intensification of farming

Traditional agricultural practices in dry lands were heavily disturbed when groundnut was introduced in Pudukkottai during the late nineteenth century. Soil erosion and nutrient decline in dry lands gained further impetus during the 1980s when new dry farming methods were introduced and land use intensified. As a result, the region has witnessed a great tension between intensification of land productivity and sustainability of the production base of the agrarian landscape.

During the pre-pump-set era, commercial crops like vegetables, banana, groundnut, green gram, and black gram were grown on garden lands with well water. The main crops on dry land were millets and groundnut. These crops were sown with a mix of crops like oil seeds, pulses, chilly, vegetables, and root crops. Mixed and multicropping systems were the rule in garden as well as in dry lands. This also spread risks of crop failure in irrigated fields when tanks went dry due to inadequate rain.

The arrival of pump sets induced a rapid shift in dry lands from multi-cropping to monocropping. Pump-set technology arrived along with other modern agricultural technologies: seed, fertilizer, and pesticides. A shift occurred towards water-intensive high-value commercial crops like sugarcane, banana, rice, and irrigated groundnut. These changes had multiple impacts on the cropping pattern in remaining rain-fed land.

Soil conservation

In 1982, soil conservation works were transferred to the Agricultural Engineering Department (AED). The AED has a top–down approach: farmers have not been consulted and sometimes not even informed about the AED work on their lands. Most bunding works were done along field boundaries, which do not match with contour lines. In heavy monsoon rains, these bunds are washed away, adding sediment to downstream tanks. Thus, the programme of soil conservation, in contrast to its objectives, has added to soil erosion and sedimentation, working against the principles supporting the functioning of interlinked tanks.

Granite mining and quarrying

Stone quarrying in Pudukkottai has long been seen by the government as a money-spinner. New additions in this regard are red soil, laterite, and granite mining for road and building works. Several mining sites—run by private contractors—are located in the upper catchment areas of tank cascades, and thousands of truck loads of red earth have been removed from these sites. Recently, since the liberalization of the economy when granite became a high-value commodity, granite mining had escalated. Earlier, stone quarries and crusher units were found mainly in uncultivated rocky areas. Now, after the granite boom, private and common lands, particularly dry lands have been targeted for mining. Earth dumped around the mining sites causes sedimentation of tanks in the valleys.

Social and legal barriers for desilting

The increasing use of chemical fertilizers since the 1970s replaced the historical practice of tank silt application to fertilize the fields. The scraping of silt from the tank beds each season kept the tank in good condition, but this practice has now almost ceased. Legal factors also discourage the removal of silt from tanks. The Government of India (GoI) has classified river sand as a minor mineral through the Mines and Minerals Act in 1957. Based on this Act, the State government issued rules controlling the utilization of minor minerals. These rules are being (mis)used by officials of the State Revenue Department to harass people wanting to remove silt or earth away from tanks for levelling their residential sites or cultivable fields (Gomathinayagam 1995). Hence, these developments in farming practices and legal measures discouraged farmers from removing silt from tanks, allowing accumulation of sediment load in tanks.

CHANGING AGRARIAN RELATIONS

A strong causal connection is found with changing agrarian relations and difficulties in the maintenance of tanks, mainly in clearing accumulated silt. During the post-independence period with the introduction of land reforms and other legislative measures in favour of labour, previously marginalized sections of society felt free from old relations of hierarchy and dominance through which social power was articulated in the operation and maintenance of tank systems (see Mosse 2003). This is part of wider socio-political changes in Tamil Nadu. The struggles waged by the Dravidian radical social reform movement and the left-wing peasant movements here created space for the emancipation of labour.

Moreover, the role of the state is crucial in dismantling the shell that kept labour in perpetual servitude. Provision of food through the public distribution system (PDS), allotment of homesteads and other pro-poor welfare measures by the state have liberated labour from older bonds. Generally, labour has become casualized, transforming tied labour relations into wage labour. New forms of labour contract have emerged with piece-rate wages for specified tasks. In addition, the growth of non-farm economic activities attracted part of the rural population. This, in turn, has brought profound changes. It has not only reduced the availability of farm labour, but also opened up space for cultural assertion of lower-caste and lower-class labour. This has challenged the ruling values in society, negating the meanings attached to labour and allowing refusal to perform socially and culturally low-status works such as desilting of tanks and channels (Anandhi *et al.* 2002; Gidwani and Sivaramakrishnan 2003).

SEDIMENTATION OF TANKS

These changing land use and agrarian relations have made the tank systems lose much of their storage capacity. A survey of tank capacity in Pudukkottai revealed that one-third of tank capacity has been lost due to sedimentation. The immediate response to reduced tank capacity was the emergence of wells and borewells in the tank command areas, as farmers adopted modern water extraction technologies. This technological solution for a socio-ecological problem resulted in the emergence of distinct landscapes, to which the discussion now returns.

THE EVOLUTION OF TWO DISTINCT TANK LANDSCAPES

This ecological crisis has worked in line with the structural features of the agrarian economy of the region, resulting in the emergence of the two distinct landscapes of wet and dry zones in an otherwise a dry region. As summarized in Table 10.1, tank cascades in wet and dry zones now exhibit variation in water management practices, cropping patterns, and degrees of commoditization. These transformations have implications for the functioning and use of tank cascade landscapes in the wet and dry zones.

CHANGING SOCIO-TECHNICAL LANDSCAPES, CHANGING CASCADES

The changing environmental and agrarian conditions described above play a crucial role in reshaping the historically recognized rights to water. These are not only encoded in the language of rules, social norms, and bureaucratic regulations of water and access, but also inscribed in the irrigation and drainage artefacts and the physical landscapes of the tank cascades.

One of the important 'signposts of struggle' (Mollinga and Bolding 1996) over rights to water along both the wet and dry zone cascades is the changing pattern of use of spillways in tanks. Calingulas are artefacts that technically connect the tanks in series. Calingulas contain elements of water rights that determine the quantum of water to be stored in a tank as well as flows of surplus water to tanks located downstream. In the context of capacity reduction of tanks and the emergence of new water-intensive cropping regimes farmers mobilize men and material to redesign the technical dimensions of calingulas. This is done by either temporarily altering the flow of surplus or by lobbying to change the basic design of calingulas into another type of spillway to retain more water in their tanks.

In this situation, any upstream tank farmer who cultivates land very close to the foreshore area of the immediate downstream tank would not support any change in the existing calingula design that raises the water level in the tank. This is one context of inter-tank conflicts centring on calingulas in tank cascades. The conflicts and struggles around calingulas illustrate the changing physical and technical features of the landscape. They are also a demonstration of the new agrarian regimes produced by broader societal, economic and environmental changes.

The necessity of retaining as much water in tanks as possible is often felt in both the wet and dry zones. It is even more crucial once tanks become understood as possible important sources of groundwater recharge for wells and borewells in this region. This is also evident in the way tank water allocation and distribution at the level of individual tanks is organized.

The intensity of the struggles over reshaping the design of calingulas is revealed in the spatial pattern of spillway use in both the wet and dry zones. As mentioned, calingulas are being extensively replaced by other types of spillways like broad-crested weirs and sharp-crested weirs. In some cases, calingulas are redesigned in combination with one of the types of spillway mentioned (for example, calingulas with broad-crested weirs and calingulas with shutter arrangements). At the same time, there is a marked difference in the spatial pattern of calingula use. For instance, in the wet zone cascade, 52 per cent of the tanks (10 out of 19) retain calingulas as spillways, in the dry zone only 28 per cent (24 out of 84). What explains these spatial differences? What are the social meanings of these different spatial patterns observed in the wet and dry zones?

Two important factors come to the fore: the social character of the wet and dry zones, and the physical character of the landscape (including groundwater availability). In both cascades, water management continues to be an important activity at the level of individual tanks. However, the intensity of management varies between tanks within each cascade, contingent upon several locally specific hydrological and structural factors. At the same time, tank use and management follow the core logic of agricultural production in each zone.

Availability of extra water through an anicut and access to groundwater shape the process of commoditization. Particularly, borewell water fuels the engine of growth. However, this growth engine is also susceptible to depreciation. The intensified groundwater-based agricultural surplus extraction is diminishing with the decline in groundwater levels over the period. In other words, commoditization processes in the wet zone, in their turn, shape the biophysical features. This factor plays a crucial role in the management of tank water. The desire to store more water in the tanks overarches the interests of well owners and non-well farmers. However, well owners exert power to control water supply at a certain level of tank storage, facilitating recharge of borewells, thereby lubricating the process of accumulation. Another important interest to keep water to a certain level in tanks up to a particular period is related to fisheries, in which the commercial farmer has a greater stake. This is more and more a feature in tanks where struggles around calingulas are intensifying. Thus, the organizing principle in the struggle over reshaping calingula design in tanks in the wet zone is profit maximization with each drop of available water.

In the dry zone, the logic is different. The principle of minimization of risk dictates strategic action in the struggles over spillway design here. This is manifested in the different strategies of tank water allocation and distribution followed under varying hydrological conditions in a given agricultural year or even within a cropping season. The agricultural practices adopted in tank ayacuts in this zone also corroborate this fact. Tanks are mainly considered as surface irrigation infrastructure. Moreover, the management of tank water in the dry zone is not only based on the irrigation requirements of crops in the ayacut but also includes water for domestic uses like bathing, washing of clothes, fisheries, livestock rearing, and recharge of wells. For all these purposes, the greatest volume of water needs to be stored in tanks up to the highest possible level.

The agrarian social interests that are a force shaping the use of spillways in tanks work not in isolation, but together with the natural qualities of the landscapes. In other words, struggles take place along the slope, gradients, contour lines, and drainage canals of both the wet and dry zones. In this way, the physical configurations of the landscapes either facilitate or constrain the choice of a particular type of artefact. For instance, the topography of the wet zone tank cascade is rather flat. The FTL marked for the downstream tank and the tank at the top of the cascade is similar, at just 50 metres at mean sea level (MSL), and there is little difference in elevation between the downstream and upstream areas of the cascade. The cascade is also located along the banks of the seasonal river, Vellar. The topography of the cascade landscape is vulnerable during floods with heavy rain (for example, in 1976, 1977, 1979, 1990, 1993, 1999, and 2005). Hence, spillways in these series of tanks mainly function as flood moderators, particularly during times of vulnerability and risk associated with abundant water.

They protect people, crops, livestock, and infrastructure, as most human settlements and irrigated fields are located within the irrigated landscape. When compared to other types of spillways, calingulas are generally considered as effective drainage infrastructure during heavy rains, as the design is adjustable to varying water levels. The topography also assists the irrigation engineers in convincing other potential actors to retain calingula designs as solutions for inter-tank conflicts over drainage flows in many cases. Therefore, calingulas in the wet zone tanks 'negotiate' (moderate) floods during heavy rains and store water in tanks at 'optimum' level during normal times. This is also one reason for the continued use of calingulas as spillways in 50 per cent of the tanks in the wet zone cascade.

In the dry zone, the slope is undulating and shows a significant variation in elevation between lower and upper parts of the cascade landscape. For example, the FTL of the downstream tank in the series is at 80 metres, whereas it is 116 metres in the most upstream tank. This difference in elevation plays a significant role in locating human settlements and other productive infrastructure. Drainage is not a problem at all. As the sources of supply for tanks in the dry zone are mainly rainfall runoff and drainage flow from nearby upstream tanks, there is little problem of flooding and drainage needs. In fact, water is a much needed resource everywhere and for everyone. In this context, the natural qualities of the dry zone landscape encourage calingula designs to be retained only in a fraction of tanks (28 per cent). In most cases people permit conversion of these historical designs (72 per cent) into other type of spillways that enable more water retention in tanks. Hence, interest in the storage function has manifested itself in the adoption of new spillways in tanks in this type of cascade.

* * *

This chapter has analysed and explained processes of change in and around tank systems, applying the concept of a socio-technical landscape to the study of interlocking tank cascades in Pudukkottai, Tamil Nadu. The analysis has shown that changes in land and water use around tanks are spatially varied and far more complex than the explanations given in the ongoing debates on the status and functioning of tanks in India. The starting challenge of this chapter can now be revisited: how should one envision the current state of irrigation tanks? How could one capture the dynamism exhibited by tank systems? The chapter has discussed that the socio-technical landscape is a dynamic outcome of contestation and negotiation between the state and agrarian society. Moreover, the tank-irrigated landscape is built through prolonged social struggle between powerful landed interests and subjugated labour. Yet the social and biophysical landscape of tanks is also shaped and reshaped through the interventions of the state. Besides, the physical landscape itself becomes a social actor in this struggle. The changing water supply, and land and water use in tank cascades in recent history indicate the shifts in power balance among social classes. All these transformations are encapsulated by the varied spatial pattern seen in the shaping and reshaping of irrigation and drainage artefacts in tanks.

Instead of perceiving tanks as static entities, the discussions on tanks could sharpen their analytical strength by recognizing that the current status of tanks is an outcome of prolonged social struggle between state and agrarian society, and among social classes within agrarian society. The future of the tank cascades requires a drastic change in the way tanks are seen today. First, current development thinking on tank irrigation needs to recognize that tank landscapes are complex, varied, interconnected, and dynamic. Second, it is important to see how the prevailing social structure and skewed power relations among social actors shape and reshape tank landscapes. Third, it requires an approach that contains principles and good practices of integrated natural resource management, incorporating new institutional structures for equitable and sustainable development.

Notes

[1] The literature on NTS is extensive; see Agarwal and Narain (1997) for 'community traditions' in water management. For a critique on ideas of community in natural resource management, see Agrawal and Gibson 1999 and Sinha *et al.* 1998.

[2] Main proponents of NIS are Palanisami (2000), Shah *et al.* (1999), and Sharma (2003).

[3] For an exhaustive bibliography, see Harriss (1980), Patnaik (1990), and Thorner (1982). For a summary of critique, see Pandian (1990).

[4] The calingula is the historical design of overflow weirs in tanks in south India. It attained currency before British rule.

[5] Nadu connotes two different but interrelated meanings: (*i*) a micro-territorial unit comprising of several villages, and (*ii*) an assembly that 'governs' the territory (Karashima 1984; Subbarayalu 1973).

[6] Property rights in Tamil society are known as the *kaniyatchi* system. In the medieval social order, caste and kaniyatchi were mutually constitutive, and caste became the cultural marker to define property. Rights in agrarian social resources were firmly embedded in the prevailing social hierarchy of caste. See Granda (1984), Shah (1985), and Sivakumar (1978). As Ludden states, 'caste boundaries, rituals of ranking, and state grants of entitlement were cultural implements for the social reproduction of rights' (1985: 167).

[7] Pre-colonial governments in India managed land mainly under two systems. Under the first, the produce of the land was divided into three shares: the government's share (*melvaram*), cultivator's share (*kudivaram*), and a share for village service holders (*suthandiram*). In the other system, the government granted tax-free land to privileged groups. In the colonial period these systems were reformulated. New land revenue assessment and collection systems were constituted, which replaced grain shares by cash as land tax and established private property on land.

[8] This research examined cases mounted in the Darbar office and courts for adjudication. Out of 594 cases studied, 156 (26 per cent) concerned issues related to calingulas alone; 501 (84 per cent) related to interlocking tank networks including redesign of calingulas (based on data from the Index of the Pudukkottai Darbar Office Records and the Pudukkottai Dewan Peishkar Office Records).

[9] Stone designs of sluices and weirs had been the work of *ulliyan* (tank–technician–artisans), until British colonial engineers came to be involved.

[10] A swing basket is a tin container with a double rope on opposite sides used by two men to lift water from a surface water body, which is quite labour-intensive.

[11] Most landowners operated bullock-driven devices to lift water from wells.

[12] For more on pannaiyal relations in Tamil Nadu, see Kumar (1965).

[13] Ambalar or ambalam was a hereditary position recognized by the previous rulers of the region. The ambalams came from the dominant Kallar caste. They had control over men and other material resources. Ambalams exercised social and political power through several means (Dirks 1989: 280–3).

[14] Social forestry—the planting of trees on village common lands—started as a pilot project in 1960 under the name of Farm Forestry and was continued as part of the third Five Year Plan from 1961. The project claimed to ensure sustained supply of fuel wood, small timber, fodder, grass, oil seeds, and other minor forest produce for rural population needs.

11 MICRO-HYDEL AND IRRIGATION: IS ANY OTHER WATER TECHNOLOGY DIFFERENT?[1]

Amreeta Regmi and *Linden Vincent*

In the hills of Nepal, ecology and terrain bring special challenges and opportunities for the provision of electrical power supply. While facing accessibility problems, there are abundant streams and rivers where the motion of flowing water can be converted into steady mechanical power for electricity generation. These sources are often already tapped for traditional technologies of water mills and farmer-managed irrigation systems (FMIS), whose design and operation have shown the capabilities of farmers to manage water. Micro-hydel energy systems[2] (MHES) have been an ongoing focus of attention in electricity supply, with a vigorous design effort, often emphasized and romanticized as the panacea for rural electrification under an 'appropriate technology' and 'small is beautiful' paradigm. The design of MHES in Nepal is related with provision of five main functions, alone or together. Lighting, agro-processing, and irrigation are most common; others are for small-scale industrial purposes supporting private sector economic activities, and tourism. However, studies have shown that these small-capacity turbines rarely operate to meet all the community's energy needs (Regmi 2004a). Performance statistics from the late 1990s showed that only 23.1 per cent of MHES in Nepal were running well; the remainder did not function optimally, and 73.4 per cent were categorized functionally as failures (Nepal 1998), contradicting support in this sector by various institutions.

This chapter presents two case studies of MHES from a larger study undertaken in the Kabhre Palanchowk district, Central Development Region. Fieldwork in 2001–2 investigated these realities in communities where MHES were developed alongside irrigation systems (Regmi 2004a). The study examined hydel power systems that reflected key MHES technologies[3] for supplying local electricity supplied by streams of the Roshi river, one of the major tributaries of the Sun Kosi river. Social and hydrological data were collected through various qualitative and quantitative methods.[4] Both case study villages, Pinthali and Katunje Besi, have MHES integrated into an irrigation system, but demonstrate different technologies and design approaches, which evolved in very different political contexts. This allows comparison of how technologies and institutions have interacted dynamically in shaping the system and in service provision to communities. The research was undertaken during a challenging period of conflict between the Maoist movement and the state. Regmi (2004a) gives more details of these power struggles in the area and their difficulties for researchers.

This chapter explores how the new MHES interfaced with existing social structures, and the roles and actions of humans as agents in getting these systems to provide their energy needs with equity, achieving what we call 'technological democracy'. In this regard, the chapter presents findings on how micro-hydel technology becomes 'adaptive', that is, able to function in its technical and social environment to meet the changing needs of the community it is supposed to serve. The chapter aims to improve understanding of joint development of MHES and irrigation, capable of supplying a viable electricity supply without affecting irrigation. It illustrates how such joint systems and development of MHES infrastructure are 'no different' from the socio-technical systems of irrigation discussed in earlier chapters.

This chapter draws on two complementary conceptual frameworks to study these MHES technologies as socio-technical systems. First, the concept of 'community-oriented' technology is seen as evolving inside a triad of three dominant forces: structures (policy, institutions, and governance), systems (designed physical system in its environment), and agents (people and knowledge). This community micro-hydel technology is then examined as an adaptive socio-technical system of water control for energy generation, looking at three levels in this adaptability. First, it is seen as a hydraulic ensemble in its early design and choices integrating technology, agro-ecology, and society. Then, as it adapts as a transformative unit, the design enables users to meet and transform both productive and consumptive uses of water (for power and irrigation). Finally, it emerges as an evolutionary system, yielding tangible and intangible benefits, with local institutions that can control and manage their own resources, showing reflexive coping with change.

Second, the study applies the framework of accountability developed for irrigation, to study how operational systems evolve to manage irrigation and electricity supply. In the case of microhydel, 'accountability' links the community that uses the technology with not only the designers and implementers, but also with the donor agencies, the state, the funding agency, and the local governing structure. Subsequently, these actors are linked with the managers, operators, and energy users' committee running the systems. The processes for accountability are built in a given 'political space' that allows designed systems to evolve, as committees build their understanding, capacities and choices. Kloezen (2002: 12) saw accountability as a 'process and an outcome of negotiations' and studied this concept in three different fields: operational accountability, financial accountability, and socio-economic and political accountability. The first field is closely interlinked with technical 'performance' as shaped also by operation and maintenance organized by management. The second field shows how users and managers can monitor, control, and influence the financial management of the system. The third field indicates the manner in which the management arrangements (for decision-making, user representation, and leadership selection) are negotiated. In addition to these fields, this study introduces the concept of 'constitutional accountability' to describe the governance mechanisms providing the supportive legislative and constitutional frameworks ensuring electricity supply for citizens. Constitutional accountability emerges as a core index (Regmi 2004b) within the different fields of accountability in making technology democratic.

The following section examines the policy space in which MHES have evolved in Nepal. The following two sections describe the case study sites and the provision of electricity services and adoption of electrical goods by villagers. The subsequent four sections look at issues of operational, financial, socio-political, and constitutional accountability. The conclusions follow thereafter.

SYSTEMS, STRUCTURES, AND AGENTS
MATERIALIZING MHES IN NEPAL

MHES in Nepal did not evolve as part of a comprehensive energy policy and strategy, but rather as a set of 'tinkering' interventions. Although modern hydropower was introduced in Nepal almost a century ago, Nepal's energy sector development has been seen as slow, costly, and urban-biased (Pandey 1996). Various parallel dichotomous structures began to take shape during this time—rural versus urban supplies, grid versus non-grid systems—linking with different possibilities for macro, small, and MHES technology, and accepting multiple influences on technology from bilateral donors.[5] The micro-hydel sector has been entirely funded by external agencies and heavily linked with soft loans and major programmes.

Four major government institutions were involved in the rural electrification sector in Nepal at the time of this study. These were the Nepal Electricity Authority (NEA), the Remote Area Development Committee (RADC), the Alternative Energy Promotion Centre (AEPC), and the King Mahendra Trust for Nature Conservation.

The NEA carries a mandate for rural electrification. However, actual installations and institutional arrangements have also developed in parallel with new local institutions. The RADC, under the Ministry of Local Development (MLD) which oversees social and economic development, has been actively involved in rural electrification, working in close cooperation with international non-governmental organizations (INGOs). The Rural Energy Development Programme (REDP), the AEPC—in which important roles have been played by the INGO Intermediate Technology[6]—the King Mahendra Trust for Nature Conservation and other influential non-governmental organizations (NGOs) have also been critically involved. These agencies and institutional arrangements interfaced with the local planning and decision-making processes, leading to the emergence of various local approaches and models for local MHES governance systems. These, in turn, shaped the formation of networks, forums, and coalition groups such as Energy Management Committees (EMC), Energy User Groups (EUG), and other functional support organizations. From this complex of agencies two community-oriented models of power supply have emerged that are relevant to this study. The first is a 'subsistence consumptive' perspective, in which technology is designed to cater to minimal needs (40–200 watt) per household. The second is the productive consumptive model, where technology aims to supply and promote both consumptive and productive demands for electricity, in domestic, agro-processing, or small-scale industry. This model is found at Pinthali and Katunje Besi.[7]

Two parallel efforts were visible in design and dissemination of MHES prototypes. The Swiss Association for Technical Assistance (SATA) supported the Kathmandu manufacturing sector, whereas the United Mission to Nepal (UMN) entered the Central-Western Region and established the decentralized Butwol Training Institute (BTI). As such, various micro-hydel prototypes such as the Francis Wheel, Pelton, cross-flow, and propeller designs gradually entered the market, with the introduction of the first locally manufactured propeller of 5 kilowatts in 1962. As with all new designs, problems associated with prototype development were not uncommon. Therefore research and development (R&D) pursued more adaptable alternative turbine designs (Aitken et al. 1991), and by the 1970s cross-flow turbines were recognized as the most suitable technology for the rural hill areas. This shaped the rise of private sector companies such as Balaju Yantra Shala (BYS) in Kathmandu, and Development and Consulting Services (DCS) in Butwol. By 1987–8, 423 installations of cross-flow turbine plants were recorded (Jantzen and Koirala 1989), and over 600 Pelton turbines.

Guided by concurrent incentive policies for certain prototype diffusion, both peltric and non-peltric turbines became popular. Peltric sets were used for demands of less than 5 kilowatt electricity generation, and non-peltric sets were used for higher demands. By 2001, on average 60–70 schemes were being installed annually and over 1990 MHES plants were recorded. During the 1998–2001 period, the REDP installations indicate rapid diffusion of the 'community model', which became a dominant approach of development. In the fiscal year 2000, over 53 per cent out of 36 MHES plants installed were undertaken by the REDP. Prior to 2000, design prototypes focused on producing small-capacity turbines. While the rise in peltric sets peaked, little attention was given to the standardization of Operation and Maintenance (O&M), mechanical electrical guidelines, and facilitation for spare parts services. In time, changes in sector and subsidy policies emphasized more add-on schemes for local production and service support at the local level.

While several agencies, networks and coalitions were created for promoting MHES in Nepal, three major interventions paved the way towards local planning and implementation. The World Bank emphasized the role of the region in energy sector development, and initiatives for monitoring and supervising energy implementation at the district level. The Danish International Development Agency (DANIDA) coordinated strategy and policy development with the national agencies, whereas the AEPC and REDP decentralized model of development gave a close interface with the district development committee (DDC) and village development committee (VDC). In an effort to promote this decentralized model, agencies and institutions were created at different levels. Most important at the national level was the Rural Energy Development Board (REDB) under the MLD. This Board is governed by a management committee (MC) comprising the MLD, the Water and Energy Commission Secretariat (WECS), the National Planning Commission (NPC), and REDP. This was an important recognition of REDP's 'decentralized' community-oriented design approach as a dominant model for Nepal. Besides, this move signalled a first indication of an effort to integrate MHES in the national planning process. However, the planning interface between the national and district level committees was not yet clearly defined at the time of this study. Submission of demand from village level and the vetting procedures in fulfilling these required demands were cumbersome processes, as described ahead and in the case studies.

At the district level, two major planning documents, a five-years' plan and annual plans, documented the district's energy needs. These included primarily rural expansion through the NEA; planning for MHES was not included as part of a comprehensive district energy strategy at the time of study.

THE CROSS-FLOW TURBINE AT PINTHALI

Electricity has transformed our village: and after I die, I will take a letter from the community to inform our ancestors so that they believe me.
—Kainla Lama, Pinthali village, June 2001

The introduction of micro-hydel technology in Pinthali began in 1997 by building and connecting the design to an irrigation system. A 12-kilowatt capacity cross-flow turbine of impulse type was installed in 1998 under a design developed by United Nations Development Programme (UNDP)/REDP. From the outset, the community planned for lighting and agro-processing to be introduced into the design, to drive an oil mill and a rice huller. All three electro-mechanical components were designed, manufactured, and installed by Mr. Akkal Man Nakarmi of Kathmandu Metal Industries (KMI) in Kathmandu, later brought to Pinthali. This project has been part of a longer-term effort of the Tamang community to re-identify itself socially and technologically from an ethnically backward group to a hardworking innovative community.

Prior to the micro-hydro intervention, there were three community-constructed irrigation canals in the Daunne catchment. The first irrigation canal was initiated to water the unbunded slopes. In 1979, a second canal was constructed to irrigate lower terraces to produce paddy, buckwheat, and maize. Water scarcity began to be felt consequent to high conveyance losses, the narrow canal width, and difficulties in maintaining canal alignment. As economic activity grew, a third canal was constructed to irrigate cash crops such as cumin, garlic, and grams on the plateau where the village lies. The canal system became a source of conflict over water rights, as there was no formal mechanism to resolve competing claims for water allocation and distribution. This water scarcity and contestation led to a new design for the irrigation systems and formalization of the local water allocation. In 1981, the community decided to consolidate three canal structures at the source, followed by increases in canal width to augment the volume of water and extending its length to irrigate the fertile plateau of Pinthali. The principles for local institutional development were established from early on.

Pinthali residents manifested their need for power in 1980 when a formal request was put forward to the Agriculture Development Bank of Nepal (ADBN) for a MHES, with subsequent requests to the MLD around 1985 and 1988. Then in 1992, another request was made to the Economic Development Council (EDC) for electricity, using the community's financial resources to initiate a feasibility study. The lengthy processes and delayed responses of the central government did not deter local leaders and entrepreneurs from taking action. A community entrepreneur, Bajra Dhoj Lama installed a *paani ghatta* (water mill) to facilitate the agro-processing needs of the community. In addition to milling services, in 1997 he generated sufficient electricity to light 35 houses in Pinthali by installing a diesel generator, providing electricity, and levying a formal tariff of NR 1 for every 25-watt bulb. Since local initiatives could not supply sufficient electricity to all households, the community approached the REDP office for electrification. A feasibility survey indicated that Daunne khola had sufficient water resources to generate 12 kilowatt of electricity. This development was soon followed by REDP's intervention in January 1997 with the introduction of a cross-flow turbine that generated 12 kilowatt of electricity. Figure 11.1 shows the hydraulic ensemble of the MHES and village and irrigation layout.

The REDP energy intervention in Pinthali was made as part of its wider holistic approach to rural livelihood enhancement through sustainable development. This model included not only rural energy development, but also environmental conservation, local economy improvement, and capacity building through institutional development, aided by a social mobilization process. This community mobilization package consisted of six basic principles: organization

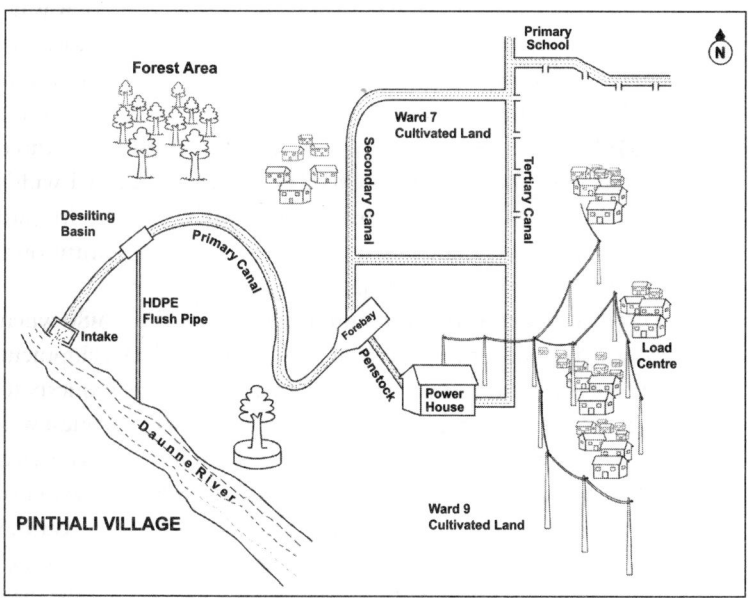

FIGURE 11.1 Pinthali Village, MHES, and Irrigation System

Source: Reprinted from Regmi (2004a).

development; skill enhancement; capital formation; technology promotion; environmental management; and women's empowerment
(Kabhre DDC/REDP 2002). With the assistance of the local NGO
partner organization Resource Management and Rural Empowerment
Centre (REMREC), a mass meeting was conducted to inform all
community members about the upcoming project. In August 1997,
REDP initiated its community mobilization programme and formed
five female and six male community organizations (COs). Different
functional groups (FGs) were then formed to undertake sectoral
activities such as the micro-hydro, forestry, health and sanitation,
adult literacy, improved stove, and small business enterprises. The
social mobilizer lived in the village for about six months assisting the
COs to mature and transform themselves into FGs. After the initial
six months, REDP staff made intermittent field visits to ensure that
the groups were functioning according to REDP's expectation. In
the course of mobilization, the final committee included a member
from each FG to represent the highest tier of the local institution.
This set-up was called the Micro Hydro Functional Group (MHFG).
Subsequently, after the implementation phase, the MHFG was converted to an executive committee known as the Daunne River Micro
Hydro Power Executive Committee (DRMHPEC) and was eventually
registered at the district level as a cooperative.

The MHFG became responsible for resource mobilization and
implementation of the project. The MHFG evolved from the COs,
each of which designated a member to constitute the eleven-member
team consisting of six males and five females. The community also
selected a chairperson and a manager. The implementation process
then secured contributions and commitments from the community
and various stakeholders for the proposed programme. Using a bid
process to identify the installer and the designer, KMI, the lowest
bidder, was selected to install the cross flow turbine to generate electricity. REDP transferred a check for NR 1,257,336 to DRMHEC
for initiation of the project.[8] The transfer of funds to the community
catalyzed the community in completing work through local participation. In addition, about 20 community members guaranteed their
land as collateral to ADBN, for the additional loan of NR 189,686
required to be pledged from the community. Other contributions
included: about NR 100,000 from DDC; NR 200,000 from VDC;

a government grant subsidy of NR 199,150; and about NR 79,500, from the village community funds, in addition to labour and local material. The cost per kilowatt production of electricity amounted to about NR 123,459. About 70 per cent of the total budget was spent on the purchase of the turbine. Extensive work on canal and other civil structures was carried out at a low cost utilizing in kind labour contributions from the village.

The project implementation was initially hampered by some disputes over erecting poles on certain private land holdings, later resolved by negotiation in a mass meeting. Then on 14 June 1998 electricity was first generated through the DRMHPEC effort. After completion of the project, the DRMHPEC did not function as smoothly as anticipated. This was partly because the police operation of 'Kilo Sierra Two'—created to eliminate all traces of the Maoist movement—generated tension within the community. Towards the end of 1998, the manager went underground[9] because of rumours circulating that he was sought by the police. Therefore, the responsibility of managing the executive committee was handed over to Gyan Bahadur Lama. Following this institutional development, two village youths were trained as operators and two as repair and maintenance in-charge workers. Consultative mass meetings had been taking place regularly and on a demand basis. Notes and records from these meetings were being maintained.[10] Thus, a formalized governing system was put in place whereby management, operational, and financial systems began to conform around an agreed institutional structure.

Pinthali shows the evolution of a 'transformative system', driven by and embedded within adaptive and accountable design processes to meet the irrigation, energy, and agro-processing needs of the community. From an early stage of technology development, the interface between systems, structure, and agents steered the development of technology, beginning from a selected innovation, to the community's participation in engineering of an effective management system. The Pinthali case study shows the suitability of technology in meeting local demands of society, and capacity to manage within a given environment. Additionally, this example shows that evolution of design alone is insufficient to promote technological democracy and accountable governance. Rather, the design process needs to create space to develop knowledge systems, leadership, and the capacity to manage these systems.

THE PROPELLER AT KATUNJE BESI

This project has become like a big bone that is stuck in our throat, we can neither spit it out not swallow it.
—Devi Shrestha, Katunje Besi village, May 2001

The 8-kilowatt propeller turbine installed and implemented at Katunje Besi in 1998, under an R&D initiative between a national designer-manufacturer KMI and REDP, triggered a contrasting story of adaptation to technology. The selected design option to supply electricity was not the community's first choice. The prototype was built in a workshop in Kathmandu and later installed in Katunje. The main objectives of the Katunje project were: to evaluate the technical reliability of a locally built propeller set running under extreme conditions of a wild Nepal river; to develop local capacity for repair and renewals; and to harness low head power potential of 1 to 5 metres (KMI 1998). Typically, a low head turbine of this sort is used for irrigating the Terai (flat plain) area of Nepal. Its efficiency and safety requires precision design, high operational skill, and good technical support and knowledge. The design and implementation stages were clouded by contestation from the community on site and technology selection. The community at the time was fragmented, diverse, and undergoing rapid social and political changes. However, the installation took place almost overnight because of the rising Maoist conflict. The technology became dysfunctional within six months, due to high management turnover, poorly trained operators, O&M problems, and the nature of the mobile community.

Katunje Besi is located about 105 kilometres east of Kathmandu, in a narrow valley within the Roshi watershed, encompassing an irrigation command area of about 11 hectares. Most inhabitants depend on agriculture, largely sharecropping maize, paddy, mustard, and wheat. Cash crops and dairy farming became increasingly popular. As with the Pinthali case study, this design was combined with an existing irrigation system. Unlike the design in Pinthali, where the forebay is designed to supply water for power and irrigation use, this plant made provision for irrigation supply by diverting water to a secondary canal for electricity use. The water conveyance structure consisted of three separate canals, the primary canal leading to the powerhouse, and the secondary and tertiary canals serving the upper and lower command

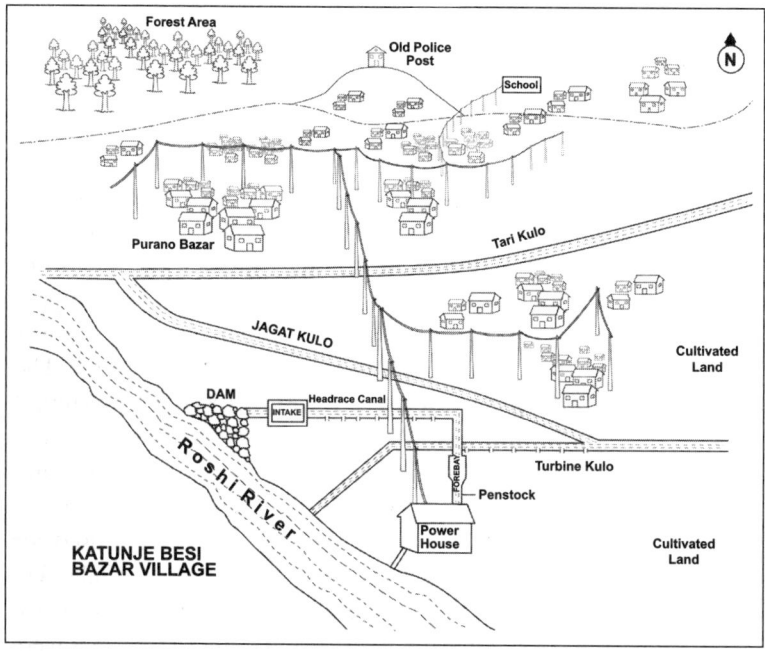

FIGURE 11.2 Katunje Besi Village Bazaar, Irrigation Canals, and MHES

Source: Reprinted from Regmi (2004a).

area (Figure 11.2). This did not affect water allocation drastically, given the abundance of water. The water supply is used for irrigation during the day and for power generation at night.

Development of Katunje's narrow floodplain began in 1979, as a temporary day base for farmers living on the mountain top who were engaged in economic and trade activities. In 1979, there was only one house where a mill eventually developed. This area expanded rapidly with road construction. Interestingly, the community's first request for electricity in 1989, to the central level authorities, was for connections from a gridline four miles away. This request led to a survey of the area. As migration from the hills to the plains increased, so did innovations introduced by local entrepreneurs. To support small-scale business activities, by 1998 various agro-processing mills, a saw mill, and other traditional technology for grinding and hulling were introduced. The demand for increased electricity to support these industrial activities became apparent. Between 1991 and 1992, the community organized itself and distributed electricity to 30 homes from a diesel-run generator. Subsequently, the original mill owner Nanda Raj Lama distributed 5 kilowatt of electricity from his private mill. By 2002, Katunje Besi had a community of over 66 households, including several landless families. However, one of the consequences of the Maoist insurgency was a steady outmigration of youths to India and elsewhere. Within a hardworking community there were coexisting dichotomies of development, divisive politics, and social unrest.

The community mobilization for Katunje Besi initially differed from regular REDP programmes, as it was considered a prototype intervention. The project agreement was signed by the donor, designer, and the VDC. Eventually, a similar REDP mobilization approach was applied in Katunje Besi with a 16-member management team that included a chairperson, manager, and an operator. However, the process to mobilize the community took over six months before COs were in place. Finally, REDP was reconciled in accepting the site in its programme and began applying its community mobilization model using organizers from REMREC.

At the local level, the village of Katunje Besi Bazaar had no earlier formal water user's group (WUG), as water was abundant. In the narrow flood plain farmers irrigated directly from canals, then channelled water from field to field. Farmers reciprocated in controlling water for irrigation, as required. Repair work was organized and conflicts resolved through mutual discussion and informal meetings in teashops, elders' homes, and the market place. However, given the recent and transitory nature of the settlement and turnover of people, social relations and incipient institutions were weaker and more vulnerable to disagreements than at Pinthali.

After the initial step of CO formation, the selection of 8 male and 8 female members from various sub-COs led to formation of an FG with 16 members. The selected group then identified a chairperson and a manager for the FG. Thereafter, the rules and formalities for project initiation were undertaken. Thus, in early 1998 a formal agreement was made between this FG and REDP to begin the Roshi River Micro Hydropower Project (RRMHP). With this agreement, REDP was responsible for providing electrical equipment, including the turbine and electro-mechanical components. The FG was responsible for mobilizing the community for construction of the canal and powerhouse, and providing poles and cables.

After formation of the FG, cash contributions were collected for the construction of the scheme. Some community members reluctantly provided support during the construction. In the Katunje Besi area, voluntary labour was difficult to get, and the members felt that they were being taken advantage of. Each household was responsible for constructing 17 metres of canal, including masonry work, carrying cement and stone, and preparing gravel. In transmission lines alone, the community invested over USD 1,000. In addition, seven members of the community deposited their land titles with the ADBN as collateral for a community loan, to meet the community's share of cash contribution. Construction of the canal and powerhouse affected the agricultural land of 11 community members, which was acquired free of cost for the project. The community members lamented that they were fooled by REDP in allowing such financial burdens, and were convinced that the electricity supply would not be a successful project. They mocked that the canal they were digging was meant to bury the REDP staff members and the social mobilizers. Though there remained differences over this project, some active members ensured that the project got completed; thus, in June 1998, the RRMHP was successfully tested. For about five months, the project seemed to be running fine. Then, in the sixth month—when the warranty period of the manufacturer KMI was over—problems began appearing. The community eventually felt that the inappropriate turbine and the temporary design of civil works indicated that both donor and manufacturer failed to foresee the significant burden the community would be exposed to.

After project completion, the FG became transformed into a management committee with an elected chairperson and a manager, and was called the Roshi River Micro Hydropower Management Committee (RRHMC). No written by-laws existed, but the RRHMC was expected to conduct monthly meetings and minute them, and disclose financial statements. However, these procedures were not complied with, and there was no documentation available for review. As no member wanted to take responsibility for collecting the electricity tariff, the REDP intervened. The project was transformed and registered as a co-operative as the Roshi Khola Micro Hydro Cooperative (RKMHC). The management committee handed over the project to the previously elected manager of the FG on a contractual basis for NR 1,500 per month; this effort failed when contracted wages were not paid.

After a few months, community discontent gained momentum as financial records were non-existent and accountability issues were raised. With REDP's presence in the district, support and supervision by the district extended beyond project completion. However, REDP was also under scrutiny from members who represented both the state police and the Maoists. To help solve diverse problems, REDP issued an additional amount of approximately USD 1,300 to the management committee to improve the temporary weir supplying the lower command area, plus about USD 2,060 to RKHMC for their other holistic activities. Yet the financial accountability of allocated sums for identified enterprises was not transparent, and there was no clear authority designated for distributing finances. The weir remained temporary; the community continued to stabilize it every monsoon. Finally, in September 2001, the community gathered together and confronted the manager of RKHMC over his financial accountability at the police post. The manager reimbursed some money not accounted for. As a result, committee members were reshuffled and the management leadership was replaced.

However, the project had faced such severe technical and management problems that it remained shut for three months. In April 2002, with renewed interest from the REDP, which coincided with visits by various scholars and potential donors to the site, additional development took place. However, despite several management changes, there was no visible progress. The REDP presented an idea to the community for a management committee comprising a 100 per cent female membership. However, the female members argued that if the community did not cooperate with men, why would they behave differently with women? Beyond management and financial issues, the fundamental problem with the project was a technical one. In all fairness to the community, the propeller design was inappropriate within the Roshi hydrology and agro-ecology. REDP finally spent approximately USD 400 to bring KMI back to the village in April 2002 to repair the machine. This was their first visit after the installation was completed in 1997.

Katunje Besi's design and development were surrounded by social, political, and environmental changes. Here, the choice of technology undermined possibilities to evolve as a transformative unit. The critical limitations were the lack of design consideration in engaging the community collectively, and the chaotic efforts in institutionalizing necessary mechanisms for O&M of the system. Katunje Besi had seen plenty of prior innovation in agro-technologies under local leadership and entrepreneurship. However, the introduction of the propeller turbine discounted this local reality with the narrow prototype approach that failed to interlink with existing knowledge systems and provide scope for further growth. The design was based on insufficient consultations and restricted ecological and technological precepts, without fully understanding the social and political dynamics or site specifics of the location. A formal governing institution like that found in Pinthali failed to establish itself easily in Katunje. However, community action was able to challenge financial accountability in its attempt to democratize the technology.

PROVISION OF ELECTRICAL SERVICES AND THE ADOPTION OF ELECTRICAL GOODS

Providing a micro-hydel service is seen as supplying electricity, while simultaneously ensuring reliability of services, repairing breakdowns and assuring safety. Most MHES supply electricity throughout the night, from dusk to dawn. Electricity allocation varies by project, from a 25-watt light bulb only, to a combination of two or three devices including television or radio to reach a higher total per household, also varying between households. The access to electricity per household was 100 watts at Pinthali and 80 watts at Katunje Besi, levels considered as standard for sufficient supply of electricity. However, the potential demand was higher in both Pinthali and Katunje Besi, with some consumers indicating the need for 300 watts per household. They recognized how limits of power supply prevented expansion into economic activities like a bakery, ice-cream factory, or milk-chilling factory. Both the Pinthali and Katunje Besi sites run additional mills and refrigerators on diesel. During holidays and special occasions like weddings, electricity is supplied during the day for television and music systems.

The adaptation of new technology has opened the community to new ideas and gadgets, but not fully replaced older devices. Rice cookers, a symbol of social status that also increase electricity demand and potential overload, may be positioned visibly alongside manual utensils like a *dekchi* or a *karai* (cooking pot) still in use. A fixed light bulb of 25 watts is considered too dim for children to study when fixed to the ceiling. Portable kerosene light bottles and other kerosene tools are still in use in households requiring more energy than the allocated wattage. Villagers still use firewood for cooking, and deforestation has not reduced.

Most male villagers are seen enjoying the benefits of electricity, while it has increased tasks and lengthened the working day for women, who often work until 11 pm, finishing up household chores. However, electric light in the kitchen has improved visibility for tasks, while television in the kitchen provides opportunities for amusement while cooking. The kitchen and the *angan* (courtyard) in most households have electric bulbs installed. The angan bulb remains on to be visible from a distance at night. The kitchen bulb normally gets relocated to bedrooms after chores are over.

COMMON OPERATIONAL PROBLEMS

The most common problems related to O&M are from lightning, loose connections, and breakdowns. The practice of routine proactive maintenance is a new technical concept; therefore, operators mostly respond to crisis or breakdown maintenance.[11] Breakdowns are common, from problems in the canal blockages and damage to diversion weirs, to mechanical failures in system components and improper wiring. Breakdowns due to lightning are also frequent, particularly during the monsoon period. However, preventive measures against lightning-related problems through adequate earthing installations are lacking in most MHES systems in Nepal. There is a lack of local capacity for repairs and frequently the community is seen transporting a broken component to Kathmandu, which is costly.

Overloading makes it common for houses to undergo a blackout at least twice monthly, but operators can handle such situations adeptly and replace fuses. As seen in Pinthali and Katunje Besi, some households use more than their allocation, stressing the system. Penalties for non-compliance—disconnection for a day—are not effective.

In particular, non-compliance increases when the light generated is dim. This occurs frequently during the wet season when sediment is carried into the canal and the flow is disrupted and blocked and when the volume of water in the canals is low, and during the hot season when the competition for irrigation use increases. When light is dim, the villagers use more electric bulbs to compensate. The turbine capacity dictates in legitimizing rules and building compliance among electricity users.

Household members are not yet habituated to turning off the electricity switch. Consequently, when the operator turns on the system at the powerhouse, electric bulbs burn out faster with power surges. Options like individual metering and house switches for disconnection when not in use do not exist.

The current status of service provision shows the importance of improving the technical capacity and social skills of operators. Various experiences indicate that, when support structures for MHES are located nearby, better installation, operation, and maintenance evolves. In this same regard, design of MHES has to be realistic about what they can supply. They may always really be for 'basic needs', but MHES introduce new demands for design consideration.

FINANCIAL ACCOUNTABILITY: LEVYING CHARGES AND MANAGING FUNDS

Allocation, access, and payment for electricity are shaped by use, rules, and fee structures that vary between MHES. These introduce new concepts of paying wages to a community member to run a community service. If fees are not collected in a timely and equitable manner, salaries get compromised, yet the manager and the operator have to perform to fulfil community expectations. Financial accountability shapes the provision of services, not only through salary structures and O&M but also in how contracts are made and funds are managed to ensure sustainability of the MHES.

The standard electricity charges adopted by most REDP programmes are NR 1 per light bulb or appliance wattage of electricity. Most hydel systems require households to pay an upfront connection fee; these are often costly and can range from free service to NR

1,500 if transmission and wiring costs are included. Exemptions of connection charges occur for households affected by the MHES infrastructure, including transmission poles on private property, relocating a mill, or a canal running through an individual's property.

At installation, a 'tariff collection card' is issued to each household. These can be viewed and written on by the operator or the manager during every monthly collection period, but are not always used. In Pinthali, the operator is responsible for fee collection. Fees are collected from individual households, and a flexible person-to-person discussion and understanding makes collection easier.

All households in Pinthali are connected to the micro-hydel. Decisions concerning connection charges and tariffs are made by the committee members and usually approved in a village mass meeting. The connection fee in Pinthali and Katunje Besi was NR 100 per household. In Pinthali, a tariff of NR 60 per 100 watt appliance wattage of electricity was also established. Katunje Besi had two different tariffs shaped by whether land and labour had been contributed in development. All houses were connected; the tariffs were: NR 1 per watt for those who contributed labour and NR 2 per watt for those who had not. There were no written rules and regulations, and practices became more ambiguous as users were covered by differing tariff structures. The donor agency recommended that the tariff be doubled for those consuming 80 watts and above. However, as the management structure remained weak, this proposal was not reinforced.

In Katunje Besi there were three different responses to tariff payment from electricity consumers. One group followed regulations and kept to the monthly allocation and payment, recognizing that the system needed running. A second group, claiming contributions through labour or political representation in the intervention process, considered itself exempt from paying tariffs. This group claimed electricity consumption above allocation, careless of responsibility for system breakdown. The third group considered water as common property, and thus felt entitled to claim electricity without payment. There were households with dues pending from over a year. As regulations for non-compliance were not formalized, a fair amount of stealing of electricity was present, including by the chairman and the technical operators.

The Management Contract and Financial Interface in Katunje Besi

As outlined earlier, to overcome numerous problems REDP tried to contract out the management of the Katunje Besi plant without community consultation. As the system became dysfunctional, the committee continued to fragment. The contract made the 'manager' accountable but he was rarely paid. Tariff collection became onerous, and compounded with intermittent electricity supply as the system was rarely working. Additionally, the local police post[12] consumed the highest monthly electricity of over 1 kilowatt, without payment, leaving a monthly revenue loss of over NR 2,000 obligated on to the community. The committee was powerless to confront the situation, given differing opinions on action. Some believed that the police were entitled to the privilege as guardians of the community. Others perceived free entitlement as unethical: tariff exemption signified taking advantage of the community's resources. The manager's lack of capacity became apparent to the community, which forced him to open the account books and recoup funds; later a new committee was formed.

OPERATIONAL ACCOUNTABILITY: DESIGNING FOR THE WATER–POWER INTERFACE

Flow measurements at the case study sites showed whether a hydraulic ensemble delivered canal discharges equivalent to the design flow initially calculated as necessary for power supply. This section compares the actual flow with the discharges required for power generation, to understand how these fluctuate together.

To ensure water supply for all activities, the interface of power generation and irrigation is an important design feature in MHES interventions. The Pinthali and Katunje Besi designs deliberately interfaced power generation and irrigation uses of water in their initial designs. Interfacing power generation and water use not only requires adequate supply of water in the canal, it is also about managing canal conveyance losses that affect releases.

In Pinthali, water supply increased after improvements to the original canal system. The newly combined Pinthali canal system maintained its initial design objective that focused on 'lighting, agro-processing, and

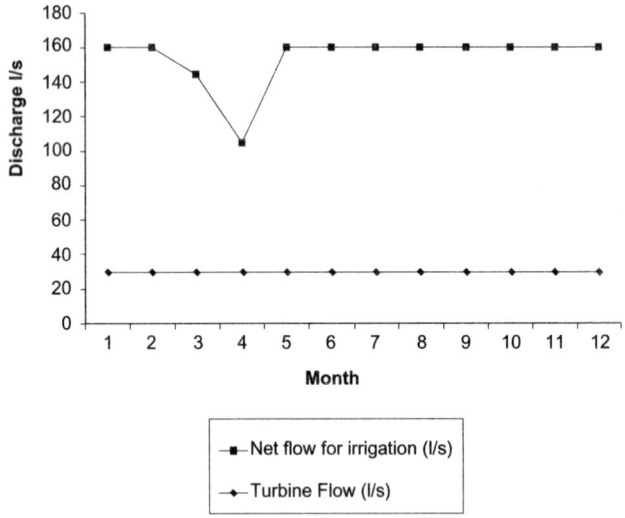

FIGURE 11.3   Water Supply for the Turbine and Other Uses in Pinthali

Source: Field measurements, January–December 2001.

irrigation' at the level of power envisaged then for core needs. However, potential village power demand exceeded the design estimates of needs given emerging additional demands during adaptation.

The hydrograph for Pinthali, shown in Figure 11.3, indicates issues around ensuring supply and managing conveyance losses in systems with long canals. The canal capacity at the Pinthali plant is designed at 200 litres per second, and the flow at the intake reached a maximum of 160 litres per second in August and a minimum of 105 litres per second in April 2001. The graph indicates a low flow period, when water is diverted from irrigation to ensure power generation and run agro-processing machines during the day. Discharge fluctuates annually, based on rainfall patterns and peaks during the monsoon months. The dip in the graph also indicates that when the flow is low during the dry season, the diversion for energy use decreases availability for other purposes. Field data indicated that, while 30 litres per second is diverted for electricity generation, some remainder is always available for irrigation but this varies over the seasons. Thus, the setting of the design discharge for electricity generation and irrigation indicates a reasonable allowance for daytime irrigation and agro-processing at this time during low flow months. However, this needs monitoring for the future. A cement and stone intake has replaced the temporary one, increasing the width of the canal to 63 centimetres and the height to 1.35 metres.

A comparison between the releases in the canal and at the source indicated conveyance losses of 44 per cent in August and 31 per cent in April. These losses result from the semi-permanent construction of the canal, leakage per length of the system, and the open outlets for irrigation.

In Katunje Besi, a net flow of 560 litres per second water is available from the canal for irrigation purposes even during the low flow periods when the plant is in operation. The turbine, when in use, utilizes a substantial amount of water, given the nature of the low head design. The canal is designed to supply 1,200 litres per second of water; after deduction of conveyance losses, about 960 litres per second is available. Figure 11.4 shows the annual discharge curve for power and other water uses, suggesting that there is no water scarcity for interfacing power and irrigation uses.

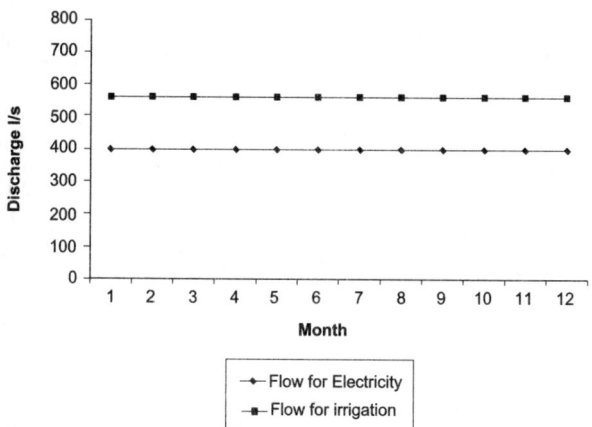

FIGURE 11.4   Water Supply for the Turbine and Other Uses in Katunje Besi

Source: Field measurements, January–December 2001.

While losses are less significant given the short conveyance system, Katunje Besi faces competition from upstream abstractions by water-consuming industries using the Roshi river. However, so far, water is sufficient year-round for domestic, electricity, and irrigation requirements, and other economic sectors.

In future monitoring of water flows and electricity generation for local planning, it is important to make allowances for efficiencies of operation and breakdowns.[13]

SOCIO-ECONOMIC AND POLITICAL ACCOUNTABILITY

The approaches and methodologies applied in different sites for forming the committees had an impact on leadership development and representation. Decentralization processes in Nepal are often linked to the creation of new institutions through group formation, where leadership and representation are built by including the politically knowledgeable agents. In this process, the political leadership is seen as sustaining initiatives, whereas the traditional technical entrepreneur is perhaps perceived as a threat. Both Katunje Besi and Pinthali have a strong history of local technical innovation. Yet in Katunje Besi, the creation of *samiti*s (committees) capitalized on local political leadership for representation that alienated the citizens, as decisions were not transparent, and made without involving them. Procedures became affected by wider political struggles and unrest. These nuances in leadership building demonstrate how representation evolves beyond the promotion of popular development models of institutions. For example, the creation of samiti and *mandal*s (groups) does not necessarily address the structural problems existing prior to group formation (Waghmore 2002). Additionally, extending the circle of social actors and groups does not necessarily ensure participation (Hennen 1999) or representation, but can question the social and political legitimacy of technology-related development as shown from Katunje Besi.

Where previously the leaders and village elders had substantial responsibility in making decisions, these powers have now shifted to a samiti form, re-formulated by the intervening agencies. The roles are now more committee-oriented, even though the power exerted is still influenced by individual membership. Therefore, in a typical REDP model, the 11 and 16-member group, comprising the FG, takes on shifting roles as seen in Katunje Besi and Pinthali. Where previously the farmers, as water users through the FMIS, congregated at the village *chautara* (a raised built structure that is an open meeting area) to resolve conflicts, now the FG convenes institutionalized mass meetings that are recorded. These groups have also introduced the offices of a manager and an operator, creating a new local 'expert system'. While the farmers previously attended and supported the cleaning and maintenance of canals, coordination of this responsibility has now shifted to the technical operator where skills-orientation creates a more individualized role in performance. However, decision-making, in terms of rules and regulations, is often still shaped collectively.

In Pinthali, the plant operates as a cooperative; therefore, the role of the farmers has changed from that of water users to that of clients or consumers of services. Prior to power use, the responsibility for the maintenance of the irrigation system was undertaken and managed collectively. However, the village elders still command influence and respect, and traditional leadership demonstrates its social capital through its reconciliation responsibilities. However, O&M as a collective responsibility is not dependent now on any formal management 'committee'. In practice, with a new hydel regime, this may remain only with the operator.

The initial composition of the Katunje Besi management committee included some members who donated land for the installation of the plant. However, the selection of the committee members was undertaken during politically volatile times. The case study discussion shows how evolution of a relevant management structure proved difficult, with experiments in contracted managers who were then confronted by users, and appointment of insufficiently trained operators to run the propeller turbine. During this period, police and Maoist campaigns affected efforts to operate the system, and there were no older institutional experiences from the irrigation systems to fall back on.

The operator's role in critically balancing power relations is visible beyond the functional tasks of operations, undertaking repairs and replacement, and resolving conflicts with farmers during paddy cultivation. The operator must maintain his established social relations despite social dilemmas. Requests for daytime electricity for individual households or political leaders cannot be refused, and operators must deal with use of electricity above quotas by those who may be their uncles and grandfathers. In most cases, the operators maintain a blind eye when the elderly or the politically powerful use electricity above their allocations. In this respect, these social relations are placed above issues of accountability.

As an externally driven and guided intervention, micro-hydel shapes the leadership in a given community. Those who are technologically aware—often already entrepreneurial agents—tend to take responsibility for directing innovation. Evidence from earlier community initiatives led by Nanda Raj Lama and Subarna Shrestha suggest that being technologically aware meant taking responsibility for an adequate management structure. The case studies show that the operator becomes a critical agent for the MHES, sometimes even replacing managers, committee members, and the committee itself during the insurgency periods. It is evident that the use and success of contracts for certain roles and functions ensures that machines are run, as shown by the agro-processing machines in Pinthali.

NEGOTIATING CONSTITUTIONAL ACCOUNTABILITY

Modern processes of intervention introduce diverse external agents and plural forms of networking. Therefore, the democratic interest of the citizens is not solely dependent on formal institutions. A larger polity controls governance processes, where power is dispersed among various networking agencies. Building leadership and representation has links with this wider local governance.

In micro-hydel intervention, governance processes can be seen as interfaces within two distinct arenas of the DDC and VDC (Kabhre DDC/REDS 2000). The DDC level primarily functions to filter financial arrangements between the centre (Kathmandu) and the local implementation and operational arrangements being made at VDC level. In reality, the case study sites show that these local-level institutions have very little influence on technology system choices. Rather, technology development issues are directly negotiated through interactions between community organization and donor agencies. Contracts are negotiated centrally (Kathmandu) in terms of manufacturing, installation, technological training, and other support structures. The social mobilization contract is undertaken at the district level between the donor agency and local networks. The role of these organizations and contracts is relevant and active only during the intervention period. After this, practical issues like availability of spare parts and troubleshooting become inconsequential to the CO, as they are not directly involved in negotiating these contracts. Therefore, the 'CO' is rather seen as a one-off activity, and not really as an institution-building exercise. Here the processes that define the relationship between DDC and the implementing or donor agency in reality appear to shape the community organization.

Group formation approaches discussed here, with their focus on representation, seem to visualize the community as an object ready to accept development, regardless of associated problems or lack of benefits to them. In most MHES, the process of site selection limits assessment of levels of cooperation and interest in the community for a given technology, and its capacity to oversee the intervention. Social mobilization and social mapping are carried out after the decision has been taken by the implementing and donor agencies to intervene in a given area. The next step, then, focuses on the process of participation, which takes the form of selection of appropriate members through the formation of community organizations. The contributions of voluntary labour, in cash and kind from the community, and the formation of committees are viewed as indicators of accountability, ownership, and users' involvement or participation. To fulfil the requirements of development models, gender and ethnic diversities had to be represented in the committee, regardless of who makes these decisions and their local social significance. Improving understanding of constitutional responsibility, in how models and processes for change are brought together, remains a critical area for thinking about the future of MHES.

* * *

This study has shown that, despite four decades of development assistance by diverse actors in the MHES sector, the sector's performance has remained marginal. Understanding of best practices for MHES promotion can be improved. Nonetheless, alternative energy continues to play an integral part in the quest for a sustainable comprehensive energy solution. This chapter has shown that adaptation of technology is driven by the interaction of systems, structures, and agents. Technological democracy—in getting MHES that work well with irrigation to provide for users' needs and equity—is only attainable if built within a framework of accountability.

The case studies of Pinthali and Katunje indicate that managing power supply and making MHES work is more than converting water into electricity. It is also about designing governance systems, building technical knowledge and managerial skills, developing rules and contracts, and reshaping a gamut of related knowledge systems. As newer governance systems get embedded in hydel technology, the interfacing of water, irrigation, and energy constructs a complex technology management system; this represents acceptance, interest, and pride in the MHES.

As with irrigation systems, MHES technology can be poorly designed just as an ensemble, without adequate attention to its real operational functionality and capacities to support livelihoods. Correspondingly, conflict emerges in the operation of systems, often leading to system breakdown, as seen in Katunje Besi. In other situations, interactions build accountable frameworks, as in Pinthali. Technology can then evolve as a transformative unit, bringing welfare and livelihood gains, as a cohesive factor bringing the community together.

A democratic MHES technology can be built from an initial hydraulic ensemble, just as an irrigation system can evolve with time. It can be adapted to evolve and transform in meeting the users' needs when systems, structures, and agents, come together, making it work within a framework of accountability. The role of constitutional accountability plays an equally important role, together with financial, operational, and socio-political accountability.

As with any other sectoral intervention and development process, the effectiveness of MHES comes from being included within an integrated comprehensive policy and planning strategy to bring about planned change. MHES development must be included within national, district, and local planning and policy processes for energy, an integrated understanding of community development, and institutions for the management of irrigation and natural resources. A good understanding of the wider technology trajectory chosen for

MHES is a prerequisite in selecting the most appropriate technology locally. This understanding is as important for irrigation alone, as it is for MHES operating with irrigation, as the chapters by Gautam, Krishnan, and Manimohan in this volume have shown.

Negotiating governance of MHES as a structure and a process at the community organization level remains a challenge. It has so far been highly dependent on skill-building from external agents. Strengthening decentralized skill building and capacity development by better interfacing the district and village levels is paramount for MHES implementation. While previous irrigation experience fosters the development of energy user committees, these new committees and their regulations trigger changes into the O&M of irrigation systems. What is clearly visible from the case study sites is that both the local customary institutions (such as the existing network of respected elders and political leaders) and the existing technological networks greatly influence the newly formed energy management structures. Therefore, negotiating governance at the community organization level is a challenge in decentralization efforts and in promoting autonomous structures. At this level, governance may be better negotiated by allowing the customary institutions and networks as a focus of power building in a local intervention.

Corroborating the above, building new governance through constitutional accountability must increase control of development at local level while ensuring access to appropriate technology and equity in service that can drive performance and progress. Re-building knowledge and enhancing local capacity within the existing ecological and political conditions are the basic tenets for implementing change with wider democratic linkages. This is as true for irrigation alone, as can be seen from the variable outcomes of management reforms documented by Nikku in this volume, as for MHES alone.

Therefore, building new technology and system designs requires not only strengthening national institutions but also strengthening local institutions, thereby bringing MHES technology and its benefits closer to users' needs and demands. This implies introducing new sets of institutions while supporting the existing ones. The incentive and policy frameworks creating these new rules, regulations, and structures must address local capacity to manage and care for the technology, and clarify and ensure the benefits associated with these responsibilities. This is critical for both MHES and irrigation systems and their joint operation.

Notes

[1] Figures 11.1 and 11.2 are reprinted from Regmi (2004a) with kind permission of Orient Blackswan, New Delhi.

[2] The definitions of MHES overlap among planners, manufacturers, and implementers in Nepal. The classification widely used in Nepal is the United Nations Industrial Development Organisation (UNIDO) 1996 Kathmandu Convention. An MHES produces up to 100 kilowatt of energy. Plants generating 100–1,000 kilowatt are considered 'mini' systems, while 'small' plants range from 1–5 MW (Pandey 1998). Only about three per cent of micro-hydel plants generate more than 20 kilowatt (Rijal 2000). Systems below 50 kilowatt are classified into four basic types of units considered appropriate for rural application (Rijal 1998): traditional water wheel, multipurpose power units (MPPU), cross-flow turbines, and peltric sets, also known as 'pico systems'.

[3] These designs are: (*i*) cross-flow turbine of 12 kilowatt in Pinthali Mangaltar, integrated with irrigation and agro-processing to supply a micro-network for lighting; (*ii*) propeller turbine of 8 kilowatt at Katunje Besi, working with an irrigation system and ghatta (water mill) to supply a micro-network for lighting; (*iii*) Pico Power Pack (PPP) of 4.2 kilowatt in Nayagaon, Kusha Devi with irrigation and no additional technologies to supply a micro-network for lighting; (*iv*) the Panuati Small Hydro Plant of 2.4 megawatt at Khopasi, supporting networks for irrigation and to supply electricity to the central grid.

[4] Qualitative techniques included structured and semi-structured interviews and follow-up discussions and informal meetings with relevant stakeholders. Quantitative data included primary information on users and non-users of electricity and irrigation, hydraulic features of the design system, and water measurements undertaken between January 2001 to August 2002 (encompassing two dry seasons and one wet season), enabling bi-monthly calculations of discharge at selected points at the intake and canal system. Interviews of households at MHES village sites included 25 per cent non-electricity users. Two local community mobilizers assisted with data collection.

[5] The MHES in particular were supported by the Swiss, Danish, American, Canadian, and German aid organizations and multilateral donors (particularly UNDP and the World Bank).

[6] Now called Practical Action.

[7] Other rural electricity supply models include a centralized and controlled NEA model feeding into the grid, and a private sector model designed to supply the grid but including a small adjunct micro-system for local supply.

[8] This amount equalled about USD 17,000 in the late 1990s (around 1997–8), an overwhelming amount for the community, of which REDP funds for micro hydro were only NR 668,336. REDP also released simultaneous funds for other activities, including promotion of the surrounding areas and environment management. The community was unclear on how these funds were spent on the proposed activities.

[9] Because of alleged increased pressure from the Maoist groups for a cash contribution of NR 50,000, the manager abandoned Pinthali in November 2002, and decided to move to Banepa.

[10] The account books and project documents were destroyed when the Maoists set fire to the VDC office in June 2002.

[11] The operator usually takes care of minor repair of nuts and bolts. For major spare parts, the operator approaches the users' committee for a decision, who may call a village mass meeting to resolve the issue. The manufacturers' warranty for electromechanical components is said to be one year from the period of operation. The lifespan of a hydel plant is normally accepted as 10 years.

[12] As the insurgency escalated in November 2001 the police post moved to the nearest town of Bhakunde Besi. Many of the young police officers earlier involved in the surveillance of Katunje bazaar were massacred in Bhakunde in February 2002.

[13] The figures given as the 'installed capacity' can be deceptive, as the efficiency of these plants is always 20–40 per cent below the installed capacity. Also, there is no means to verify actual electricity production and turbine efficiency, or record breakdowns. In the research sites observed, electricity disruption of over three weeks was a frequent occurrence.

12 BOUNDARY CONCEPTS FOR THE INTERDISCIPLINARY ANALYSIS OF IRRIGATION WATER MANAGEMENT IN SOUTH ASIA

Peter P. Mollinga

On the basis of the rich empirical material provided by the chapters in this volume, the case for natural-social science interdisciplinarity in water resources management analysis is easily made. It derives from, first, the complexity or multidimensionality of water resources management as a concrete phenomenon and, second, the perceived urgency and intractability of water resources management problems and conflicts, urging decision makers to look for 'integrated', 'adaptive', or otherwise comprehensive approaches. The latter has led to an upsurge in water resources management studies and policy approaches that seek to capture and cut through the socio-material complexity of water systems dynamics, and their contested and negotiated transformation. The relevance and need for this is as apparent in South Asia as elsewhere (see Ballabh 2008; Baviskar 2005; Joy *et al.* 2008; Shah 2009).

However, the intellectual and institutional odds are still largely against practicing science in the interdisciplinary mode. Academic specialization, no matter how useful for some purposes, makes 'integration', on any definition, a tall order (Pohl and Hadorn 2007). The organizational division of labour in government and administration means that ministries and sectors experience great difficulties to interact and collaborate effectively. Finally, research remains uncomfortably related to policy, expert knowledge, and scientific knowledge separated from lay and local knowledge. Cutting through all this are different world views and political standpoints, within science as much as outside it (Lele and Norgaard 2005). Nevertheless, the poignancy of contemporary natural resources management controversies increasingly forces interdisciplinary and transdisciplinary research. Klein's (1996) seminal book on interdisciplinarity is titled *Crossing Boundaries*. There is no scarcity of boundaries to cross for those who attempt to develop 'integrated perspectives' on the complex system that water resources management is.[1] In the United States (US), the term 'boundary work' has been coined for sustainability science to refer to the concerted and systematic effort that boundary crossing involves (Cash *et al.* 2003). This work primarily focuses on the research–policy boundary, and what can be done for a more productive interface. The boundary work concept can, however, also be used in a broader sense, referring to the variety of boundaries that exist in natural resources management research and practice, and need to be 'managed'.

This chapter is a reflection around the main conceptual, theoretical, and methodological messages of this book. It seeks new ways of approaching interdisciplinarity in what is an ongoing process of conceptual, theoretical and methodological refinement. Chapter 1 focused on the socio-technical approach as the framework to study water control in most of the chapters in this volume. While acknowledging this work I move forward from it towards new ways to theorize the materiality of social change in water management.

Elsewhere (as in Mollinga 2010a) I have suggested that boundary work in interdisciplinary and transdisciplinary research on natural resources management has three components: boundary concepts, boundary objects, and boundary settings. Boundary concepts allow us to think (conceptually communicate) about the multidimensionality of the phenomena studied and addressed. Boundary objects are devices and methods that allow us to act in situations of incomplete knowledge, non-linearity, and divergent interests (the category includes models, frameworks, and participatory processes for decision-making and planning, among other things). Boundary settings are the institutional arrangements within which these concepts, devices, and methods can be fruitfully developed and effectively put to work.

This chapter concentrates on the first of these three elements—the theoretical constructs required to think across the boundaries of the natural and the social sciences, leaving the acting and enabling of the second and third elements aside. By exploring the conceptual terrain at this interface, I aim to contribute to a 'critical interdisciplinarity' (Klein 1996) in the domain of water resources.

Perhaps the major analytical boundary challenge is binarism as a form of analytical reductionism (Castree 2002). At the heart of interdisciplinary analysis of water resources management lies the material–social binary. This binary is 'real' within the water resources domain as a division between disciplines and professions that are labelled 'technical' and 'social'. The division is also institutional (as embodied in government and university departments, professional associations, and academic journals), cultural (as embodied in professional identities), and cognitive (by considering the material and social as distinct objects). The fallacy of binarism is that the world is ontologically complex—structured, stratified, and heterogeneous (Sayer 1984). Irrigation systems provide an archetypical example. The networked system of dams, weirs, canals, and other *ouvrages d'art* structures both the physical landscape and institutional and economic life at various levels of scale.[2]

The collection of boundary concepts presented in the following four sections is informed by an understanding of sociology as a 'science of connections' or 'science of associations' (Latour 2005). Social-ness exists in the act of creating and maintaining networks of heterogeneous objects and relationships. It is in that sense that human beings are the prime movers in 'socially constructing' water resources management systems. The second starting point is that 'putting and holding things together' in configurations that have some endurance, that is structures (Archer 1995), involves such configurations having 'internal relations' and exhibiting 'emergent properties' (Sayer 1984). In the inventory of attempts at capturing emergence related to irrigation technical artefacts and irrigation landscapes in the following sections, the term 'institutions' is used for what are conventionally called social dimensions, social factors, or social aspects—formulations that treat the social as a distinct object. 'Social' in this chapter is reserved for the activity of making, reproducing, and transforming the hybrid and complex sets of 'associations' and 'connections' in water resources management configurations. This adds new space for thinking beyond the institutional–organizational and political–social processes of water control introduced in Chapter 1, and for challenging the limitations of global institutional thinking discussed there.

The following four sections give an overview of the different ways in which conceptual hybridization has been attempted in the South Asian, particularly Indian, context, with a focus on irrigation. The second section looks at how water rights and entitlements can be understood in an interdisciplinary way. It discusses the concept of hydraulic property and, as an extension of that, how ecological relations are part of rights and entitlements concepts. The third section looks at the interdisciplinary analysis of water use; it explores conceptualizations of design–management relations and the social construction of irrigation technology, producing management 'scripts'. The fourth section addresses the impacts of water use in developmental terms. Boundary concepts discussed are 'landesque capital' and the value or valuation of ecosystem goods and services (EG&S). The fifth section looks at interdisciplinary conceptualizations of the embeddedness of irrigation and water resources management processes, by discussing time and space relations in irrigation and the 'cultural politics of water' perspective. Then, looking ahead, section six discusses how these boundary concepts can help to further develop interdisciplinary social theory on water, and identifies some formal and substantive theoretical avenues for that. The final section concludes the chapter by outlining potential further research activities.

WATER AND RIGHTS: HYDRAULIC PROPERTY AND ECOLOGICAL INTEGRITY

HYDRAULIC PROPERTY

In his analysis of farmer-managed irrigation systems (FMISs) in Thailand, Indonesia, and the Indian Himalayas, Coward (1986a, 1986b, 1990) analyses the intimate relations between the social relations of water management and the technical infrastructure. His basic argument is that 'creation of irrigation facilities establishes among the creators property relations' (Coward 1986b: 227). Naturally, '[n]one of this property can be sustained over time without frequent renewal through the investment of labour and capital' (1986b: 225). Therefore, 'the basis for [the] social action [of the community irrigation group] is the common relationship they have with regard to property objects which they have created' (1986b: 225). This means that the creation and upkeep of irrigation infrastructure go hand in hand with the (transformation of the) social relations through which that infrastructure is used; they co-evolve and are each other's expression as 'hydraulic property'.

On property rights arrangements in FMIS in the Kangra valley of Himachal Pradesh, Coward (1990) shows that the consolidation of land rights in the colonial period included a specification of the materials to be used for diversion weir construction and a description of the (proportional) division structures for water distribution. 'The width of the openings created by the thelu [division structure] is measured in "fingers" depending upon the area of land to be served by a given turnout ... the thelu is a simple but effective device by which the abstract water rights of individuals can be translated into calibrated water flows' (Coward 1990: 83). The land and water rights thus defined had to be reproduced through the contribution of labour to maintenance and repair. Coward also shows that distribution of rights and access in local irrigation systems are not necessarily equitable. 'The lower zone people ... reproduce their water rights in the Bharul network even though the costs to them are considerably higher than those incurred by the upper groups' (1990: 84).

The concept of hydraulic property thus captures two theoretical ideas: first, that water rights take on a material form in the characteristics of the infrastructure of the systems in and for which they exist; second, that the activity of infrastructure creation and upkeep is a process of property rights creation and upkeep. Though the initial formulation of the concept is partly India-based, it has attracted little follow-up work in South Asia outside Nepal (but more in Latin America; for example, Gerbrandy and Hoogendam 1996). However, the chapter by Parajuli in this volume clearly stands in this research tradition. Parajuli's description of the design and operation of proportioning weirs are comparable with the *thelu* described by Coward, as a materialization of rights. Parajuli also analyses the choice of division structures and operational approaches in relation to the agro-ecology of the hills. Chapters by Khanal and Parajuli both describe FMIS where water rights are related with original involvement in system construction, and reproduced through responsibilities in maintenance and operation, although they also show how trading of rights may be allowed. They contribute to our understanding of hydraulic property and the interrelationships between technology and institutions, and of processes of change in irrigation systems where water rights definitions evolve with agrarian and economic transformations.

PROPERTY RIGHTS AND ECOLOGICAL INTEGRITY

The scope of the hydraulic property concept can be broadened to the ecology or the landscape. Case studies in this book show the linkages between property rights, ecological relations, and landscape. In her chapter on the interaction of pond (small tank) and canal water management in a watershed in Kerala, Krishnan links the ecological characteristics of the landscape to the (land and) water rights that govern its use. In another publication (Krishnan 2009) she details how ecological relations were historically part of the definition of land and water rights in ways that achieved ecological sustainability. The land used to be owned by landlords (*janmis*), who rented it out to tenants through intermediaries (managers). Those cultivating land in the command area of a pond had a water right attached to it, involving rights to pond water and to the upland forested area for forest products for their own use. There were also arrangements for pond (tank) maintenance. The janmi undertook regular desilting of the pond, through the supervisor appointed by him. Day-to-day activities like cleaning runoff channels in the catchment (necessary to fill the tank) and other regular tasks were undertaken by permanent labourers who worked for the concerned tenant(s).

The Kerala government decided to implement a land reform. Around the same time the vesting of privately owned forests with the government was implemented. The time lag between the promulgation of the Forest Act and its final implementation enabled landlords to dispose of the valuable trees, resulting in deforestation of the uplands. Landlords also maintained access to the valuable valley lands by strategic registration of plots. When the uplands and lowlands were redistributed under land reform, only land rights were consciously redistributed: many former tenants who obtained small plots of land remained without water rights.

As described in this volume, government irrigation systems were implanted on the landscape without reference to existing pond or tank systems, and the canal water used to fill ponds and tanks. This public water provided by government canals was privatized when it entered the ponds and tanks available. The chapter by Manimohan on tank cascades in Tamil Nadu shows how water users had a sense of rights and responsibilities over water in the landscape, extending to the guiding of catchment runoff into the tank as well as maintenance of the tanks themselves. Tank users protested when government forestry plantations disrupted these runoff patterns, but to no avail.

Notwithstanding the depressing outcome of the rights reform and development process described by Krishnan,[3] the theoretically interesting point is that ecological relations were part of the definition of land and water rights. How rights are defined shapes the landscape, and the reproduction of certain landscapes assumes specific property rights arrangements. The case studies in this book have contributed to the analysis of such interrelationships. However, more work needs to be done on this theme.

THE USE OF WATER: DESIGN-MANAGEMENT RELATIONS AND THE SOCIAL CONSTRUCTION OF TECHNOLOGY

DESIGN-MANAGEMENT RELATIONS

Notwithstanding images and ideologies of western technical sciences being standardized and universal, irrigation engineering has a strongly regionally specific history. Technical concepts and approaches developed in, say, British, French, and Dutch colonial engineering are quite distinct. Within these there are variations reflecting contexts of development—in British colonial engineering, for instance, for Egypt and India. When to this are added irrigation engineering traditions from other regions like China, Japan, the USA, and Europe, and local engineering knowledge of, for instance, hill irrigation in the Andes and Himalayas, it is clear that there is a wide variety of cultures of engineering.

Apart from physical differences in terrain, rainfall, and runoff regimes, available construction material and the like, the variation is due to a series of institutional factors. As shown in this and other chapters, designs materialize property rights. Within government-owned and managed irrigation systems the variation lies in the state–irrigator relationships incorporated in technical design; that is, the form of organization for management and governance of the system materialized in technical design principles.

The significance and specificity of design-management relations has been innovatively explored by Horst (1996, 1998; also see Levine 1980). The thrust of Horst's analysis is that different types of division or offtake structures (fixed, open–closed, gradually adjustable) associate with different forms and principles of operation and management (Horst 1998: 84). Moreover, each of these offers different opportunities for manipulation, adding a second layer to the analysis: the discrepancy between on-paper and real operation and management practices.

Bolding *et al.* (1995) discuss how colonial governments attempted to implement the rationing of irrigation supply in canal irrigation through varying combinations of technologies and institutions differently in the northern, western and southern parts of India. Early nineteenth century efforts to introduce the so-called 'block system' in present-day Maharashtra show the two levels of analysis: the search for an appropriate combination of technical and institutional features, and the undermining of this in the realities of irrigation system use. This allows an analysis of the features and contradictions of colonial rule as well as the dynamics of peasant social differentiation in an emerging capitalist rural economy. Narain (this volume) has provided an analysis of the contemporary relevance of division structures and canal design principles in farmer organization in Haryana and Maharashtra. He shows how organizational concepts of local water users' association (WUA) as promoted under government participatory irrigation management (PIM) policies and programmes do not fit with the technical reality of India's canal irrigation systems. Blueprint models of WUAs and global discourses on markets and pricing have been pursued without an understanding of the various water allocation systems in these states, and the different technologies and possibilities of operational control. Khanal's case study of the West Gandak irrigation system in Nepal documents the diverse range of control structures brought in by different intervention programmes. These were too numerous, costly, and badly installed for farmers' groups to manage when donor support withdrew. Also for Nepal, Regmi and Vincent have discussed in their chapter how water system designs evolve over time in their joint evolution of technology and institutions. The authors show how systems evolve from a simple 'hydraulic ensemble' to 'evolutionary systems' providing diverse benefits and maturing institutions showing reflexive coping with change. The interaction of structures, systems, and agents shapes accountability mechanisms of various kinds, local and constitutional, which are critical in these material transformations.

SOCIAL CONSTRUCTION OF TECHNOLOGY

The inquiry into design-management relations as described before, pursued from a civil engineering starting point, could be more comprehensively theorized using the 'social construction of technology' perspective that became available in the 1980s.[4] From this perspective, Shah (2003) investigates the 'social designs' of tank irrigation in Karnataka, and her chapter in this volume has given a cogent review of how tank designs have transformed under a recursive state–society relationship. Many tanks were built in pre-colonial times. She suggests that 'the design principle of a labour intensive construction method of embankments carries the imprint of the historical era that rested on a rigidly built, hierarchical social order which exerted a considerable degree of control over labour' (Shah 2003: 261). When this order changed, maintaining the infrastructure in a good state became difficult. Presently, with expanded market relations, decentralization policies, and a general loosening of social rigidities, rural elites find it increasingly difficult to mobilize labour for tasks like canal cleaning, sluice operation, and field-to-field irrigation from lower caste labourers. They turn to the state for investment in maintenance and management (Shah 2003: 262–3).

These observations mean that the technical features of water infrastructure must be understood as historical products, fitting a particular context but potentially inadequate in others. The historical literature on irrigation and flood control in India and South Asia profiles this statement with rich and fascinating accounts (for example, Gilmartin 1995; Stone 1984; Weil 2006). The broader theme is the role of water resources development and technology in colonization and nation-building.

The social constructivist analysis of technology and technological systems (see Hughes 1987) can also be reversed. With particular societal objectives and forms of organization in mind, efforts can be made to consciously design technological systems in such a way that they fit these objectives and forms. A perspective of design and planning as self-conscious social construction has been developed for irrigation in certain parts of the world, notably Latin America and Africa. The relative absence of participatory approaches to irrigation design and planning in India can perhaps be explained by the hierarchical and prescriptive style of government intervention on one side, and the glorification of 'traditional' irrigation by civil society organizations on the other. An innovative effort at redesigning the Sardar Sarovar dam, part of the Narmada project, and its downstream irrigated area is by Paranjape and Joy (1995). It is telling that the proposal did not spark anyone's imagination (see Mollinga 2010a). Differences in irrigation management reform approaches between India and Nepal are shown by Nikku and Khanal in this volume. Nepalese policies did allow for participatory development processes between engineers and farmers, aimed at building local organizations and helping them identify and implement infrastructural and institutional priorities.

The growing importance of groundwater development in economic transformation of the region can also be studied through these social constructivist analyses (also see the Introduction). Prakash (this volume) has shown the interlocking agrarian networks that have helped drive the unsustainable expansion of well technology in Gujarat. He has also demonstrated the inequalities emerging in access to water and land under emergent groundwater markets supplied by these wells, and in the sharecropping systems now linked with their productive use.

The upshot of the discussion is that the designs of irrigation infrastructure carry, in social constructivist language, management 'scripts' or, as outlined in the Introduction and Chapter 1, have 'social requirements for use'. Technology is not neutral; its contextuality can be revealed through careful observation and analysis of irrigation designs and water management practices, and by documenting the life histories of artefacts, as in this volume.

WATER, LANDSCAPES, AND LIVELIHOODS: LANDESQUE CAPITAL AND THE VALUE OF ECOSYSTEM GOODS AND SERVICES

LANDESQUE CAPITAL: CULTURAL AND ENGINEERED LANDSCAPES

Amartya Sen is credited with first using the term 'landesque capital', in a conceptual combine with 'labouresque capital' (Sen 1968; Widgren 2007). The term refers to human alterations to the landscape designed to yield long-term gains in productivity, and is mostly used in analyses of agricultural intensification, particularly in smallholder subsistence agriculture. The building of irrigation systems, and water infrastructure generally, is a form of creating landesque capital (see Cosgrove and Petts 1990; Earle and Doyel 2008; Lansing 1991; and Trawick 2008). Apart from agricultural productivity, the term has also been used in relation to the symbolic (identity) dimensions of landscapes. Harrower suggests for ancient irrigation in Southwest Arabia that '[i]rrigation structures not only served as landesque capital ... improvement that established economic investments in landscape infrastructure, but more importantly ... irrigation structures and tombs served as *symbolic landesque capital* investments that proclaimed people-land relations' (2008: 505). Greider and Garkovich (1994) speak of 'landscapes as definitions of ourselves'. Landscapes are as relational as technology, and as contested (Strang 2001).

The richness of these perspectives is still to be appropriated fully in South Asian water and irrigation studies. For India, the historical literature of irrigation, floods, and landscapes comes closest to it (D'Souza 2006b; Ludden 1978). In this volume, Manimohan reviews the changing and different investments in water technologies in 'wet' versus 'dry' tank cascades under changing agrarian politics including differences in groundwater uptake and tank renovation strategies. He gives a powerful sense of the dynamism, present in commoditization and social differentiation, that controls the productivity gains in paddy and sugarcane cropping in wet cascades and in diverse crops with lower water demands in dry cascades. This reality is quite different from the discourses of neglect and decay in tank management, which legitimize diverse development plans for tanks.

ECOSYSTEM GOODS AND SERVICES: VALUE AND VALUATION

The new millennium saw the global consolidation of the 'ecosystem goods and services' concept through the Millennium Ecosystem Assessment (MEA 2005). The notion of 'ecosystem goods and services' is a boundary concept because the globally shared typology distinguishing between supporting, provisioning, regulating, and cultural ecosystem services allows ecologists, economists, sociologists, and other disciplines to have a common language on the usefulness of ecosystems to human society (Slootweg and Mollinga 2009). In India, the epistemic community most closely associated with the analysis of the valuation of EG&S is the Indian Society for Ecological Economics. The literature on (agricultural) water from this perspective is still very small (Kerr 2002; Puttaswamaiah and Raju 2009). In the international sphere, the interdisciplinary analysis of wetlands has contributed much to the development of the perspective—a field scarcely developed in India (Narayanan and Venot 2009). However, an 'ecosystem goods and services' notion is implicit in much debate on local irrigation systems, as these usually have multiple uses and users (Agarwal and Narain 1997). It would seem relevant to explore more fully the value and valuation of water, and its commensuration, beyond the existing economistic thematics of water pricing and payment/markets for ecosystem services (Espeland and Stevens 1998).

EG&S is both a richer and a poorer concept than the Marxist categories of goods/commodities having use value and exchange value. The plurality of goods and services identified by the EGS category is a plurality of meanings of water and of interest groups. EGS is, however, also primarily a descriptive concept that lists and maps, without much analytical 'punch'. The use/exchange value pair has a lot of such 'punch' (through the labour theory of value), but theorizes away materiality of commodities by focusing upon exchange value. Neoclassical and neo-institutional economics tend to be totally agnostic about the material specificities of commodities, as became dramatically clear in neoliberal economists' advocacy of 'tradeable water rights' and 'water markets' in the early 1990s (see Rosegrant and Binswanger 1994). In this volume, particularly the studies on FMIS in Nepal open possibilities to understand how local water organizations account for water rights under diverse land and water access arrangements, and may or may not allow controlled trading or sale. Gautam's chapter illustrates how the Chattis Mauja irrigation organization allows village groups to leave and rejoin the system in relation to their choice to use other water sources. The study is not formulated around ecosystem services, but provides a different view of how local groups can assess the ecosystem's capacity to supply services in return for diverse 'payments'.

IRRIGATION AND SOCIETY: SPACE–TIME RELATIONS AND A CULTURAL POLITICS OF WATER

The broader the canvas, the more tentative the conceptualizations of the multidimensionality of irrigation are. Geography and political ecology-inspired perspectives are making significant contributions to a more broadly-based interdisciplinary analysis of irrigation.

SPACE–TIME RELATIONS

An early and perceptive account of the connection between the physical characteristics of irrigation systems and the social processes that take place in them is VanderMeer's (1971) historical analysis of water thievery in Taiwanese rice irrigation. In his explanation of types and frequency of thievery, the spatial organization of the canals and the land play an important role. Very few analyses of Indian irrigation address the complexity of spatial relations with such nuance, even when the 'head–tail' problematic of unequal water distribution along canals has been a central theme in Indian, and global, irrigation management studies since the 1980s (Chambers 1988).

The 'head–tail' problematic is usually stated in the apparently straightforward terms of locational advantage: farmers with land on the upstream 'head' of the canal have better access to water than those further down at the 'tail-end'. Head-enders thus have better options for productive farming and become rich farmers, while tail-enders remain or become poor. Mollinga's (2003) analysis of the spatial form of the social differentiation in the Tungabhadra Left Bank Canal irrigation system in Karnataka, India, shows that locational advantage is not a 'given'. When analysis is done over a period of time, of whose land is located where in the canal system, and how the canal system itself is partly remodelled in the process of emerging head–tail patterns, the dynamic nature of locational advantage can be shown. In this particular case that dynamism involved land acquisition by (experienced) migrant rice farmers from a neighbouring state, patterned by government crop zoning (called localization) and strategic settlement (on canal-road crossings), and the institutional mechanisms of loans, indebtedness, lease agreements, and land registration. Responses to emerging inequalities in access were both physical and institutional (including a lift irrigation boom in the 1990s and forms of political and administrative lobbying and corruption).

While the causes of these inequalities are known, remedial action is rare. In particular, programmes of system rehabilitation brought in alongside policies for Irrigation Management Transfer could have addressed them. In this volume, Khanal has outlined how a participatory technology development approach enabled engineers and farmers to reform the water conveyance system in the Nepalese Khagheri system: farmers themselves renegotiated water allocations and management representation between head- and tail-end areas. However, such studies are rare, and the results of such attempts to practice alternative approaches are mixed. Thus, the chapter by Nikku on the Andhra Pradesh (AP) reforms showed how proposals to empower representatives with new local procedures to improve maintenance and supply were captured by elites and bureaucrats.

The construction of irrigation canals creates linkages between individuals, settlements, and communities by organizing them in a complexly bifurcated and queued sequence for access to water. Once constructed, there are inflexibilities that provide strategic advantage to some and disadvantage to others. Thus, a spatial pattern of social differentiation is configured. However, the implanted canal system grid of the social processes of irrigated agriculture is not immutable. Over time, the grid is partly remodelled and spatially reorganized through a variety of means and mechanisms, making time an important dimension (see Carlstein 1982). The chapter by Gautam shows how successive programmes promoting well technology in areas of the Tinau basin in the Nepalese Terai, that also had surface irrigation systems, left farmers with a choice of water sources. They developed diverse water use complexes to combine use of these sources at different times of the year. Transformation in designs of deep tube wells (DTWs) changed their management requirements and farmers' interests in them. These interests changed over time and space as costs of water sources and cropping options changed.

A Cultural Politics of Water

Political ecology has been strongly interested in transcending the society–ecology binary, critiquing and complementing political economy perspectives exclusively focused on access and distribution. In

India, a political ecology of agricultural water approach is still largely to be developed (Mollinga 2010b). The closest to it is Baviskar's 'cultural politics' approach to natural resources management, which aims to combine political economy and the epistemological variant of political ecology (Baviskar 2003; Mollinga 2010b). It emphasizes that '[s]truggles over water are simultaneously struggles for power over symbolic representations and material resources' (Baviskar 2007: 1) The perspective aims at 'incorporation of *ecological specificity* into the cultural politics of water' (2007: 7; emphasis in original) and wants to 'emphasize the distinctive bio-physical properties of water which shape its modes of appropriation' (2007: 5). Along similar lines Mehta (2005) has analysed the social construction of concepts of scarcity and droughts. This perspective allows analysis of local knowledges and cosmologies, as well as policy and other public discourses, within a single framework, enriching the understanding of human agency and the meaning of landscapes in water resources management. The chapter by Manimohan, discussing the diverse economic and cultural trajectories of wet and dry tank cascades under transforming agrarian politics, and the dynamism present beneath the development rhetoric of decline, is a good example from irrigation. When combined with a more 'materialist' understanding of irrigation, political ecology perspectives can make a significant contribution to interdisciplinary irrigation studies.

DISCUSSION: TOWARDS INTERDISCIPLINARY SOCIAL THEORY ON WATER

In critical realism's stratified ontology of structures, mechanisms, and events, the previous four sections have mapped different structures and some associated mechanisms as their emergent properties. The structures are the structure of property rights and entitlements; of techno-institutional arrangements for everyday water use, management, and governance; of water resource-based livelihoods; and the social structure in general.

For all of these structures, some mechanisms were identified. The objective was to show the materiality of the social process of water management, particularly that related to technology/infrastructure and ecology/landscape. There is, thus, no claim to comprehensive mapping, more so because materiality as embodiment of agency, and of water as substance, was left aside. The purpose is conceptual. The four sections together suggest that boundary concepts that allow us to think across the boundaries of the natural and social sciences can and have come into existence for irrigation/water resources management.

Among these boundary concepts, some are neatly articulated as concise and precise concepts, like 'hydraulic property' and 'landesque capital'. Others are more metaphorical in nature, like the management 'scripts' of socially constructed technology. Some are descriptively valid but lack social 'punch', like 'ecosystem goods and services', though this is perhaps compensated by the force of the concept of 'value'. Some are indications of areas of inquiry that need further conceptual articulation specific to water resources situations, like 'space–time relations' and 'cultural politics'/'political ecology'. The conceptual boundary work is work in progress, but sufficient evidence exists, at least in the view of this author, that the multidimensionality of irrigation/water resources management and the ontological complexity of its internal relations can be fruitfully captured, and understanding of it moved forward, by adopting an interdisciplinary perspective.

The 'water control' concept discussed and used throughout this volume can be seen as a boundary concept and also a 'loose concept', what (Löwy 1992) identified as a conceptual space in which the human metabolic engagement with nature in the form of irrigated agriculture can be studied in an interdisciplinary way, by unravelling the hybrid connections that water management involves. The boundary concepts discussed in the previous sections can be read as specifications (but by no means derivations) of this general theoretical idea. With this conceptual space now being well populated, an effort at subsequent general, and more precise, theorization may be in order.

The structures and their emergent properties identified in the previous four sections can be seen as a non-reductionist dissection of the 'concentration of many determinations' that water resources management is. All mechanisms or emergent properties identified derive from hybrid structures, against reductionism, which involves positing ontologically singular structures.[5] This image of dissected determinations begs the question of their 'concentration', that is, how they fit together. For the concrete situations discussed earlier, the fit is relatively easily suggested. Property rights and entitlement arrangements are one of the structuring forces of water use, management, and governance practices (by shaping access and control), while the practices in their turn shape the resource-based livelihoods they support, which shape and are shaped by the broader societal structure and dynamics they are part of. This is neither a time sequence, nor a hierarchy of causality—all this shaping happens simultaneously. However, the possibility to identify related but nevertheless distinct 'determinations' suggests that, indeed, reality is structured and stratified as critical realist philosophy would have it.

The increasing complexity, ontologically as well as societally, of natural or water resources management in combination with the increased scientific legitimacy of notions of complexity and emergence (see Scheffer 2009; Trosper 2005) seems to induce new theorizations of the 'concentration of determinations'. In the rest of this section I sketch some research avenues that seem to me particularly promising for 'substantiating' the water control concept. The first of these sketches focuses on formal theory, the second on substantive theory.

FORMAL THEORY: THE HYDROSOCIAL CYCLE AND MORPHOGENESIS

Within the field of geography, a political ecology perspective on water resources is engaged in developing the concept of the 'hydrosocial cycle', as an interdisciplinary counterpoint to the physically reductionist renderings of the hydrological cycle found in hydrological textbooks (Linton 2008). 'In a sustained attempt to transcend the modernist nature-society binaries, hydro-social research envisions the circulation of water as a combined physical and social process, as a hybridized socio-natural flow that fuses together nature and society in inseparable manners' (Swyngedouw 2009: 56).

This perspective potentially brings together several theoretical components that allow a formal social theory of water resources in the sense of Latour's sociology of connections or associations. Human intervention in the physical cycle of water behaviour is about rearranging flow, availability, and quality of water in time and space. These rearrangements are consolidated in technologies (for example, storage reservoirs), landscapes (for example, polders), and a variety of institutional arrangements for water use, management, and governance; made, reproduced and transformed by various categories of human actors embodying various forms and dimensions of human agency, thus constituting different 'logics of structuration' (Kontopoulos 1993).

The link with formal social theory about structure–agency dynamics seems relatively straightforward, at least at this level of abstraction. Archer's (1995) morphogenetic approach to structural elaboration (the change of structures through episodes of the deployment of human agency, requiring time, that is, in recursive cycles) strongly resonates with the hydro-social cycle perspective. Archer's approach is not explicitly interdisciplinary in the sense of this chapter, but her approach applies as much to hybrid structures and their emergent properties as to social structures as conventionally understood.[6]

Formal theoretical resources for thinking through the time and space dimensions of morphogenesis for water resources management, and implicit in the issue of (multiple) scale(s), can be found in recent work on the politics of space (Jessop et al. 2008) and older work on time-geography (Carlstein 1982). Both provide typologies as heuristics for exploring structured diversity. Jessop et al. (2008) distinguish four key spatial dimensions of social relations: territory, place, scale, and network. Carlstein considers that '[h]uman time is a resource, since *all* activities necessarily require it as input and since we have limited capacity to act in relation to time' (1982: 27). This gives three types of time–space constraints: capacity constraints, coupling constraints, and regulatory/authority constraints (Carlstein 1982: 260).

A second formal theoretical thematic is the concepts of change (and power) to be adopted. In the Marxist tradition the dynamics of change are often conceived as 'dialectics' (see Swyngedouw 1999). Other critical perspectives would use the terminology of 'technologies of rule' (Lansing 1991) for the concentrations (in the plural) of determinations at different levels and in different domains. Such perspectives would take the critical realist notion of society as an 'open system' further than most Marxist theory might. The way forward, I suggest, is concrete research on mechanisms/emergent properties/ logics of structuration in water management transformation processes. Events are easily described and structures are easily labelled, but unravelling the mechanisms/emergent properties/logics of structuration is hard empirical and analytical work.

TOWARDS SUBSTANTIVE THEORY: THINKING MATERIALITY

In terms of substantive theory I confine myself to listing what I consider prospective concrete thematics for advancing theorization of the materiality of social change in water resources management. I want to suggest that rethinking of the commodity form, a materialist institutionalism, and the embodiment of agency are useful entry points for advancing 'hybrid' social theory on water. Given the era in which this chapter is written, the necessary setting of such exploration is, in this author's view, the process of (neo-liberal) capitalist globalization (Brenner *et al.* 2010; Conca 2006; Moore 2010).

Commodities: The social life of things

An emotive controversy in contemporary water disputations is whether water should be considered as an 'economic good' or as a 'social good'. The former is strongly associated with the 1990s neo-liberal development paradigm of market fundamentalism; the latter is the core strategic essentialism of much alternative water politics. The Marxist binary mapping onto this is that of the exchange value and use value of commodities, with exchange value taking on 'a life of its own' under capitalism. Most Marxist theory has focused on the exchange value dynamics, to the detriment of the use value component. 'Social good' perspectives on water can be understood as giving primacy to the use value dimension, and the plurality and diversity associated with that, wanting to keep the exchange value dimension at bay. I suggest that a richer conceptualization of water as a commodity and other commodities implicated in its use can help avoid the 'oversocialization' that stands in the way of capturing hybridity.

Appadurai's attempt to develop 'a new perspective on the circulation of commodities in social life' (1986: 3) understands 'the creation of value [as] a politically mediated process' (1986: 6). It is an effort 'to restore the cultural dimension to societies that are too often represented simply as economies writ large, and to restore the calculative dimension to societies that are too often simply portrayed as solidarity writ small' (1986: 12). He proposes that 'the commodity situation in the social life of any "thing" be defined as the situation in which its exchangeability (past, present, or future) for some other thing is its socially relevant feature' (1986: 13). Whether and how things move in and out of commodity status is a matter of social (including cultural) regulation, that is, it can be done in structurally different ways in different 'regimes of value' (1986: 15). The 'commodity-hood' of a thing can be regarded as an emergent property, constituted by the qualities of the thing itself and the configuration of which it is part. As Appadurai observes, the 'formal truth' that things have no meaning other than that humans endow them with 'does not illuminate the concrete, historical circulation of things. For that we have to follow the things themselves, for their meanings are inscribed in their forms, their uses, their trajectories' (1986: 5).

It is difficult to imagine a water resources management process of some scale that does not have an 'exchange' dimension to it, as securing of the capture, distribution, and use of water requires socially organized investments of labour and materials (see Bakker 2003). This means that rather than 'commodification yes or no?', the question has to be one of *forms* and *patterns* of commodification, perhaps differentiated by phase or compartment of the circulation process, type of water use, and a variety of contextual factors. This applies to 'modern' and 'traditional', 'state', 'corporate' and 'community' water resources management alike. In this way the economic versus social good simplification can be transcended, facilitating more refined theorizations of water resources management and concomitant logics of accumulation in global(izing) capitalism.

Materialist institutionalism

A 'materialist institutionalism' as proposed here involves an analysis of institutional arrangements and forms of organization that takes into account the material mediations of water's circulation. Sections two to five above provided several concrete starting points for 'materialist institutionalist' analysis. The hydraulic property concept (and its extension to ecology/landscape) neatly captures the materialization of rights. The emergent properties of property rights arrangements can, of course, be seen in the way power is exerted in water allocation and distribution. It is clear that technological structure and spatial extent/organization recursively structure management and governance regimes, but how exactly remains to be theorized more precisely (see Wade 1995). The emergent properties of 'locational advantage' and 'queuing' in canal irrigation point in the direction of a (to be developed) typology of 'system characteristics' of canal infrastructure. The connection between water resources development and nation building suggests the need for an interdisciplinary analysis of institutions as technologies of rule (Lansing 1991).

To link a 'materialist institutionalist' analysis of irrigation/agricultural water management with the process of commodification discussed above, it seems useful to me to explore something in between Loftus's 'production of everyday environments' and 'the waterscape as an accumulation strategy' (2009: 964ff).[7]

The embodiment of agency[8]

Human beings have a direct material experience of water through their senses that is meaningful, remarkably consistent over time, and which shapes our engagements with and views about water (Strang 2005). The strong cultural values (and taboos) attached to water have a lot to do with the multifarious personal encounters with water that are part of human life.

A second form of embodiment of agency in water resources management is that use, management, and governance are work; labour processes performed by persons, individually and collectively, by sexed humans, with physical bodies. These persons and their bodies are gendered and of a certain age. This affects water needs as well as capacity to perform water work, individually and collectively as 'materially situated selves'. The body is the repository of specific water resources management knowledge, skills, and experience. Dramatic examples of this can be observed in some forms of spate irrigation, where the diverting of flash floods may require dangerous acts of management in fast flowing streams; or for drinking water supply in water scarce areas where force and agility are needed to carry water safely over large distances. The performance of water work also has bodily effects, for example health effects of headloading and working in paddy fields. Social power is thus partly bodily defined; analysis of water work as labour, and the labour processes in which water is governed, managed, and used seem to me promising entry points for addressing the embodiment of human agency.

As discussed, there remain challenges to non-reductionist theorising of the social in studying the transformations of complex associations and connections that make up water resource management, and to analyse what Archer (1995) describes as the agential properties of emergent water developments. The question remains to avoid the over-socialization of analysis (and also the over-technicality of the past) and keep attention to the material relationships that shape the power of resources in interdisciplinary studies. Chapter 1 referred to the concept of agency used in research for the chapters in this volume, particularly explored through the concepts and frameworks of water (actor) networks and development arenas. One emergent framework is a typology of water politics: the everyday politics of water, the day-to-day contestations of water use; inter-state water politics and negotiations between states on water allocation and distribution; and global water politics studying the relatively new phenomenon of global discourses on policies and regulation.

The capacity to shape water control across South Asia differently in the future for better human development outcomes will at least partly depend on the capacity to analyse and explain more rigorously the hybrid nature of the hydrosocial dynamics of water control.

WHAT NEXT?

I started this chapter by stating that the case for interdisciplinarity is easily made on the grounds of complexity, as also illustrated by this volume. With the growing interest in complexity as a scientific puzzle, the disciplinary–interdisciplinary dichotomy can become a caricature. More relevant seems to be distinguishing the different ways in which complexity can be approached and addressed. These differences are more 'paradigmatic' than having to do with disciplines, which are, according to Lele and Norgaard (2005) better seen as 'academic administrative artefacts'. It is for this reason that this chapter has devoted considerable attention to the ontological premises that I find useful for interdisciplinary analysis of water resources management.

I conclude by listing five research activities that could lift the idiosyncratic focus on irrigation and South Asia of this chapter to a more generic approach to the analysis of hybrid and contested water resources management:

1. A geographically, historically, and sub-sectorally broad-based review of each of the boundary concepts identified in sections two to five, and potential additional ones, and the structures and mechanisms they seek to capture. This may

systematically consolidate existing conceptual framings of the diverse 'determinations' operating in irrigation and water resources management situations.

2. To deploy the existing collection of boundary concepts in single, intensive case studies, to explore the complexity of internal relations in water resources management situations, and to develop theoretical capacity to capture the 'concentration' part of the determinations.

3. Subsequently and in parallel, undertake systematic comparative analysis of the structurally diverse dynamics of water resources management situations.

4. Develop the formal theoretical base of an interdisciplinary political sociology of water resources (see Mollinga 2008a) by elaborating the formal theorization of structure–agency dynamics and water circulation (with a suggested focus on the concepts of hydro-social cycle and morphogenesis).

5. Develop substantive theorization of the materiality of social change in water resource management by elaborating the suggested water-specific rethinking of the commodity form, of different varieties of materialist institutionalism, and of the embodiment of agency.

The chapters in this volume have made a start to these debates, both through their studies of the materiality of social processes shaping water resources development, as well as the socio-technical processes creating contemporary irrigation management. Like the 'Matching Technology and Institutions' research programme from which this book originates, the suggested agenda and collective undertaking of critical interdisciplinary water research will undoubtedly go unexpected, exciting, and complex ways.

Notes

[1] On the emergence of the boundary vocabulary, see Gieryn (1983) and Star and Griesemer (1989). Mollinga (2010a) distinguishes three forms of complexity of natural resources management: ontological (heterogeneity in components and relations), societal (its contested nature), and analytical (difficult to understand). Ontological and analytical complexity constitute the case for interdisciplinarity; societal complexity for transdisciplinarity. Adjectives I use for these three complexities are hybrid, contested, and complicated.

[2] Tanks in south India are an example of village-scale structuring (Shah 2003); canal irrigation (Mollinga 2003), and interlinked system tanks (see

Manimohan, this volume) an example at district level; the interconnected Indus plain irrigation system of Pakistan an example at country level (Merrey 1983).

[3] Including the irony that ecological sustainability existed under a system with feudal characteristics, while ecological degradation ensued when land reform was implemented on welfarist principles driven by a communist party political agenda.

[4] Seminal papers include Pinch and Bijker (1984) and Winner (1986). For further application, see Ertsen (2010), and Bolding (2004) for 'technography' as a methodology for 'following the artefacts'.

[5] My understanding of reductionism originates from Rose (1987). Reductionism is of at least two kinds: 'true' specialization, as in hydraulics exclusively theorising the mechanics of physical water flow, and 'imperial' forms of reductionism that impose a single metric or frame on plurality and diversity, like reducing value to price.

[6] Archer's approach distinguishes three types of emergent properties: structural, cultural, and agential. Archer's perspective that '… structural emergent properties …, irreducible to people and relatively enduring, as with all incidences of emergence, are specifically defined as those internal and necessary relationships which entail material resources, whether physical or human, and which generate causal powers proper to the relation itself' (1995: 177) allows, if not calls for, an interdisciplinarity as explored in this chapter. This means that hybrid phenomena like irrigation systems, practices, and situations have properties that are constituted by their physical and meaning/institutional dimensions *simultaneously*, resulting from the precise way they have been put together (rather than one reflecting or being instantiated by the other).

[7] See Swyngedouw's (2007) analysis of the reconstructing of the complete hydraulic landscape in Franco's fascist Spain as part of a socio-environmental and socio-spatial project of nation building and capitalist accumulation.

[8] This section is strongly shaped by discussions with Frances Cleaver, whose contribution I gratefully acknowledge; the usual disclaimers apply.

GLOSSARY

akhtiyar	chairman in Nepalese FMIS
ambalar/ambalan	hereditary political position
angan	courtyard
anicut	diversion weir (India)
annaikal	dam stones: stones crenulating the crest of a tank spillway
attha coolie	day labourer
atchakat	paddy-growing command area of a tank
ayacut	command area
ayacutdhar	irrigators in the command area
azmoish	see joint azmoish
bhaichaara	sense of brotherhood and friendship
bhaichaara panchayat	local informal congregation of village elders
bhagiyo	sharecropper
bigha	0.68 hectare
bighatti	in proportion to bigha
birta land	land grants in Nepal awarded by the king to private individuals
burmeli	Nepali people who lived in Burma (Myanmar)
calingula	tank spillway
cess	tax or fee (collected on crop type and area)
chak	outlet command
chak hadi	an opening over into the watercourse
chattis	thirty-six
Chaudhary	local Tharu landlord
chautara	raised platform used as a meeting place
cholam	variety of millet
cumbu	variety of millet
darbar	former feudal lords
dhani	landlord

dekchi	cooking pot
elaka/ilaka	subdivision of a district, and electoral unit (divided into wards)
gaon	village
ghatta	mill
haath	local measure of distance (around 0.45 metres)
hankalu	dry land
hawdi	a circular embankment into which water flows from the outlet before flowing into the watercourse
hunda	contract farming
jagir land	land grants by the Nepalese kingdom to government officials and members of the army
janmi	landlord
jharan	water flow from sub-surface springs and drains
jharan-in-charge	person responsible for managing the infrastructure conveying jharan water
Joint *azmoish*	joint supervision; survey of command area in a hydraulic unit
jowar	sorghum
kadu	forest land
kalam	main cropping season for rice in tanks, July–January
kalayi	double or triple cropped paddy lands located in valley bottoms
kallar	caste name
kallar ambalar	official supervising water rights, turns and tank maintenance
kaniyatchi	system of property rights in Tamil society
karai	cooking pot
karais	important lineages in a village
kattaru	jungle streams
kattha	water share unit
kharif crop	summer or monsoon crop
khola	small stream
khola khet	foothill terrace
kila	acre
kisan	farmer

kodai	summer season cultivation in a tank, February–July
kos	irrigation device operated manually using draught animals
kudivaram	cultivator's share of land
kulahi	canal cleaning operations
kulam	tank or pond
kulara	water entitlement from the main irrigation system; obligation to mobilize resources
kulo	canal; farmer-managed surface irrigation system
kulo samiti	irrigation committee
kumili	sluice of a tank
kuzhi	shallow pit
lingajat	non-Brahmin agricultural caste in Karnataka, south India
luskar	lowest level of irrigation agency staff
mandal	a group (of villages)
manjoor shuda	approved, having been sanctioned by law
marukal	surplus discharge
mato muri	traditional measure of land and water shares
mauja	village irrigated by a common village canal of a traditional surface irrigation system
melvaram	government's share of land
mogha	outlet
murai	rotational turn of irrigation in a tank
nadu	a micro-territorial unit comprising of several villages, and an assembly that 'governs' the territory
nanchei	command area (ayacut) of a tank
neerganti	worker who distributes water in a tank command, operates the system and cleans canals
nirani	tank sluice operator
numbari land	originally developed land in an irrigation system
odai	minor ephemeral streams
ovu	pond sluices; Mele ovu: at higher level; Kizhe ovu: at lower level
paani ghatta	water mill
paani jimmawal	irrigation system representative

paatkari	a waterman in charge of water distribution in an area of about 800 to 1,000 hectares in the shejpali system
pali patrak	final water distribution schedule
panchanama	additional irrigation above that sanctioned
panchayat	local, village-level administrative body; village council
pangu	water share in a distribution system of a tank command
pani book	land register
pannaiyal	attached farm labourer
parmar	cobbler caste
parambu	upstream lands which cannot be irrigated
Patel	agriculturalist caste (Gujarat) and village-level revenue officer (Karnataka)
potta	single cropped paddy land on the valley slopes
prajapati	potter caste
pucka warabandi	schedule for water distribution below the outlet in warabandi systems
puja	worship, devotional observances, offerings
punchei	dryland
rabi crop	winter crop
ragi	millet
saacho	wooden proportioning device
saajedaari agreement	sharecropping agreement
samiti	committee
saru jameen	lower or seepage land in a tank-irrigated area
sarkar vellam	water to which property rights are claimed by the government
shanbhoga	village official in Karnataka
shejpali	water allocation system based on farmer demand
sirbaar	manager (over sharecropping arrangements)
sorha	sixteen
suthandiram	land share for village service holders System representative paani jimmawal
taluka	lowest level of government administration above the village level
talwar	village servant
tar	slope-hill terraces

tawan	registration of water theft and penalty note
terai	a flat plain, geographically a zone of southern Nepal and northern India
thakore	agriculturalist caste
tharu	original inhabitant of the Terai plain in Nepal
theka	a tenancy arrangement which includes water rights and land agreements
Thola	ancestral family unit
tole/tola	village or settlement
tol samiti	section committee (for part of an irrigation system)
thottam	garden lands supported by wells, growing vegetables
tuki	light bottle
vaghari	vegetable vendor caste; double or triple cropped paddy land in the valleys
varambu	main bund of a pond
vari jameen	upper land in a tank-irrigated area
vokkaligas	non-Brahmin agricultural caste in Karnataka, South India
warabandi	form of protective irrigation based on a rotational system of fixed turns
Zilla Parishad	elected district-level political and administrative unit of a three-tier structure of local government

BIBLIOGRAPHY

Abeyratne, S.D. 1990. 'Rice, Rehabilitation and Rural Change: Social Organization and State Intervention in Small-scale Irrigation Systems in Sri Lanka', PhD dissertation, Cornell University, Ithaca, NY.

ADB (Asian Development Bank). 1995. 'Project Administration Memorandum, Irrigation Management Transfer Project (IMTP)', Loan No. 1311–NEP(SF).

Agrawal, A. 2003. 'Sustainable Governance of Common-pool Resources: Context, Methods, and Politics', *Annual Review of Anthropology*, 32: 243–62.

Agrawal, A. and C.C. Gibson. 1999. 'Enchantment and Disenchantment: The Role of Community in Natural Resource Conservation', *World Development*, 27(4): 629–49.

Agrawal, A. and K. Sivaramakrishnan (eds). 2001. 'Introduction: Agrarian Environments', in *Social Nature: Resources, Representations, and Rule in India*. New Delhi: Oxford University Press.

Agarwal, A. and S. Narain. 1997. *Dying Wisdom: Rise, Fall and Potential of India's Traditional Water Harvesting Systems*. New Delhi: Centre for Science and the Environment.

Aitken, J.M., G. Cromwell, and G. Wishart. 1991. 'Mini and Micro-hydropower in Nepal', ICIMOD Occasional Paper No. 16. Kathmandu: International Centre for Integrated Mountain Development.

Ambler, J. 1989. 'Adat and Aid: Management of Small-scale Irrigation in West Sumatra, Indonesia', PhD dissertation, Cornell University, Ithaca, NY.

———. 1990. 'The Influence of Farmer Water Rights on the Design of Water-proportioning Devices', in R. Yoder and J. Thurston (eds), *Design Issues in Farmer Managed Irrigation Systems*. Colombo: International Irrigation Management Institute.

———. 1994. 'Small-scale Surface Irrigation in Asia: Technologies, Institutions and Emerging Issues', *Land Use Policy*, 11(4): 262–74.

Anandhi, S., J. Jayaranjan, and R. Krishnan. 2002. 'Work, Caste and Competing Masculinities: Notes from a Tamil Village', *Economic and Political Weekly*, 37(43): 4397–406.

Anderson, T.L. and P. Snyder. 1997. *Water Markets: Priming the Invisible Pump*. Washington, DC: Cato Institute.

Appadurai, A. (ed.). 1986. 'Introduction: Commodities and the Politics of Value', in *The Social Life of Things: Commodities in Cultural Perspective*. Cambridge: Cambridge University Press.

APROSC (Agriculture Project Services Centre). 1978. 'Irrigation Impact Evaluation Study; Gajuri, Manusmara and Tika Bhairab Irrigation Project'. Kathmandu: APROSC.

Archer, M.S. 1995. *Realist Social Theory: The Morphogenetic Approach*. Cambridge: Cambridge University Press.

Arnold, M., A. Bergman, and G. Djurveldt. 1988. *Forestry for the Poor: An Evaluation of the SIDA Supported Social Forestry Project in Tamil Nadu, India*. Stockholm: Swedish International Development Report.

Ashby, J.A. and L. Spurling. 1994. 'Institutionalizing Participatory, Client-driven Research and Technology Development in Agriculture', Agriculture Administration Network Paper 49. London: Overseas Development Institute.

Assadi, M.H. 1997. *Peasant Movement in Karnataka: 1980–94*. New Delhi: Shipra Publications.

Athreya, V.B., G. Djurfeldt, and S. Lindberg. 1990. *Barriers Broken: Production Relations and Agrarian Change in Tamil Nadu*. New Delhi: Sage Publications.

Attwood, D.W. 1985. 'Peasants versus Capitalists in the Indian Sugar Industry: The Impact of the Irrigation Frontier', *Journal of Asian Studies*, 45(1): 59–80.

———. 1987. 'Irrigation and Imperialism: The Causes and Consequences of a Shift from Subsistence to Cash Cropping', *Journal of Development Studies*, 23(3): 341–66.

Bakker, K.J. 2003. *An Uncooperative Commodity: Privatizing Water in England and Wales*. Oxford: Oxford University Press.

Ballabh, V. 2005. 'Emerging Water Crisis and Political Economy of Irrigation Reforms in India', in G.P. Shivakoti, D.L. Vermillion, W.F. Lam, E. Ostrom, U. Pradhan, and R. Yoder (eds), *Asian Irrigation in Transition: Responding to Challenges*. New Delhi: IWMI/Sage Publications.

——— (ed.). 2008. *Governance of Water: Institutional Alternatives and Political Economy*. New Delhi: Sage Publications.

Ballabh, V., C. Kameshwar, S. Pandey, and S. Mishra. 2008. 'Groundwater Governance in Eastern India', in V. Ballabh (ed.), *Governance of Water: Institutional Alternatives and Political Economy*. New Delhi: Sage Publications.

Bandaragoda, D.J. 1998. 'Design and Practice of Water Allocation Rules: Lessons from Warabandi in Pakistan's Punjab', Research Report 17. Colombo: International Irrigation Management Institute.

Bandyopadhyay, J. 2009. *Water, Ecosystems and Society: A Confluence of Disciplines*. London and New Delhi: Sage Publications.

Bardhan, P. and I. Ray (eds). 2008. *The Contested Commons: Conversations between Economists and Anthropologists*. Oxford: Blackwell.

Barker, R. and F. Molle. 2005. 'Perspectives on Asian Irrigation', in G.P. Shivakoti, D.F. Vermillion, W.F. Lam, E. Ostrom, U. Pradhan, and R. Yoder (eds), *Asian Irrigation in Transition: Responding to Challenges*. New Delhi: IWMI/Sage Publications.

Baviskar, A. 2003. 'For a Cultural Politics of Natural Resources', *Economic and Political Weekly*, 38(49): 5051–5.

———. 2005. *In the Belly of the River: Tribal Conflicts over Development in the Narmada Valley*. New Delhi: Oxford University Press.

———. (ed.). 2007. *Waterscapes: The Cultural Politics of a Natural Resource*. New Delhi: Orient Longman.

Bayliss-Smith, T. and S. Wanmali (eds). 1984. *Understanding Green Revolutions: Agrarian Change and Development Planning in South Asia*. Cambridge: Cambridge University Press.

Benda-Beckmann, F. von. 2001. 'Between Free Riders and Free Raiders: Property Rights and Soil Degradation in Context', in N. Heerink, H. van Keulen, and M. Kuiper (eds), *Economic Policy and Sustainable Land Use: Recent Advances in Quantitative Analysis for Developing Countries*. Heidelberg and New York: Physica-Verlag.

Benda-Beckmann, F. von, K. von Benda-Beckmann, and M.G. Wiber. 2006. 'The Properties of Property', in F. von Benda-Beckmann, K. von Benda-Beckmann, and M.G. Wiber (eds), *Changing Properties of Property*. New York/Oxford: Berghahn.

Benton, T. 1992. 'Ecology, Socialism and the Mastery of Nature', *New Left Review*, I(194): 55–74.

Berkes, F. (ed.). 1989. *Common Property Resources: Ecology and Community-based Sustainable Development*. London: Belhaven.

Berkoff, D.J. 1990. 'Irrigation Management on the Indo-Gangetic Plain,' World Bank Technical Paper No. 129. Washington, DC: World Bank.

Bhalerao, C.N. 1973. 'Bureaucracy as an Instrument of Modernisation in India: Some Issues', in Ramesh Arora (ed.), *Administrative Change in India*. New Delhi: Aalekh Publishers.

Bhatia, B. 1992. *Lush Fields and Parched Throats: The Political Economy of Groundwater in Gujarat*. Helsinki: UNU World Institute for Development Economics Research (UNU/WIDER).

Bijoy, C.R. 2006. 'Kerala's Plachimada Struggle: A Narrative on Water and Governance Rights', *Economic and Political Weekly*, 41(41): 4332–9.

Biswas, A., K. Rangachari, and C. Tortajada. 2009. *Water Resources of the Indian Subcontinent*. USA: Oxford University Press.

Boelens, R. and G. Davila (eds). 1998. *Searching for Equity: Conceptions of Justice and Equity in Peasant Irrigation*. Assen: Van Gorcum.

Boelens, R., M. Zwarteveen, and D. Roth. 2005. 'Legal Complexity in the Analysis of Water Rights and Water Resources Management', in D. Roth, R. Boelens, and M. Zwarteveen (eds), *Liquid Relations: Contested Water Rights and Legal Complexity*. Piscataway: Rutgers University Press.

Bolding, A. 2004. 'In Hot Water: A Study on Sociotechnical Intervention Models and Practices of Water Use in Smallholder Agriculture, Nyanyadzi catchment, Zimbabwe', PhD dissertation, Wageningen University, The Netherlands.

Bolding, A., P. Mollinga, and K. van Straaten. 1995. 'Modules for Modernisation: Colonial Irrigation in India and the Technological Dimension of Agrarian Change', *Journal of Development Studies*, 31(6): 805–44.

Bottrall, A. 1978. 'Technology and Management in Irrigated Agriculture', *Development Policy Review*, A11: 22–50.

———. 1992. 'Fits and Misfits Over Time and Space: Technologies and Institutions of Water Development in South Asian Agriculture', *Contemporary South Asian Studies*, 1(2): 227–47.

Brass, T. (ed.). 1995. *New Farmers' Movement in India*. Essex: Frank Cass.

Brenner, N., J. Peck, and N. Theodore. 2010. 'Variegated Neoliberalization: Geographies, Modalities, Pathways', *Global Networks*, 10(2): 182–222.

Brewer, J., S. Kolavalli, A. Kalro, G. Naik, S. Ramnarayan, K.V. Raju, and R. Sakthivadivel. 1999. *Irrigation Management Transfer in India: Policies, Processes and Performance*. New Delhi and Calcutta: Oxford & IBH.

Briscoe, J. and R.P.S. Malik. 2007. *Handbook of Water Resources in India*. New Delhi: Oxford University Press.

Bruns, B. and R. Meinzen-Dick (eds). 2000. *Negotiating Water Rights*. London: Intermediate Technology Publications/International Food Policy Research Institute.

Bruns, B., C. Ringler, and R. Meinzen-Dick (eds). 2005. *Water Rights Reform: Lessons for Institutional Design*. Washington, DC: International food Policy Research Institute.

Burke, J., C. Sauveplanne, and M. Moench. 1999. 'Groundwater Management and Socio-economic Responses', *Natural Resources Forum*, 23(4): 303–13.

Burke, J. and M. Moench. 2000. *Groundwater and Society: Resources, Tensions and Opportunities*. New York: United Nations Publications.

Burt, C. and S. Styles. 1999. 'Modern Water Control and Management Practices: Impacts on Performance', FAO Water Report 19. Rome: FAO.

Carlstein, T. 1982. *Time Resources, Society and Ecology: On the Capacity for Human Interaction in Space and Time*, vol. 1, *Preindustrial Societies*. London: George Allen and Unwin.

Cash, D., W. Clark, F. Alcock, N.M. Dickson, N. Eckley, D. Guston, J. Jäger, and R. Mitchell. 2003. 'Knowledge Systems for Sustainable Development', Proceedings of the National Academy of Sciences of the USA. Available at http://www.pnas.org/content/100/14/8086. full (accessed 6 November 2011).

Castree, N. 2002. 'False Antitheses? Marxism, Nature and Actor-networks', *Antipode*, 34(1): 111–46.

Census of India. 1991. *District Census Handbook Palghat*. Trivandrum: Department of Information and Public Relations, Government of Kerala.

Chakravarti, A. 2001. *Social Power and Everyday Class Relations: Agrarian Transformation in North Bihar*. New Delhi: Sage Publications.

Chambers, R. 1986. 'Irrigation against Rural Poverty', Paper for the INSA National Seminar on Water Management—The Key to the Development of Agriculture, Indian National Science Academy, New Delhi, 27–9 January.

———. 1988. *Managing Canal Irrigation: Practical Analysis from South Asia*. New Delhi: Oxford & IBH.

Chambers, R., N. Saxena, and T. Shah. 1989. *To the Hands of the Poor: Water and Trees*. India: Oxford University Press/IBH, and London: Intermediate Technology Publications.

Chandrakanth, M., B. Shivakumaraswamy, and K. Ananda. 1998. *Economic Implications of Unsustainable Use of Groundwater in Hard Rock Areas of Karnataka*. Bangalore: Department of Agricultural Economics, University of Agricultural Sciences.

Chandrashekar, B. 1984. 'Panchayati Raj Law in Karnataka: Janata Initiative in Decentralisation', *Economic and Political Weekly*, 19(16): 683–92.

Cheung, S. 1969. *The Theory of Share Tenancy*. Chicago: University of Chicago Press.

Chow, L. 1960. 'Development of Rotational Irrigation in Taiwan', *Journal of the Irrigation and Drainage Division*, 86(3): 1–12.

Clay, E. and B. Schaffer (eds). 1984. *Room for Manoeuvre: An Exploration of Public Policy in Agriculture and Rural Development*. London: Heinemann.

Cleaver, F. 2002. 'Reinventing Institutions: Bricolage and the Social Embeddedness of Natural Resource Management', *The European Journal of Development Research*, 14(2): 11–30.

Conca, K. 2006. *Governing Water: Contentious Transnational Policies and Global Institution Building*. Cambridge, MA: MIT Press.

Cooper, D. 1998. *Governing Out of Order: Space, Law and the Politics of Belonging*. London: Rivers Oram Press.

Cosgrove, D. and G. Petts. 1990. *Water, Engineering, and Landscape: Water Control and Landscape Transformation in the Modern Period*. London and New York: Belhaven Press.

Cotula, L. 2006. 'Land and Water Rights in the Sahel: Tenure Challenges of Improving Access to Water for Agriculture', IIED Issue Paper No. 139. London: IIED.

Coward, E.W. 1979. 'Principles of Social Organization in Indigenous Irrigation Systems', *Human Organization*, 38(1): 28–36.

———. 1986a. 'State and Locality in Asian Irrigation Development: The Property Factor', in K.C. Nobe and and R.K. Sampath (eds), *Irrigation Management in Developing Countries: Current Issues and Approaches*. Boulder: Westview Press.

———. 1986b. 'Direct or Indirect Alternatives for Irrigation Investment and the Creation of Property', in K.W. Easter (ed.), *Irrigation Investment, Technology and Management Strategies for Development*. Boulder: Westview Press.

———. 1990. 'Property Rights and Network Order: The Case of Irrigation Works in the Western Himalayas', *Human Organization*, 49(1): 78–88.

Coward, E.W. and G. Levine. 1987. 'Studies of Farmer Managed Irrigation Systems: Ten Years of Cumulative Knowledge and Changing Research Priorities', in International Irrigation Management Institute (IIMI)/ Water and Energy Commission Secretariat (WECS) (eds), *Public Intervention in Farmer-Managed Irrigation Systems*. IIMI and WECS of the Ministry of Water Resources, Government of Nepal, Sri Lanka: International Irrigation Management Institute.

Cullet, P. and J. Gupta. 2009. 'India: Evolution of Water Law and Policy', in J.W. Dellapenna and J. Gupta (eds), *The Evolution of the Law and Politics of Water*. New York: Springer Verlag.

Dahal, N. 1997. 'A Review of Nepal's First Conference on Agriculture', *Water Nepal*, 5(2): 149–64.

Dahl, R.A. 1961. *Who Governs? Democracy and Power in an American City*. New Haven: Yale University Press.

Dani, A. and N. Siddiqi. 1989. *Institutional Innovation in Irrigation Management: A Case Study from Northern Pakistan*. Colombo: International Irrigation Management Institute.

Dev, M. and J. Mooij. 2002. 'Social Sector Expenditures and Budgeting: An Analysis of Patterns and the Budget Making Process in India in the 1990s', IDS Working Paper 164. Brighton: Institute of Development Studies.

Dhawan, B.D. 1982. *The Development of Tubewell Irrigation in India*. New Delhi: Agricole Publishing Academy.

———. 1988. *Irrigation in India's Agricultural Development: Productivity, Stability, Equity*. New Delhi: Sage Publications.

———. 1993. *Trends and New Tendencies in Indian Irrigated Agriculture*. New Delhi: Commonwealth Publishers.

Dhungel, D.N. and S.B. Pun. 2009. *The Nepal-India Water Resources Relationship: Challenges*. New York: Springer Verlag.

Diemer, G. and F. Huibers (eds). 1996. *Crops, People and Irrigation*. London: Intermediate Technology Publications.

Diemer, G. and J. Slabbers (eds). 1992. *Irrigators and Engineers*. Amsterdam: Thesis Publishers.

Dirks, N.B. 1989. *The Hollow Crown: Ethnohistory of an Indian Kingdom*. Cambridge: Cambridge University Press; and Bombay: Orient Longman.

Donahue, J.M. and B.R. Johnston (eds). 1998. *Water, Culture, and Power: Local Struggles in a Global Context*. Washington, DC: Island Press.

D'Souza, D. 2002. *The Narmada Dammed: An Inquiry into the Politics of Development*. New Delhi: Penguin Books.

D'Souza, R. 2006a. 'Water in British India: The Making of a Colonial Hydrology', *History Compass*, 4(4): 621–8.

———. 2006b. *Drowned and Dammed: Colonial Capitalism and Flood Control in Eastern India*. New Delhi: Oxford University Press.

———. 2008. 'River-linking and its Discontents: The Final Plunge for Supply-side Hydrology in India', in K. Lahiri-Dutt and R. Wasson (eds), *Water First: Issues and Challenges for Nations and Communities in South Asia*. New Delhi: Sage Publications.

Dubash, N.K. 2002. *Tubewell Capitalism: Groundwater Development and Agrarian Change in Gujarat*. New Delhi: Oxford University Press.

Dye, T.R. 2001. *Top Down Policymaking*. New York and London: Chatham House Publishers.

Earle, T. and D.E. Doyel. 2008. 'The Engineered Landscapes of Irrigation', in L. Cliggett and C.A. Pool (eds), *Economies and the Transformation of Landscape*. Plymouth: AltaMira Press.

EDI (Economic Development Institute). 1998. *Handbook on Participatory Irrigation Management*, compiled by D. Groenfeldt. Washington, DC: Economic Development Institute of the World Bank.

Egeberg, M. 1999. 'The Impact of Bureaucratic Structure on Policy Making', *Public Administration*, 77(1): 155–70.

Eggink, J.W. and J. Ubels. 1984. *Irrigation, Peasants and Development*. Wageningen: Department of Irrigation and Civil Engineering and Department of Rural Sociology, Wageningen University.

Ertsen, M. 2010. *Locales of Happiness: Colonial Irrigation in the Netherlands East Indies and its Remains, 1830–1980*. Delft: VSSD.

Espeland, W.N. and M.L. Stevens. 1998. 'Commensuration as a Social Process', *Annual Review of Sociology*, 24: 313–43.

Farmer, B.H. 1986. 'Perspectives on the "Green Revolution" in South Asia', *Modern South Asian Studies*, 20(1): 175–99.

Frankel, F.R. 1972. *India's Green Revolution: Economic Gains and Political Costs*. Princeton: Princeton University Press.

Freeman, D.M. and J. Wilkins-Wells. 1989. *Local Organization for Social Development: Concepts and Cases of Irrigation Organization*. Boulder: Westview Press.

Freeman, D. with V. Bhandarkar, E. Shinn, J. Wilkin-Wells, and P. Wilkin-Wells. 1989. *Local Organizations for Social Development: Concepts and Cases for Irrigation*. Boulder: Westview Press.

Freyfogle, E.T. 1996. 'Ethics, Community, and Private Land', *Ecology Law Quarterly*, 23: 631–61.

Gandhi, S.G. 1983. *Tamil Nadu District Gazetteers: Pudukkottai*. Madras: Government Press.

Gautam, S.R. 2006. *Incorporating Groundwater Irrigation: Technology Dynamics and Conjunctive Water Management in the Nepal Terai*. Wageningen University Water Resources Series 8. Hyderabad: Orient Longman.

Geijer, J.C.M.A. 1995. 'Irrigation Management in Asia: Papers from the Expert Consultation on Irrigation Management Transfer in Asia', Bangkok and Chaingmai, 25–9 September.

Geijer, J.C., M. Svendsen, and D.L. Vermilion. 1996. 'Transferring Irrigation Management Responsibility in Asia: Results of a Workshop—FAO/IIMI Expert Consultation on Irrigation Management Transfer in Asia, Bangkok and Chiang Mai, 25–9 September 1995'. Colombo: IIMI.

Gerbrandy, G. and P. Hoogendam. 1996. 'The Materialization of Water Rights: Hydraulic Property in the Extension and Rehabilitation of Two Irrigation Systems in Bolivia', in G. Diemer and F. Huibers (eds), *Crops, People and Irrigation*. London: Intermediate Technology Publications.

Gidwani, V. and K. Sivaramakrishnan. 2003. 'Circular Migration and the Spaces of Cultural Assertion', *Annals of the Association of American Geographers*, 93(1): 186–213.

Gieryn, T. 1983. 'Boundary Work and the Demarcation of Science from Non-science: Strains and Interests in Professional Ideology of Scientists', *American Sociological Review*, 48(6): 781–95.

Gilmartin, D. 1994. 'Scientific Empire and Imperial Science: Colonialism and Irrigation Technology in the Indus Basin', *Journal of Asian Studies*, 53(4): 1127–48.

———. 1995. 'Models of the Hydraulic Environment: Colonial Irrigation, State Power and Community in the Indus basin', in D. Arnold and R. Guha (eds), *Nature Culture Imperialism: Essays on the Environmental History of South Asia*. New Delhi: Oxford University Press.

Giordano, M. and K.G. Villholth (eds). 2007. *The Agricultural Groundwater Revolution: Opportunities and Threats to Development*. Wallingford: CAB International.

Glennon, R. 2002. *Water Follies: Groundwater Pumping and the Fate of America's Fresh Waters*. Washington, DC: Island Press.

GoAP (Government of Andhra Pradesh). 1996. *Irrigation Sector: A Factual Note*. Hyderabad: Irrigation and Command Area Development Department, GoAP.

GoI (Government of India). 1972. 'Report of the Irrigation Commission', New Delhi: Ministry of Irrigation and Power.

GoI/MoWR (Government of India/Ministry of Water Resources). 1987. *National Water Policy*. New Delhi: MoWR.

Goldman, M. 1998. 'Inventing the Commons: Theories and Practices of the Commons' Professional', in Michael Goldman (ed.), *Privatizing Nature: Political Struggles for the Global Commons*. London: Pluto Press in association with Transnational Institute.

Gomathinayagam, P. 1995. 'Tank Irrigation in Ancient Tamil Nadu', in N.V. Pundarikanthan and L. Jayasekar (eds), *Proceedings of the National Workshop on Traditional Water Management for Tanks and Ponds*. Madras: Anna University.

Gorter, P. 1989. 'Canal Irrigation and Agrarian Transformation: A Case of Kerala', *Economic and Political Weekly*, 24(39): A94–105.

Government of Kerala. 1971 and 2005. *Economic Review, Agricultural Census*, Annual series. Trivandrum: Department of Economics and Statistics.

———. 2001. *Statistics for Planning 2001*. Trivandrum: Department of Economics and Statistics.

———. 2003. *District Handbooks of Kerala: Palakkad*. Trivandrum: Department of Information and Public Relations.

Government of Kerala. 2006. *Panchayat Level Statistics 2006*. Trivandrum: Department of Economics and Statistics.

———. 2007a. *District Level Database 2006, Palakkad*. Trivandrum: Department of Local Self-government and KILA.

———. 2007b. *Guidelines for the Preparation of Annual Plan (2007–2008) and Eleventh Five-Year Plan (2007–2012)*. Trivandrum: Department of Local Self-government.

Granda, P. 1984. 'Property Rights and Land Control in Tamil Nadu: 1350–1600', PhD dissertation, University of Michigan, Ann Arbor, MI, USA.

Greider, T. and L. Garkovich. 1994. 'Landscapes: The Social Construction of Nature and the Environment', *Rural Sociology*, 59(1): 1–24.

Grindle, M.S. 1977. *Bureaucrats, Politicians and Peasants in Mexico: A Case Study in Public Policy*. Berkeley: University of California Press.

Grindle, M.S. and J.W. Thomas. 1990. 'Policy Makers, Policy Choices, and Policy Outcomes: The Political Economy of Reform in Developing Countries', *Policy Sciences*, 22(3/4): 213–48.

Groenfeldt, D. (ed.). 1999. 'Volume 1: Handbook', in *Capacity Building for Participatory Irrigation Management (PIM)*, Training Course, 28–9 September. Bari, Italy: CIHEAM, IAM-B, and World Bank.

Groenfeldt, D. and M. Svendsen (eds). 1997. *Case Studies in Participatory Irrigation Management*. Washington, DC: World Bank.

Groot, A.E. 2002. 'Demystifying Facilitation of Multi-actor Learning Processes', PhD dissertation, Wageningen University, The Netherlands.

Gulati, A. and R. Chadha. 1999. 'India Poised for Take-off? Reforms in Trade, Industry and Agriculture: Stock-taking and Future Challenges', in U. Kapila (ed.), *Indian Economy since Independence*. Ghaziabad: Academic Foundation.

Gulati, A., R. Meinzen-Dick, and K.V. Raju. 1999. *From Top Down to Bottom Up: Institutional Reforms in Indian Canal Irrigation*, A Collaborative Study of IEGNCAER–ISEC–IFPRI. New Delhi: Institute of Economic Growth.

———. 2005. *Institutional Reforms in Indian Irrigation*. New Delhi: International Food Policy Research Institute/Sage Publications.

Gunawardana, R.L.A.H. 1984. 'Intersocietal Transfer of Hydraulic Technology in Precolonial South Asia: Some Reflections on Preliminary Investigation', *Southeast Asian Studies*, 22(2): 115–42.

Gupta, A. 1998. *Postcolonial Developments: Agriculture in the Making of Modern India*. London: Duke University Press.

———. 2008. 'The Monsoon Rivers of South Asia: A Geomorphological Perspective on Managing Monsoon Rivers', in K. Lahiri-Dutt and

R. Wasson (eds), *Water First: Issues and Challenges for Nations and Communities in South Asia*. New Delhi: Sage Publications.

Gurukkal, R. 1986. 'Aspects of the Reservoir System of Irrigation in the Early Pandyan State', *Studies in History*, 2(2): 155–64.

Guttierrez Pérez, J. 2005. 'Appropriate Designs and Appropriating Irrigation Systems: Irrigation Infrastructure Development and Users' Management in Bolivia', PhD dissertation, Wageningen University, The Netherlands.

Gyawali, D. 2001. *Water in Nepal*. Kathmandu: Himal Books and Panos South Asia with Nepal Water Conservation Foundation.

Gyawali, D. and A. Dixit. 1999. 'Fractured Institutions and Physical Interdependence: Challenges to Local Water Management in the Tinau River Basin, Nepal', in M. Moench, E. Caspari, and A. Dixit (eds), *Rethinking the Mosaic: Investigations into Local Water Management*. Kathmandu: Nepal Water Conservation Foundation.

Ham, C. and M.J. Hill. 1984. *The Policy Process in the Modern Capitalist State*. New York: St. Martin's Press.

Hann, C. 2007. 'A New Double Movement? Anthropological Perspectives on Property in the Age of Neoliberalism', *Socio-Economic Review*, 5: 287–318.

Hanna, S., C. Folke, and K.G. Maler (eds). 1996. 'Property Rights and the Natural Environment', in *Rights to Nature: Ecological, Economic, Cultural, and Political Principles of Institutions for the Environment*. Washington, DC: Island Press.

Hardiman, D. 1998. 'Well Irrigation in Gujarat: Systems of Use, Hierarchies of Control', *Economic and Political Weekly*, 33(25): 1533–44.

Harriss, J. 1980. 'Contemporary Marxist Analysis of the Agrarian Question in India', Working Paper 14. Madras: Madras Institute of Development Studies.

———. 2002. *Depoliticizing Development: The World Bank and Social Capital*. London: Anthem Press.

Harrower, M.J. 2008. 'Hydrology, Ideology, and the Origins of Irrigation in Ancient Southwest Arabia', *Current Anthropology*, 49(3): 497–510.

Hasnip, N., L. Vincent, and K. Hussein. 1999. 'Poverty Reduction and Irrigated Agriculture', IPTRID Issue Paper 1. Rome: FAO.

Heitzman, J. 2001. *Gifts of Power: Lordship in an Early Indian State*. New Delhi: Oxford University Press.

Hennen, L. 1999. 'Uncertainty and Modernity: Participatory Technology Assessment: A Response to Technical Modernity?' *Science and Public Policy*, 26(5): 303–12.

Hodgson, S. 2004. *Land and Water: The Rights Interface*. Rome: FAO.

Hoogendam, P. 1994. *Lively Practice and Hardened History: A Research Program on Irrigation Technology, Engineers and Artifacts*. Wageningen: Wageningen Agricultural University.

Horst, L. 1996. 'Intervention in Irrigation Water Division in Bali, Indonesia: A Case of Farmers' Circumvention of Modern Technology', in G. Diemer and F.P. Huibers (eds), *Crops, People and Irrigation: Water Allocation Practices of Farmers and Engineers*. London: Intermediate Technology Publications.

———. 1998. *The Dilemmas of Water Division: Considerations and Criteria for Irrigation System Design*. Colombo/Wageningen: International Irrigation Management Institute/Wageningen Agricultural University.

HR Wallingford. 2001. 'Sustainable Irrigation Turnover. Report on Infrastructure, Field Investigation Results', KAR Project R7389.

Hughes, T.P. 1987. 'The Evolution of Large Technological Systems', in W.E. Bijker, T.P. Hughes, and T.J. Pinch (eds), *The Social Construction of Technological Systems*. Cambridge, MA: MIT Press.

Hunt, R. and E. Hunt. 1976. 'Canal Organization and Local Social Organization', *Current Anthropology*, 17(3): 389–411.

Hussain, I. and E. Biltonen (eds). 2001. 'Pro-poor Irrigation Intervention Strategies in Irrigated Agriculture in Asia: Developing the Project Framework', in *Managing Water for the Poor: Proceedings of the Regional Workshop on Pro-poor Intervention Strategies in Irrigated Agriculture in Asia*. Colombo: International Water Management Institute.

Hussain, I., K. Yokoyama, and I. Hunzai. 2001. 'Irrigation Against Rural Poverty: An Overview of Issues and Pro-poor Intervention Strategies in Irrigated Agriculture in Asia', in I. Hussain and E. Biltonen (eds), *Irrigation against Rural Poverty: An Overview of Issues and Pro-poor Intervention Strategies in Irrigated Agriculture in Asia*. Colombo: International Water Management Institute.

IIMI/WECS (International Irrigation Management Institute/Water and Energy Commission Secretariat) (ed.). 1987. *Public Intervention in Farmer-Managed Irrigation Systems*. IIMI and WECS of the Ministry of Water Resources, Government of Nepal, Sri Lanka: International irrigation Management Institute.

IIMA–IWMI (Indian Institute of Management Ahmedabad–International Water Management Institute). 1999. *Maharashtra IMT Impact Assessment. A Collaborative Study*. Ahmedabad and Colombo: IIMA and IWMI.

Iyer, R. 2008. 'National and Regional Water Concerns: Setting the Scene', in K. Lahiri-Dutt and R. Wasson (eds), *Water First: Issues and*

Challenges for Nations and Communities in South Asia. New Delhi: Sage Publications.

Jacob, S.B. (ed.). 1995. *Hill Irrigation Engineering*. Kathmandu: Tribhuvan University, Institute of Engineering.

Jain, S.K., P.K. Agarwal, and V.P. Singh. 2007. *Hydrology and Water Resources of India*. New York: Springer Verlag.

Jairath, J. 1985. 'Private Tubewell Utilisation in Punjab', *Economic and Political Weekly*, 20(40): 1703–12.

Jairath, J. and V. Ballabh (eds). 2008. *Droughts and Integrated Water Management in South Asia*. UK: Sage Publications.

Janakarajan, S. 1989. 'Characteristics and Functioning of Traditional Irrigation Institutions', *Management of Renewable Resources*, 19(12): 81–101.

———. 1993. 'In Search of Tanks: Some Hidden Facts', *Economic and Political Weekly*, 28(26): A53–60.

———. 1994. 'Trading in Groundwater: A Source of Power and Accumulation', in M. Moench (ed.), *Selling Water: Conceptual and Policy Debates over Groundwater Markets in India*. Ahmedabad: VIKSAT, Pacific Institute, Natural Heritage Institute.

———. 1997. 'Consequences of Aquifer Overexploitation: Prosperity and Deprivation', *Review of Development and Change*, 2(1): 52–71.

———. 1999. 'Conflicts over the Invisible Resource in Tamil Nadu: Is There a Way Out?' in M. Moench, E. Caspari, and A. Dixit (eds), *Rethinking the Mosaic: Investigations into Local Water Management*. Kathmandu and Colorado: Nepal Water Conservation Foundation and Institute for Social and Environmental Transition.

———. 2004. 'Irrigation: The Development of an Agro-ecological Crisis', in B. Harriss-White and S. Janakarajan (eds), *Rural India Facing the 21st Century*. London: Anthem Press.

Jantzen, D. and K. Koirala. 1989. *Micro-hydro Power in Nepal*. Kathmandu: Bikash Enterprises.

Jentoft, S. 2004. 'Institutions in Fisheries: What They Are, What They Do, and How They Change', *Marine Policy*, 28: 137–49.

Jessop, B., N. Brenner, and M. Jones. 2008. 'Theorizing Sociospatial Relations', *Environment and Planning D: Society and Space*, 26: 389–401.

Johnson, J., D.L. Vermillion, and J.A. Sagardoy (eds). 1995. 'Irrigation Management Transfer. Selected Papers from the International Conference in Irrigation Management', Water Report No. 5. Rome: FAO/IIMI.

Jorgensen, U. and O.H. Sorensen. 1999. 'Arenas of Development: A Space Populated by Actor-worlds, Artefacts and Surprises', *Technology Analysis and Strategic Management*, 11(3): 409–29.

Joy, K.J., B. Gujja, S. Paranjape, V. Goud, and S. Vispute (eds). 2008. *Water Conflicts in India: A Million Revolts in the Making*. New Delhi: Routledge.

Jurriëns, R. 1993. 'Protective Irrigation: Essence and Implications. Water Management in the Next Century', 15th Congress on Irrigation and Drainage, Volume 1-A. The Hague: International Commission on Irrigation and Drainage.

Jurriëns, R., P. Mollinga, and P. Wester. 1996. 'Scarcity by Design: Protective Irrigation in India and Pakistan', Liquid Gold Paper 1. Wageningen: Wageningen University.

Kabhre DDC/REDS (District Development Committee/Rural Energy Development Section). 2000. *District Development Committee/Rural Energy Development Section. Kabhre Achievements 1997–2000*. Kabhre Palanchowk: Kabhre DDC and REDS.

———. 2002. *REDP 5 Years in Kabhre, 1997–2001: A Document of Achievement*. Nepal: DDC and REDP.

Kahnert, F. and G. Levine. 1993. *Groundwater Irrigation and the Rural Poor: Options for Development in the Gangetic Basin*. Washington, DC: World Bank.

Kansakar, D.R. 2005. 'Understanding Groundwater for Proper Utilization and Management in Nepal', in B.R. Sharma, K.G. Villholth, and K.D. Sharma (eds), *Groundwater Research and Management: Integrating Science into Management Decisions. Proceedings of IWMI-ITP-NIH International Workshop on Creating Synergy between Groundwater Research and Management in South and Southeast Asia, 8–9 February 2005*. Colombo: IWMI.

Karashima, N. 1984. *South Indian History and Society: Studies from Inscriptions A.D. 850–1800*. New Delhi: Oxford University Press.

Keeley, J. 1997. 'Reconceptualising Policy Processes, The Dynamics of Natural Resource Management and Agricultural Intensification Policy-making in Ethiopia, 1984–97', MPhil dissertation, IDS, University of Sussex, Brighton, UK.

Kerala State Land Use Board. 2001. 'Vibhava Bhoopada Nirmana Report on Resource Map Preparation, Kollengode and Elavenchery Grama Panchayats, Palakkad Zilla' (in Malayalam).

Kerr, J. 2002. 'Watershed Development, Environmental Services, and Poverty Alleviation in India', *World Development*, 30(8): 1387–1400.

Khanal, P.R. 2003. *Engineering Participation: The Processes and Outcomes of Irrigation Management Transfer in the Terai of Nepal*. Wageningen University Water Resources Series 2. Hyderabad: Orient Longman.

Kishore, A. 2004. 'Understanding Agrarian Impasse in Bihar', *Economic and Political Weekly*, 39(31): 3484–91.

Klein, J.T. 1996. *Crossing Boundaries: Knowledge, Disciplinarities, and Interdisciplinarities*. Charlottesville/London: University Press of Virginia.

Kloezen, W.H. 2002. 'Accounting for Water: Institutional Viability and Impact of Market-oriented Irrigation Intervention in Central Mexico', PhD dissertation, Wageningen University, The Netherlands.

Kloezen, W.H. and M. Samad. 1995. 'Synthesis of Issues Discussed at the International Conference on Irrigation Management Transfer, Wuhan, China, 20–24 September 1994', Short series on locally managed irrigation. Colombo: IIMI.

Kloezen, W.H. and P.P. Mollinga. 1992. 'Opening Closed Gates: Recognizing the Social Nature of Irrigation Artefacts', in G. Diemer and J. Slabbers (eds), *Irrigators and Engineers*. Amsterdam: Thesis Publishers.

Klug, H. 2002. 'Straining the Law: Conflicting Legal Premises and the Governance of Aquatic Resources', *Society and Natural Resources*, 15: 693–707.

KMI (Kathmandu Metal Industries). 1998. *Final Report of R and D Propeller Turbine Installed at Katunje Kavree*. Kathmandu: Kathmandu Metal Industries (P) Ltd.

Knegt, J.W. and L. Vincent. 2001. 'From Open Access to Access by All: Restating Challenges for Groundwater Management in Andhra Pradesh', *Natural Resources Forum*, 25(4): 321–31.

Kohli, A. 1987. 'Karnataka: Populism, Patronage and Piecemeal Reform', in A. Kohli (ed.), *The State and Poverty in India*. Cambridge: Cambridge University Press.

Kolavalli, S. and D.L. Chicoine. 1987. *Groundwater Markets in Gujarat*. Ahmedabad: Indian Institute of Management.

Kontopoulos, K. 1993. *The Logics of Social Structure: Structural Analysis in the Social Sciences*. Cambridge: Cambridge University Press.

Kothari, S. and R. Roy. 1969. *Relations between Politicians and Administrators at the District Level*. New Delhi: Indian Institute of Public Administration.

Kripa, A.P. 1992. 'Farmers' Movement in Karnataka', *Economic and Political Weekly*, 27(23): 1182–3.

Krishna Murthy, A.N. 1975. 'Shimoga, Mysore', in International Rice Research Institute (IRRI) (ed.), *Changes in Rice Farming in Selected Area of Asia*. Manila: IRRI.

Krishnan, J. 2009. *Enclosed Waters: Property Rights, Technology and Ecology in the Management of Water Resources in Palakkad, Kerala*. Wageningen University Water Resources Series 9. Hyderabad: Orient Blackswan.

Kumar, D. 1965. *Land and Caste in South India: Agricultural Labour in the Madras Presidency during the Nineteenth Century*. Cambridge: Cambridge University Press.

Kumar, M.D. 2007. *Groundwater Management in India: Physical, Institutional and Policy Alternatives*. New Delhi: Sage Publications.

Kumar, R., R.D. Singh, and K.D. Sharma. 2005. 'Water Resources of India', *Current Science*, 89(5): 794–811.

Lahiri-Dutt, K. and R.J. Wasson (eds). 2008. *Water First: Issues and Challenges for Nations and Communities in South Asia*. New Delhi: Sage Publications.

Lam, W.F. 1996. 'Improving the Performance of Small-scale Irrigation Systems: The Effects of Technology Investments and Government Structure on Irrigation Performance in Nepal', *World Development*, 24(6): 1301–13.

Lansing, J.S. 1991. *Priests and Programmers: Technologies of Power in the Engineered Landscape of Bali*. Princeton: Princeton University Press.

———. 2006. *Perfect Order: Recognising Complexity in Bali*. Princeton: Princeton University Press.

Latour, B. 2005. *Reassembling the Social: An Introduction to Actor-network Theory*. Oxford: Oxford University Press.

Leach, E.R. 1959. 'Hydraulic Society in Ceylon', *Past and Present*, 15: 2–26.

———. 1980. 'Village Irrigation in the Dry Zone of Sri Lanka', in E.W. Coward Jr. (ed.), *Irrigation and Agricultural Development in Asia: Perspectives from Social Science*. Ithaca, New York: Cornell University Press.

Leach, M., R. Mearns, and I. Scoones. 1999. 'Environmental Entitlements: Dynamics and Institutions in Community-Based Natural Resource Management', *World Development*, 27(2): 225–47.

Lele, S. and R.B. Norgaard. 2005. 'Practicing Interdisciplinarity', *BioScience*, 55: 967–75.

Lele, S.N. and R.K. Patil. 1994. *Farmer Participation in Irrigation Management: A Case study of Maharashtra*. Pune: Society for People's Participation in Ecosystem Management and New Delhi: Horizon India Books.

Levine, G. 1980. 'Hardware and Software: An Engineering Perspective on the Mix for Irrigation Management', in *Report of a Planning Workshop on Irrigation Water Management*. Philippines: IRRI.

Linton, J. 2008. 'Is the Hydrologic Cycle Sustainable? A Historical-geographical Critique of a Modern Concept', *Annals of the Association of American Geographers*, 98(3): 630–49.

Lipsky, M. 1980. *Street-level Bureaucracy: Dilemmas of the Individual in Public Services*. New York: Russell Sage Foundation.

Loftus, A. 2009. 'Rethinking Political Ecologies of Water', *Third World Quarterly*, 30(5): 953–68.

Löwy, I. 1992. 'The Strength of Loose Concepts: Boundary Concepts, Federative Experimental Strategies and Disciplinary Growth: The Case of Immunology', *History of Science*, 30–4(90): 371–96.

Ludden, D. 1978. 'Ecological Zones and the Cultural Economy of Irrigation in Southern Tamil Nadu', *South Asia*, New Series, I(1): 1–13.

———. 1979. 'Patronage and Irrigation in Tamil Nadu: A Long-term View', *Indian Economic and Social History Review*, 63: 347–65.

———. 1985. *Peasant History in South India*. Princeton: Princeton University Press.

———. 1992. 'India's Development Regime', in N.B. Dirks (ed.), *Colonialism and Culture*. Ann Arbor: University of Michigan Press.

Mabry, J. (ed.). 1996. *Canals and Communities: Small-scale Irrigation Systems*. Tucson: University of Arizona Press.

MacDonald and Hunting Technical Services. 1982. 'Medium Irrigation Projects', A Design Manual. Nepal: MacDonald and Partners Ltd in association with Hunting Technical Service Ltd for Department of Irrigation.

Malhotra, S.P. 1988. 'The Warabandi System and its Infrastructure.' CBIP Publication No. 157. New Delhi: Central Board of Irrigation and Power.

Malhotra, S.P., S.K. Raheja, and D. Seckler. 1984. 'Performance Monitoring in the *Warabandi* System of Irrigation Management', in N. Pant (ed.), *Productivity and Equity in Irrigation Systems*. New Delhi: Ashish Publishing House.

Manor, J. 1989. 'Karnataka: Caste, Class, Dominance and Politics in a Cohesive Society', in F.R. Frankel and M.S.A. Rao (eds), *Dominance and State Power in Modern India: Deadline of a Social Order*, vol. 1. New Delhi: Oxford University Press.

Martin, E.D. and R. Yoder. 1987. 'Institutions for Irrigation Management in Farmer-managed Irrigation Systems: Examples from the Hills of Nepal', IIMI Research Paper 5. Colombo: International Irrigation Management Institute.

Marx, K. 1974. *Capital: A Critical Analysis of Capitalist Production*, vol. I. Moscow: Progress Publishers.

Massey, D. 1984. *Spatial Divisions of Labour: Social Structures and the Geography of Production*. London: Macmillan.

McCay, B.J. 2002. 'Emergence of Institutions for the Commons: Contexts, Situations, and Events', in E. Ostrom, T. Dietz, N. Dolsak, P.C. Stern, S. Stonich, and E.U. Weber (eds), *The Drama of the Commons*. Washington, DC: National Academy Press.

McCay, B.J. and J.M. Acheson (eds). 1987. 'Human Ecology of the Commons', in *The Question of the Commons: The Culture and Ecology of Communal Resources*. Tucson: The University of Arizona.

McKean, M.A. 2000. 'Common Property: What Is It, What Is It Good for, and What Makes It Work?' in C. Gibson, M. McKean, and E. Ostrom (eds), *People and Forests: Communities, Institutions, and Governance*, Cambridge, MA: MIT Press.

MEA (Millennium Ecosystem Assessment). 2005. *Ecosystems and Human Well-being: Synthesis Report*. Washington, DC: Island Press.

Mehta, L. 2005. *The Politics and Poetics of Water: Naturalising Scarcity in Western India*. New Delhi: Orient Longman.

Mehta, L., M. Leach, P. Newell, I. Scoones, K. Sivaramakrishnan, and S.A. Way. 1999. 'Exploring Understandings of Institutions and Uncertainty: New Directions in Natural Resource Management', IDS Discussion Paper 372. Brighton: Environment Group, Institute of Development Studies, University of Sussex.

Meinzen-Dick, R.S. 1989. 'Water in a Thirsty Land: Irrigation Development and Agrarian Structure in South India', PhD dissertation, Cornell University, Ithaca, NY.

———. 1997. 'Farmer Participation in Irrigation: 20 Years of Experience and Lessons for the Future', *Irrigation and Drainage Systems*, 11(2): 103–88.

Meinzen-Dick, R.S. and R. Pradhan. 2001. 'Implications of Legal pluralism for Natural Resource Management', *IDS Bulletin*, 32(4): 10–17.

———. 2005. 'Recognizing Multiple Water Uses in Intersectoral Water Transfers', in G.P. Shivakoti, D.L. Vermillion, W.F. Lam, E. Ostrom, U. Pradhan, and R. Yoder (eds), *Asian Irrigation in Transition: Responding to Challenges*. New Delhi: IWMI/Sage Publications.

Merrey, D.J. 1983. 'Irrigation, Poverty and Social Change in a Village of Pakistani Punjab: An Historical and Cultural Ecological Analysis', PhD dissertation, University of Pennsylvania, Philadelphia, PA.

Merrey, D.J., R. Meinzen-Dick, P.P. Mollinga, and E. Karar. 2007. 'Policy and Institutional Reform: The Art of the Possible', in D. Molden (ed.), *Water for Food, Water for Life: A Comprehensive Assessment of Water Management in Agriculture*. London: International Water Management Institute/Earthscan.

Migdal, J.S., A. Kohli, and V. Shue (eds). 1994. *State Power and Social Forces: Domination and Transformation in the Third World*. Cambridge: Cambridge University Press.

Mitra, A.K. 1992. 'Joint Management of Irrigation Systems in India', *Economic and Political Weekly*, 27(26): A75–82.

Moench, M. 1992. 'Drawing down the Buffer', *Economic and Political Weekly*, 27(13): A7–14.

———. 1994. 'Approaches to Groundwater Management: To Control or Enable?' *Economic and Political Weekly*, 33(26): A46–53.

———. 1999. 'Addressing Constraints in Complex Systems: Meeting the Water Management Needs of South Asia in the 21st Century', in M. Moench., E. Caspari, and Ajaya Dixit (eds), *Rethinking the Mosaic: Investigations into Local Water Management*. Kathmandu and Colorado: Nepal Water Conservation Foundation and Institute for Social and Environmental Transition.

———. 2000. 'India's Groundwater Challenge', *Seminar*, 486(February). Available at http://www.india-seminar.com/2000/486/486%20moench.htm (accessed 12 February 2012).

———. 2002. 'Water and the Potential for Social Instability: Livelihoods, Migration and the Building of Society', *Natural Resources Forum*, 26: 195–204.

Moench, Marcus, A. Dixit, S. Janakarajan, M.S. Rathore, and S. Mudrakartha (eds). 2003. *The Fluid Mosaic: Water Governance in the Context of Variability, Uncertainty and Change. A Synthesis Paper*. Kathmandu and Boulder: Nepal Water Conservation Foundation and the Institute for Social and Environmental Transition.

Moench, M., E. Caspari, and A. Dixit (eds). 1999. *Rethinking the Mosaic: Investigations into Local Water Management*. Kathmandu: Nepal Water Conservation Foundation.

Molden, D. (ed.). 2007. *Water for Food, Water for Life: A Comprehensive Assessment of Water Management in Agriculture*. London: International Water Management Institute/Earthscan.

Molden, D., R. Saktivadivel, C. Perry, C. de la Fraiture, and W. Kloezen. 1998. 'Indicators for Comparing Performance of Irrigated Agricultural Systems', Research Report 20. Colombo: IWMI.

Molle, F. 2008. 'Nirvana Concepts, Narratives, and Policy Models: Insights from the Water Sector', *Water Alternatives*, 1(1): 131–56.

Molle, F., P.P. Mollinga, and P.H. Wester. 2009. 'Hydraulic Bureaucracies: Flows of Water, Flows of Power', *Water Alternatives*, 2(3): 328–49.

Mollinga, P.P. 2000. 'The Inevitability of Reform: Towards Alternative Approaches for Canal Irrigation Development in India', in L.K. Joshi and R. Hooja (eds), *Participatory Irrigation Management. Paradigm for the 21st Century*, vol. 1. Jaipur and New Delhi: Rawat Publications.

———. 2001. 'Water and Politics: Levels, Rational Choice and South Indian Canal Irrigation', *Futures*, 33: 733–52.

Mollinga, P.P. 2003. *On the Waterfront: Water Distribution, Technology and Agrarian Change in a South Indian Canal Irrigation System*. Wageningen University Water Resources Series. Hyderabad: Orient Longman.

———. 2007. 'Water Policy—Water Politics: Social Engineering and Strategic Action in Water Sector Reform', Working Paper Series 19. Bonn: GEF.

———. 2008a. 'Water, Politics and Development: Framing a Political Sociology of Water Resources Management', *Water Alternatives*, 1(1): 7–23.

———. 2008b. 'The Water Resources Policy Process in India: Centralisation, Polarisation and New Demands on Governance', in V. Ballabh (ed.), *Governance of Water: Institutional Alternatives and Political Economy*. New Delhi: Sage Publications.

———. 2010a. 'Boundary Work and the Complexity of Natural Resources Management', *Crop Science*, 50: S1–9.

———. 2010b. 'The Material Conditions for a Polarised Discourse: Clamours and Silences in Critical Analysis of Agricultural Water Use in India', *Journal of Agrarian Change*, 10(3): 414–36.

Mollinga, P.P. and A. Bolding. 1996. 'Signposts of Struggle: Pipe Outlets as the Material Interface between Water Users and the State in a Large-scale Irrigation System in South India', in G. Diemer and F. Huibers (eds), *Crops, Irrigation and People: Water Allocation Practices of Farmers and Engineers*. London: Intermediate Technology Publications.

——— (eds). 2004. *The Politics of Irrigation Reform: Contested Policy Formulation and Implementation in Asia, Africa and Latin America*. Aldershot: Ashgate.

Mooij, J. 1998 *Food Policy and the Indian State: The Public Distribution System in South India*. New Delhi: Oxford University Press.

———. 2003. 'Smart Governance? Politics in the Policy Process in Andhra Pradesh', ODI Working Paper No. 228. London: Overseas Development Institute.

Mooij, J. and V. de Vos. 2003. 'Policy Processes: An Annotated Bibliography on Policy Processes, with Particular Emphasis on India', ODI Working Paper No. 221. London: Overseas Development Institute.

Moore, J.W. 2010. 'The End of the Road?: Agricultural Revolutions in the Capitalist World-ecology, 1450–2010', *Journal of Agrarian Change*, 10(3): 389–413.

Moss, T. 2006. 'Solving Problems of "fit" at the Expense of Problems of "Interplay"? The Spatial Reorganisation of Water Management Following the EU Water Framework Directive', in P.P. Mollinga, A. Dixit, and K. Athukorala (eds), *Integrated Water Resources*

Management: Global Theory, Emerging Practice and Local Needs. New Delhi: Sage Publications.

Mosse, D. 1997a. 'Ecological Zones and the Culture of Collective Action: The History and Social Organization of a Tank Irrigation System in Tamil Nadu', *South Indian Studies*, 3: 1–88.

———. 1997b. 'The Symbolic Making of a Common Property Resource: History, Ecology and Locality in a Tank-irrigated Landscape in South India', *Development and Change*, 28(3): 467–504.

———. 1999. 'Colonial and Contemporary Ideologies of "Community Management": The Case of Tank Irrigation Development in South India', *Modern Asian Studies*, 33(2): 303–38.

———. 2003. *The Rule of Water: Statecraft, Ecology and Collective Action in South India*. New Delhi: Oxford University Press.

MoWR (Ministry of Water Resources of Nepal). 1997. *Water Resources Act 2049 (1992)*. Kathmandu: MoWR.

Mukherji, A., K.G. Villholth, B.R. Sharma, and J. Wang (eds). 2009. *Groundwater Governance in the Indo-Gangetic and Yellow Rivers Basins: Results and Challenges*. Leiden: CRC Press/Balkema.

Mukundan, T.M. 1988. 'The Ery Systems of South India', *PPST Bulletin*, 16: 38.

Munoz, G., C. Garces-Restrepo, D.L. Vermillion, D. Renault, and M. Samad. 2007. 'Irrigation Management Transfer: Worldwide Efforts and Results', Paper for the 4th Asian Regional Conference and the 10th International Seminar on Participatory Irrigation Management, 2–5 May, Tehran, Iran.

Musch, A. 2001. 'The Small Gods of Participation', PhD dissertation, University of Twente.

Nadkarni, M.V. 1987. *Farmers' Movement in India*. Ahmedabad: Allied Publishers.

———. 1996. 'Accelerating Commercialisation of Agriculture: Dynamic Agriculture and Stagnating Peasants?' *Economic and Political Weekly*, 31(26): A63–A73.

Nagaraj, K. and J. Jayaranjan. 2004. 'Tamilnadu Economy: Contours of Change: A Secondary Data Exploration', mimeo, Chennai: Madras Institute of Development Studies.

Nair, S.C. 1991. 'The Southern Western Ghats: a Biodiversity Conservation Plan', *Studies in Ecology and Sustainable Development—4*. New Delhi: INTACH.

Narain, B. 1922. *Indian Economic Problems, Part II: Source Book for the Study of Indian Economic Problems*. Lahore: The Punjab Printing Works.

Narain, V. 1998. 'Towards a New Groundwater Institution for India', *Water Policy*, 1(3): 357–66.

Narain, V. 2000. 'India's Water Crisis: The Challenges of Governance', *Water Policy*, 2(6): 433–44.

———. 2003a. *Institutions, Technology and Water Control: Water Users' Associations and Irrigation Management Reform in Two Large-scale Irrigation Systems in India*. Wageningen University Water Resources Series 1. Hyderabad: Orient Longman.

———. 2003b. 'Mediating Scarcity by Design: Water Rights and Legal Pluralism in Protective Irrigation', in *Environmental Threats, Vulnerability and Adaptation: Case Studies from India*. New Delhi: TERI.

———. 2008. 'Reform in Indian Canal Irrigation: Does Technology Matter?' *Water International*, 33(1): 33–42.

Narayanamoorthy, A. 2001. 'Irrigation and Rural Poverty Nexus: A Statewise Analysis', *Indian Journal of Agricultural Economics*, 56(1): 40–56.

———. 2007. 'Tank Irrigation in India: A Time Series Analysis', *Water Policy*, 9(2): 193–216.

Narayanan, N.C. and P. Venot. 2009. 'Drivers of Change in Fragile Environments: Challenges to Governance in Indian Wetlands', *Natural Resources Forum*, 33(4): 320–33.

Nepal, G. 1998. 'A Report of Random Sample Survey to Determine Actual Status of Private Micro-hydropower Plants in Nepal. Submitted to ICIMOD and ITDG'. Kathmandu: Earth Consult.

Nikku, B.R. 2002. 'Water Users Associations in Irrigation Management: Case of Andhra Pradesh, South India. Opportunities and challenges for collective Action', Paper accepted at the 9th Biennial Conference of the International Association for the Study of Common Property, Victoria Falls, Zimbabwe, 17–22 June. Available at http://dlc.dlib.indiana.edu/dlc/handle/10535/1035 (accessed 24 May 2012).

———. 2003. 'Irrigation Reforms, Institutions and Livelihoods: The Case of Andhra Pradesh, South India', in K. Chopra, C.H. Rao, and R. Sengupta (eds), *Water Resources Sustainable Livelihoods and Ecosystem Services*. Delhi: Institute of Economic Growth.

———. 2006. 'The Politics of Policy: Participatory Irrigation Management in Andhra Pradesh', PhD dissertation, Wageningen University, The Netherlands.

Nikku, B.R. and I. van der Molen. 2008. 'Conflict, Resistance and Alliances in a Multi-governance Setting: Reshaping Realities in the Andhra Pradesh Irrigation Reforms', *Energy & Environment*, 19(6): 861–75.

Nuijten, M., G. Anders, J. van Gastel, G. van der Haar, C. Nijnatten, and J. Warner. 2004. 'Governance in Action: Some Theoretical and Practical Reflections on a Key Concept', in D. Kalb, W. Pansters, and H. Siebers (eds), *Globalization and Development: Themes and Concepts in Current Research*. Dordrecht: Kluwer Academic Publishers.

Oblitas, K. and R. Peter. 1999. 'Transferring Irrigation Management to Farmers in Andhra Pradesh, India', World Bank Technical Paper 449. Washington, DC: World Bank.

Ojwang, J.B. and C. Juma. 1996. 'Towards Ecological Jurisprudence', in C. Juma and J.B. Ojwang (eds), *In Land We Trust: Environment, Private Property and Constitutional Change*. Nairobi: Initiatives Publishers.

Olin, M. 1994. 'Transfer of Management to Water Users in Stages I & II of the Bhairahawa Lumbini Groundwater Irrigation Project in Nepal', Paper presented at the International Conference on Irrigation Management Transfer, IIMI and Wuhan University of Hydraulic and Electrical Engineering, China, 20–24 September.

Omvedt, G. 1993. *Reinventing Revolution: New Social Movements and the Socialist Tradition in India*. New York: M.E. Sharp.

Oorthuizen, J. 2003. *Water, Works and Wages: The Everyday Politics of Irrigation Management Reform in the Philippines*. Wageningen University Water Resources Series 3. Hyderabad: Orient Longman.

Orindi, V. and C. Huggins. 2005. 'The Dynamic Relationship Between Property Rights, Water Resource Management and Poverty in Lake Victoria Basin', Paper presented at the International Workshop on African Water Laws: Plural Legislative Frameworks for Rural Water management in Africa, 26–28 January 2005, Johannesburg, South Africa.

Ostrom, E. 1992. *Crafting Institutions for Self-governing Irrigation Systems*. San Francisco: ICS Press.

Ostrom, E., P. Benjamin, and G. Shivakoti. 1992. 'Institutions, Incentives, and Irrigation in Nepal'. Nepal Irrigation Institutions and Systems Project, Workshop in Political Theory and Policy Analysis. USA: Indiana University.

Pahl-Wostl, C., J. Gupta, and D. Petry. 2008. 'Governance and the Global Water System: A Theoretical Exploration', *Global Governance*, 14(4): 419–35.

Palanisami, K. 2000. *Tank Irrigation: Revival for Prosperity*. New Delhi: Asian Publication Services.

Palmer-Jones, R. 1994. 'Groundwater Markets in South Asia: A Discussion of Theory and Evidence', in Marcus Moench (ed.), *Selling Water: Conceptual and Policy Debates over Groundwater Markets in India*. Ahmedabad: VIKSAT, Pacific Institute, Natural Heritage Institute.

Pandey, B. 1996. 'Local Benefits from Hydro Development', *Studies in Nepali History and Society*, 1(2): 314–44.

Pandey, K. 1998. 'Hydropower Development in Post-1990 Nepal', *Water Nepal*, 6(1): 145–69.

Pandian, M.S. 1990. *The Political Economy of Agrarian Change: Nanchilnadu 1880–1939*. New Delhi: Sage Publications.

Pani, N. 1997. 'Towards Decentralized Agrarian Reform: Lessons from Karnataka's 1974 Experience', in A. Aziz and S. Krishna (eds), *Karnataka: Promises Kept and Missed*. New Delhi: Sage Publications.

Pant, D. 2000. 'Intervention Processes and Irrigation Institutions: Sustainability of Farmer-managed Irrigation Systems in Nepal', PhD dissertation, Wageningen University.

Pant, N. 2003. *Key Trends in Groundwater Irrigation in the Eastern and Western Regions of Uttar Pradesh*. Anand: IWMI-Tata Water Policy Programme.

———. 2004. 'Trends in Groundwater Irrigation in Eastern and Western UP', *Economic and Political Weekly*, 39(31): 3463–8.

Pant, N. and R.P. Rai. 1985. *Community Tubewell and Agricultural Development*. New Delhi: Ashish Publishing House.

Papanek, V. 1985. *Design for the Real World: Human Ecology and Social Change*. London: Thames and Hudson.

Parajuli, U.N. 1999. 'Agro-ecology and Irrigation Technology: Comparative Research on Farmer Managed Irrigation Systems in the Mid-hills of Nepal', PhD dissertation, Wageningen Agricultural University, The Netherlands.

Paranjape, S. and K.J. Joy. 1995. *Sustainable Technology: Making the Sardar Sarovar Project Viable. A Comprehensive Proposal to Modify the Project for Greater Equity and Ecological Sustainability*. Ahmedabad: Centre for Environment Education.

Patil, R.K. and S.N. Lele. 1995. 'Irrigation Management Transfer: Problems in Implementation', in S.H. Johnson, D.L. Vermillion, and J.A. Sagardoy (eds), *Selected Papers from the International Conference on Irrigation Management Transfer, Wuhan, China, 20–24 September 1994*. Water Report No. 5. Rome: FAO/IIMI.

Patnaik, U. (ed.). 1990. *Agrarian Relations and Accumulation: The Mode of Production Debate in India*. New Delhi: Oxford University Press.

Peter, J.R. 2002. 'Case Study on Andhra Pradesh', Paper presented at the Sixth International Seminar on PIM: Institutional Options for User Participation, Beijing, China, 21–6 April.

Pinch, T.J. and W.E. Bijker. 1984. 'The Social Construction of Facts and Artefacts: Or How the Sociology of Science and the Sociology of Technology Might Benefit Each Other', *Social Studies of Science*, 14(3): 399–441.

Plusquellec, H. 2002. 'Is the Daunting Challenge of Irrigation Achievable?' *Irrigation and Drainage*, 51(3): 185–98.

Pohl, C. and G.H. Hadorn. 2007. *Principles for Designing Transdisciplinary Research*. München: Oekom Verlag.

Pradhan, P. 1989. *Patterns of Irrigation Organization in Nepal: A Comparative Study of 21 Farmer-managed Irrigation Systems*. Colombo: International Irrigation management Institute.

Pradhan, R., F. von Benda-Beckmann, and K. von Benda-Beckmann (eds). 2000. *Water, Land and Law: Changing Rights to Land and Water in Nepal*. Kathmandu/Wageningen/Rotterdam: Freedeal/Wageningen Agricultural University, Erasmus University Rotterdam.

Pradhan, R. and R. Meinzen-Dick. 2003. 'Which Rights are Rights? Water Rights, Culture and Underlying Values', *Water Nepal*, 9/10(1/2): 37–61.

Pradhan, T.M.S. 1996. 'Gated or Ungated Water Control in Government-built Irrigation Systems: Comparative Research in Nepal', PhD dissertation, Wageningen University, The Netherlands.

Pradhan, U. 1988. 'Local Resource Mobilization and Government Intervention in Hill Irrigation Systems', mimeo, Cornell University Water Management Synthesis Project.

Prahladachar, M. 1994. 'Innovations in the Use and Management of Groundwater in Hardrock Regions in India', *Ecological Economics*, 9(3): 267–72.

Prakash, A. 2005. *The Dark Zone. Groundwater Irrigation, Politics and Social Power in North Gujarat*, Wageningen University Water Resources Series 7. New Delhi: Orient Longman.

Prakash, A. and V. Ballabh. 2005. 'A Win-some Lose-all Game: Social Differentiation and Politics of Groundwater Markets in North Gujarat', in D. Roth, R. Boelens, and M. Zwarteveen (eds), *Liquid Relations: Contested Water Rights and Legal Complexity*. Piscataway: Rutgers University Press.

Puttaswamaiah, S. and K.V. Raju. 2009. 'Compensation and Reward for Ecosystem Services: A New Approach for Natural Resource Management', Paper for the 5th Biennial Conference of INSEE, Ahmedabad, January.

Rajan, M.A.S. 1981. *Land Reforms in Karnataka: An Account by a Participant Observer*. New Delhi: Hindustan Publishing Corporation.

Raju, K.V., A. Narayanamoorthy, G. Gopakumar, and H.K. Amarnath. 2004. *State of the Indian Farmer: A Millennium Study*, vol. 3. New Delhi: Academic Foundation.

Ramamurthy, P. 1995. 'The Political Economy of Canal Irrigation in South India', PhD dissertation, Syracuse University, Syracuse, NY.

Rao, C. and A. Gulati. 1994. 'Indian Agriculture: Emerging Perspective and Policy Issues', *Economic and Political Weekly*, 29(55): A58–69.

Rao, P.N. and K.C. Suri. 2006. 'Dimensions of Agrarian Distress in Andhra Pradesh', *Economic and Political Weekly*, 41(6): 1546–52.

Rao, V.M. and D.V. Gopalappa. 2004. 'Agricultural Growth and Farmers' Distress: Tentative Perspectives from Karnataka', *Economic and Political Weekly*, 39(52): 5591–7.

Rap, E. 2004. 'The Success of a Policy Model: Irrigation Management Transfer in Mexico', PhD dissertation, Wageningen University, The Netherlands.

Ratnavel, S.M. and P. Gomathinayagam. 2006. *In Search of Ancient Wisdom: Irrigation Tanks*. Madurai: DHAN Foundation.

Reddy, N.D. 2002. 'Designer Participation: Politics of Irrigation Management in Andhra Pradesh', in R. Hooja, G. Pangare, and K.V. Raju (eds), *Users in Water Management: The Andhra Model and its Replicability in India*. Jaipur and New Delhi: Rawat Publications.

Reddy, V.R. 2003. 'Irrigation: Development and Reforms', *Economic and Political Weekly*, 38(12/13): 1179–89.

Reddy, V.R. and S. Mahendra Dev. 2006. *Managing Water Resources: Policies, Institutions and Technologies*. New Delhi: Oxford University Press.

Regmi, A. 2004a. *Democratising Micro-hydel: Structures, Systems and Agents in Adaptive Technology in the Hills of Nepal*. Wageningen University Water Resources Series 6. New Delhi: Orient Blackswan.

———. 2004b. 'Indexing Constitutional Accountability in Local Governance: The Search for the Water-power Interface', *Resources, Energy and Development*, 1(1/2): 91–4.

Regmi, M.C. 1978. 'Land Tenure and Taxation in Nepal', in H.K. Kuloy (ed.) *Bibliotheca Himalaya*, vol. 26, no. 1. Kathmandu: Ratna Pustak Bhandar.

Renault, D., T. Facon, and R. Wahaj. 2007. 'Modernizing Irrigation Management—The MASSCOTE Approach: Mapping System and Services for Canal Operation Techniques', FAO Irrigation and Drainage Paper 63. Rome: FAO.

Ribot, J. and N.L. Peluso. 2003. 'A Theory of Access', *Rural Sociology*, 68(2): 153–81.

Rijal, K. (ed.). 1998. *Renewable Energy Technologies: A Brighter Future*. Kathmandu: International Centre for Mountain Development.

———. 2000. 'Mini and Micro-hydro Power Development: Status, Issues and Strategies for the Hindu Kush Himalayan Region', *NEES Journal of Engineering*, 9: 32–7.

Rogers, P. and A.W. Hall. 2003. 'Effective Water Governance', TEC Background Paper No. 7. Sweden: GWP.

Rose, S. 1987. *Molecules and Minds. Essays on Biology and the Social Order*. Milton Keynes and Philadelphia: Open University Press.

Rosegrant, M.W. and H.P. Binswanger. 1994. 'Markets in Tradable Water Rights: Potential for Efficiency Gains in Developing Country Water Resource Allocation', *World Development*, 22(11): 1613–25.

Roy, A.D. and T. Shah. 2003. 'Socio-ecology of Groundwater Irrigation in India', in R. Llamas and E.A. Custodio (eds), *Intensive Use of Groundwater: Challenges and Opportunities*. Rotterdam: Balkema Publishers.

Roth, D. 2003. 'Ambition, Regulation and Reality. Complex Use of Land and Water Resources in Luwu, South Sulawesi, Indonesia', PhD dissertation, Wageningen University, The Netherlands.

———. 2006. 'Which Order? Whose Order? Balinese Irrigation Management in Sulawesi, Indonesia', *Oxford Development Studies*, 34(1): 33–46.

Roth, D., R. Boelens, and M. Zwarteveen (eds). 2005. *Liquid Relations: Contested Water Rights and Legal Complexity*. Piscataway: Rutgers University Press.

Rudolph, L.I. and S.H. Rudolph. 1987. *In Pursuit of Lakshmi: The Political Economy of the Indian State*. New Delhi: Orient Longman.

Sakthivadivel, R. and P. Gomathinayagam. 2006. *Rehabilitation of Tanks in India*. The Philippines: Asian Development Bank.

Sakthivadivel, R., N. Fernando, R. Panabokke, and C.M. Wijayaratna. 1996. 'Nature of Small Tank Cascade Systems and a Framework for Rehabilitation of Tanks Within Them', Sri Lanka Country Report 13. Sri Lanka: IIMI.

Sakthivadivel, R., P. Gomathinayagam, and T. Shah. 2004. 'Rejuvenating Irrigation Tanks through Local Institutions', *Economic and Political Weekly*, 39(31): 3521–6.

Saleth, R.M. 1998. 'Water Markets in India: Economic and Institutional Aspects', in K.W. Easter, M.W. Rosegrant, and A. Dinar (eds), *Markets for Water: Potential and Performance*. Connecticut: Kluwer Academic Press.

———. 2005. 'Water Institutions in India: Structure, Performance, and Change', in C. Gopalakrishnan, C. Tortajada, and A.K. Biswas (eds), *Water Institutions: Policies, Performance and Prospects*. Berlin and Heidelberg: Springer-Verlag.

Satyasai, K.J.S. and K.U. Viswanathan. 1997. 'Terms of Transaction in Groundwater Markets: A Study in Anantapur District of Andhra Pradesh', *Journal of Agricultural Economics*, 52(4): 751–60.

Sayer, A. 1984. *Method in Social Science: A Realist Approach*. London: Hutchinson.

Scheffer, M. 2009. *Critical Transitions in Nature and Society*, Princeton Studies in Complexity. Princeton and Oxford: Princeton University Press.

Scott, J.C. 1998. *Seeing Like a State: How Certain Schemes to Improve the Human Condition have Failed*. New Haven: Yale University Press.

Scott, W.R. 1995. *Institutions and Organizations*. London: Sage Publications.

Sen, A. 1968. *Choice of Techniques: An Aspect of Theory of Planned Economic Development*. Oxford: Blackwell.

Sengupta, N. 1980. 'Indigenous Irrigation and Social Organization in South Bihar', *Indian Economic and Social History Review*, 17(2): 157–90.

———. 1985. 'Irrigation: Traditional vs. Modern', *Economic and Political Weekly*, 20(45/46/47): 1919–38.

———. 1991. *Managing Common Property: Irrigation in India and the Philippines*. New Delhi: Sage Publications.

———. 1993. *User Friendly Irrigation Designs*. New Delhi: Sage Publications.

———. 1997. 'The Rise of the Bureaucracy in Tamil Nadu: Water Control vs. Management', *Water Nepal*, 5(2): 125–35.

Shah, E. 2003. *Social Designs: Tank Irrigation Technology and Agrarian Transformation in Karnataka, South India*, Wageningen University Water Resources Series 4. Hyderabad: Orient Longman.

———. 2008a. 'Telling Otherwise: Historical Anthropology of Tank Irrigation Technology of South India', *Technology and Culture*, 29(3): 652–74.

———. 2008b. 'What Makes Crop Biotechnology Find its Roots? The Technological Culture of Bt Cotton in Gujarat', *European Journal of Development Research*, 20(3): 432–47.

———. 2008c. 'Resource, Rules and Technology: Ethnography of Building a Water Users' Association', in V. Ballabh (ed.), *Governance of Water: Institutional Alternatives and Political Economy*. New Delhi: Sage Publications.

Shah, M. 1985. 'The Kaniatchi Form of Labour', *Economic and Political Weekly*, 20(30): 65–78.

Shah, T. 1985. 'Transforming Groundwater Markets into Powerful Instruments of Small Farmer Development', ODI Irrigation Management Network Paper No. 11d. London: Overseas Development Institute.

———. 1986. *Groundwater Markets in Water Scarce Regions: Fieldnotes from Karimnagar District (Telangana), Andhra Pradesh*. Anand: Institute of Rural Management.

———. 1993. *Groundwater Markets and Irrigation Development: Political Economy and Practical Policy*. Bombay: Oxford University Press.

———. 2003. Framing the Rules of the Game: Preparing for the First Irrigation Season in the Sardar Sarovar Project, Anand: IWMI-Tata

Research Team. Available at http://www.iwmi.org/iwmi-tata (accessed September 2004).

Shah, T. 2005. 'Groundwater and Human Development: Challenges and Opportunities in Livelihoods and Environment', in B.R. Sharma, K.G. Villholth, and K.D. Sharma (eds), *Growndwater Research and Management: Integrating Science into Management Decisions*, Proceedings of IWMI–ITP–NIH international workshop on 'Creating Synergy between Groundwater Research and Management in South and Southeast Asia', Roorkee, India, 8–9 February. Colombo, Sri Lanka: IWMI.

———. 2009. *Taming the Anarchy: Groundwater Governance in South Asia*. Washington, DC and Colombo: Resources for the Future and International Water Management Institute.

Shah, T. and K.V. Raju. 1987. 'Working of Groundwater Markets in Andhra Pradesh and Gujarat: Results of Two Village Studies', *Economic and Political Weekly*, 26(3): A23–28.

———. 2001. 'Rethinking Rehabilitation: The Socio-ecology of Tanks and Water Harvesting in Rajasthan, NW India', CAPRI Working Paper 18. Washington, DC: International Food Policy Research Institute.

Shah, T. and V. Ballabh. 1997. 'Water Markets in North Bihar: Six Village Studies in Muzaffarpur District', *Economic and Political Weekly*, 32(52): A183–90.

Shah, T., R. Seenivasan, C.R. Shanmugam, and M.P. Vasimalai. 1999. *Sustaining Tamilnadu's Tanks: Fieldnotes on PRADAN's Work in Madurai and Ramnad*. Madurai: DHAN Foundation.

Shankari, U. 1991. 'Tanks: Major Problems in Minor Irrigation', *Economic and Political Weekly*, 26(39): A115–25.

Shankari, U. and E. Shah. 1993. *Water Management Traditions in India*. Chennai: PPST foundation.

Sharma, A. 2003. 'Rethinking Tanks: Opportunities for Revitalizing Irrigation Tanks—Empirical Findings from Ananthapur District, Andhra Pradesh, India', Working Paper 62. Colombo: International Water Management Institute.

Shivakoti, G. and E. Ostrom. 2002. *Improving Irrigation Governance and Management in Nepal*. Oakland: ICS Press.

Shivakoti, G.P., D.L. Vermillion, W.F. Lam, E. Ostrom, U. Pradhan, and R. Yoder (eds). 2005. *Asian Irrigation in Transition: Responding to Challenges*. New Delhi: IWMI/Sage Publications.

Sick, D. 2007. 'Yours, Mine and Ours: Managing Water Resources in the Mexican-US Borderlands', in Amita Baviskar (ed.), *Waterscapes: The Cultural Politics of a Natural Resource*. Uttaranchal: Permanent Black.

Sikor, T. and C. Lund. 2009. 'Access and Property: A Question of Power and Authority', *Development and Change*, 40(1): 1–22.

Singh, C. 1992. *Water Law in India*. New Delhi: Indian Law Institute.

Singh, S. 2002. *Taming the Waters. The Political Economy of Large Dams in India*. New Delhi: Oxford University Press.

Sinha, S., S. Gururani, and B. Greenberg. 1998. 'The "New Traditionalist" Discourse of Indian Environmentalism', *Journal of Peasant Studies*, 24(3): 65–99.

Sivakumar, S.S. 1978. 'The Transformation of the Agrarian Economy in Tondaimandalam, 1760–1900', *Social Scientist*, 6(70): 18–39.

Slootweg, R. and P.P. Mollinga. 2009. 'The Impact Assessment Framework', in R. Slootweg, A. Rajvanshi, V.B. Mathur, and A. Kolhoff (eds), *Biodiversity in Environmental Assessment: Enhancing Ecosystem Services for Human Well-being*. Cambridge: Cambridge University Press.

Sooryamoorthy, R. 2003. 'Fast Dying River Systems: Periyar and Bharathapuzha Basins', in R. Sooryamoothy and Antony Palackal (eds), *Managing Water and Water Users: Experiences from Kerala*. Maryland: University Press of America.

Sowerine, J., G. Shivakoti, U. Pradhan, A. Shukla, and E. Ostrom (eds). 1994. 'From Farmer's Fields to Data Fields and Back: A Synthesis of Participatory Irrigation Systems and Other Resources'. Colombo and Rampur: International Irrigation Management Institute and Institute of Agriculture and Animal Sciences.

Star, S.L. and J.R. Griesemer. 1989. 'Institutional Ecology, "Translations" and Boundary Objects: Amateurs and Professionals in Berkeley's Museum of Vertebrate Zoology, 1907–39', *Social Studies of Science*, 19(3): 387–420.

State Planning Board. 1976. *The Scheme for the Free Supply of Pumpsets to Panchayats: An Evaluation Study*. Trivandrum: State Planning Board.

Stein, B. 1980. *Peasant State and Society in Medieval South India*. New Delhi: Oxford University Press.

Stone, I. 1984. *Canal irrigation in British India: Perspectives on Technological Change in a Peasant Society*. Cambridge: Cambridge University Press.

Strang, V. 2001. 'Negotiating the River: Cultural Tributaries in Far North Queensland', in B. Bender and M. Winer (eds), *Contested Landscapes: Movement, Exile and Place*. Oxford: Berg.

———. 2005. 'Common Senses: Water, Sensory Experience and the Generation of Meaning', *Journal of Material Culture*, 10(1): 92–120.

Subbarayalu, Y. 1973. *The Political Geography of the Chola Country*. Madras: State Department of Archaeology, Government of Tamil Nadu.

Svendsen, M. and W. Huppert. 2000. 'Incentive Creation for Irrigation System Maintenance and Water Delivery: The Case of Recent Reforms in Andhra Pradesh', Maintain Case Study No. 5. Germany: GTZ.

Swyngedouw, E.A. 1999. 'Marxism and Historical-geographical Materialism: A Spectre is Haunting Geography', *Scottish Geographical Magazine*, 115(2): 91–102.

————. 2007. 'Technonatural Revolutions—The Scalar Politics of Franco's Hydro-social Dream for Spain, 1939–1975', *Transactions of the Institute of British Geographers*, 32(1): 9–28.

————. 2009. 'The Political Economy and Political Ecology of the Hydro-social Cycle', *Journal of Contemporary Water Research and Education*, 142: 56–60.

Tendler, J. 1997. *Good Government in the Tropics*. Baltimore and London: The Johns Hopkins University Press.

Thorner, A. 1982. 'Semi-feudalism or Capitalism? Contemporary Debate on Classes and Modes of Production in India', *Economic and Political Weekly*, 17(49/50): 1961–8 and 1993–9.

Tippaiah, P. 1997. *Study of Causes for the Shrinkage in Tank Irrigated Area in Karnataka*. Bangalore: Institute for Social and Economic Change.

Torori, C.O., A.O. Mumma, and A. Field-Juma. 1996. 'Land Tenure and Water Resources', in C. Juma and J.B. Ojwang (eds), *In Land We Trust: Environment, Private Property and Constitutional Change*. Nairobi: Initiatives Publishers.

Trawick, P. 2008. 'Reading History in an Irrigated Landscape: The Drama of the Commons in the Andes', in L. Cliggett and C.A. Pool (eds), *Economies and the Transformation of Landscape*. Plymouth: AltaMira Press.

Trosper, R.L. 2005. 'Emergence Unites Ecology and Society', *Ecology and Society*, 10(1): 14.

Turral, H. 1995. 'Devolution of Management in Public Irrigation Systems: Cost Shedding, Empowerment and Performance. A Review', Working Paper 80. London: Overseas Development Institute.

Uphoff, N. 1986. 'Getting the Process Right: Improving Water Management with Farmer Organization and Participation.' Water Management Synthesis Working Paper, Ithaca: Cornell University.

Vaidyanathan, A. 1996. 'Depletion of Groundwater: Some Issues', *Indian Journal of Agricultural Economics*, 51(1/2): 184–92.

————. 1999. *Water Resource Management: Institutions and Irrigation Development in India*. New Delhi: Oxford University Press.

————. (ed.). 2001. *Tanks of South India*. New Delhi: Centre for Science and Environment.

VanderMeer, C. 1971. 'Water Thievery in a Rice Irrigation System in Taiwan', *Annals of the American Association of Geographers*, 61(1): 156–79.

van Halsema, G.E. 2002. 'Trial and Retrial: The Evolution of Irrigation Modernisation in NWFP, Pakistan', PhD dissertation, Wageningen University, The Netherlands.

van Halsema, G.E. and L. Vincent. 2006. 'Of Flumes, Modules and Barrels: The Failure of Irrigation Institutions and Technology to Achieve Equitable Water Control in the Indus Basin', in T. Tvedt and E. Jakobsson (eds), *A History of Water, vol. 1. Water Control and River Biographies*. London: I.B. Tauris.

van Koppen, B., C.S. Sokile, N. Hatibu, B.A. Lankford, H. Mahoo, and P.Z. Yanda. 2004. 'Formal Water Rights in Rural Tanzania: Deepening the Dichotomy?' Working Paper No. 71. Colombo: International Water Management Institute.

von Oppen, M. and K.V. Subba Rao. 1980. 'Tank Irrigation in Semi-arid Tropical India. Part II: Technical Features and Economic Performance', Economic Programme Progress Report 5. Hyderabad: ICRISAT.

van Steenbergen, F. and T. Shah. 2003. 'Rules Rather than Rights: Self-Regulation in Intensively Used Groundwater Systems', in R. Llamas and E. Custodio (eds), *Intensive Use of Groundwater: Challenges and Opportunities*. Lisse: Zwets and Zeitlinger.

Vani, M.S. 2009. 'Groundwater Law in India: A New Approach', in R.R. Iyer (ed.), *Water and the Laws in India*. New Delhi: Sage Publications.

Varshney, A. 1998. *Democracy, Development and the Countryside: Urban Rural Struggles in India*. Cambridge: Cambridge University Press.

Venkateswarulu, D. 1999. 'Politics of Irrigation Management Reforms in Andhra Pradesh', Paper presented at the researcher's conference 'The Long Road to Commitment: A Socio-political Perspective on Irrigation Reform', Hyderabad, 11–14 December.

Vermillion, D. 1992. 'Irrigation Management Turnover: Structural Adjustment or Strategic Evolution?' *IIMI Review*, 6(2): 3–12.

———. 1997. 'Impacts of Irrigation Management Transfer: A Review of the Evidence', Research Report 11. Colombo: IWMI.

Vermillion, D. and J. Sagardoy. 1999. 'Transfer of Irrigation Management Services'. FAO Irrigation and Drainage Paper 58. Rome: FAO.

Vincent, L. 1995. *Hill Irrigation: Water and Development in Mountain Agriculture*. London: Intermediate Technology Publications.

———. 1997. 'Irrigation as a Technology, Irrigation as Resource: A Sociotechnical Approach to Irrigation', Inaugural address, Wageningen Agricultural University, The Netherlands.

Vincent, L. 2001. 'Struggles at the Social Interface: Developing Sociotechnical Research in Irrigation and Water Management', in P. Hebinck and G. Verschoor (eds), *Resonances and Dissonances in Development: Actors, Networks and Cultural Repertoires*. Assen: Van Gorcum.

———. 2003. 'Towards a Smallholder Hydrology for Equitable and Sustainable Water Management', *Natural Resources Forum*, 27(3): 108–16.

Wade, R. 1982. 'The System of Administrative and Political Corruption: Canal Irrigation in South India', *Journal of Development Studies*, 18(3): 287–328.

———. 1988. *Village Republics: Economic Conditions for Collective Action in South India*. Cambridge: Cambridge University Press.

———. 1995. 'The Ecological Basis of Irrigation Institutions: East and South Asia', *World Development*, 23(12): 2041–9.

Waghmore, S. 2002. 'Rural Development: Role of State', *Economic and Political Weekly*, 37(29): 3001–3.

Wahaj, R. 2001. 'Farmers Actions and Improvements in Irrigation Performance below the Mogha. How Farmers Manage Water Scarcity and Abundance in a Large Scale Irrigation System in South Eastern Punjab, Pakistan', PhD dissertation, Wageningen University, The Netherlands.

WALMI (Water and Land Management Institute). 1998a. 'Water Distribution Practices in Maharashtra State', Publication No. 22. Aurangabad: WALMI.

———. 1998b. 'Operation and Management of Irrigation Systems in Maharashtra State', Publication No. 20. Aurangabad: WALMI.

WECS (Water and Energy Commission Secretariat). 1988. 'Water Use Inventory Study of Kabhre Palanchowk District'. Kathmandu: WECS, Multi Disciplinary Consultants.

Weil, B. 2006. 'The Rivers Come: Colonial Flood Control and Knowledge Systems in the Indus Basin, 1840s–1930s', *Environment and History*, 12(1): 3–29.

Wester, P., E. Rap, and S. Vargas-Velazquez. 2009. 'The Hydraulic Mission and the Mexican Hydrocracy: Regulating and Reforming the Flows of Water and Power', *Water Alternatives*, 2(3): 395–415.

Whitcombe, E. 1972. *Agrarian Conditions in Northern India: The United Provinces under British Rule, 1860–1900*. Berkeley: California University Press.

Whitehead, J. 2008. 'Submerged Voices and Transnational Environmentalism: The Movement against the Sardar Sarovar Dam', in K. Lahiri-Dutt and R. Wasson (eds), *Water First: Issues and Challenges for Nations and Communities in South Asia*. New Delhi: Sage Publications.

Widgren, M. 2007. 'Pre-colonial Landesque Capital: A Global Perspective', in A. Hornborg, J.R. McNeill, and J. Martinez-Alier (eds), *Rethinking Environmental History. World System History and Global Environmental Change*. Walnut Creek: AltaMira Press.

Wilson, K. 2002. 'Small Cultivators in Bihar and "New" Technology. Choice or Compulsion?' *Economic and Political Weekly*, 37(13, March): 1229–38.

Winner, L. 1986. *The Whale and the Reactor: A Search for Limits in the Age of High Technology*. Chicago: Chicago University Press.

Wittfogel, K. 1957. *Oriental Despotism: A Comparative Study of Local Power*. New Haven: Yale University Press.

Wood, G. 1999. 'Private Provision After Public Neglect: Bending Irrigation Markets in North Bihar', *Development and Change*, 30(4): 775–94.

World Bank. 1981. 'Karnataka Tank Irrigation Project'. Washington, DC: World Bank.

———. 1994. 'A Review of World Bank Experience in Irrigation', Report No. 13676. Washington, DC: World Bank.

World Bank and GoI. 1999. *The Irrigation Sector*. South Asia Rural Development Series, in collaboration with Ministry of Water Resources, the GoI. Washington, DC and New Delhi: World Bank and Allied Publishers.

Yabes, R.A. 1990. 'Indigenous Proportional Weirs and Modern Agency Turnouts: Design Alternatives in the Philippines', in R. Yoder and J. Thurston (eds), *Design Issues in Farmer-managed Irrigation Systems*. Colombo, Sri Lanka: IIMI.

Yoder, R. and S.B. Upadhyaya. 1987. 'Reconnaissance/Inventory Study of Irrigation Systems in the Indrawati Basin of Nepal', in IIMI, IAAS, and WINROCK (eds), *Irrigation Management in Nepal. Research Papers from a National Seminar*. Bharatpur, Nepal.

Yoder, R., E. Martin, R. Barker, and T.S. Steenhuis. 1987. 'Variations in Irrigation Management Intensity: Farmer-managed Hill Irrigation Systems in Nepal', Water Management Synthesis Report 67. Kathmandu: United States Agency for International Development (USAID).

CONTRIBUTORS

SUMAN RIMAL GAUTAM is a trained agricultural engineer with specialization in irrigation engineering and management, and works from Washington, DC.

PUSPA RAJ KHANAL is a water resources development and conservation expert in the Food and Agriculture Organization, Bangkok, Thailand.

JYOTHI KRISHNAN is an independent researcher in water and natural resources management, Kerala, India.

R. MANIMOHAN is a PhD researcher at Wageningen University, The Netherlands, and Research Associate at the Institute for Ocean Management, Anna University, Chennai, India.

PETER P. MOLLINGA is Professor of Development Studies at the School of Oriental and African Studies, London, United Kingdom.

VISHAL NARAIN is Associate Professor at the School of Public Policy and Governance at the Management Development Institute, Gurgaon, India.

BALA RAJU NIKKU is on the executive board of the Asian and Pacific Association of Social Work Education, visiting lecturer at the School of Social Sciences, Universiti Sains Malaysia, and General Secretary of the South Asia Consortium of Interdisciplinary Water Resource Studies.

UMESH NATH PARAJULI is a freelance international consultant in several countries in Southeast and Central Asia.

ANJAL PRAKASH is the Executive Director of the South Asia Consortium for Interdisciplinary Water Resources Studies, Hyderabad, India.

AMREETA REGMI is Election Poll Manager for Gwinnett County, Georgia.

DIK ROTH is Assistant Professor of the Rural Development Sociology Group, Social Sciences Department, Wageningen University, The Netherlands.

ESHA SHAH is Assistant Professor in the Department of Technology and Society Studies, Faculty of Arts and Social Sciences, Maastricht University, The Netherlands.

LINDEN VINCENT is Professor of the Irrigation and Water Engineering Group, Environmental Sciences Department, Wageningen University, The Netherlands.